The Study of Time IV

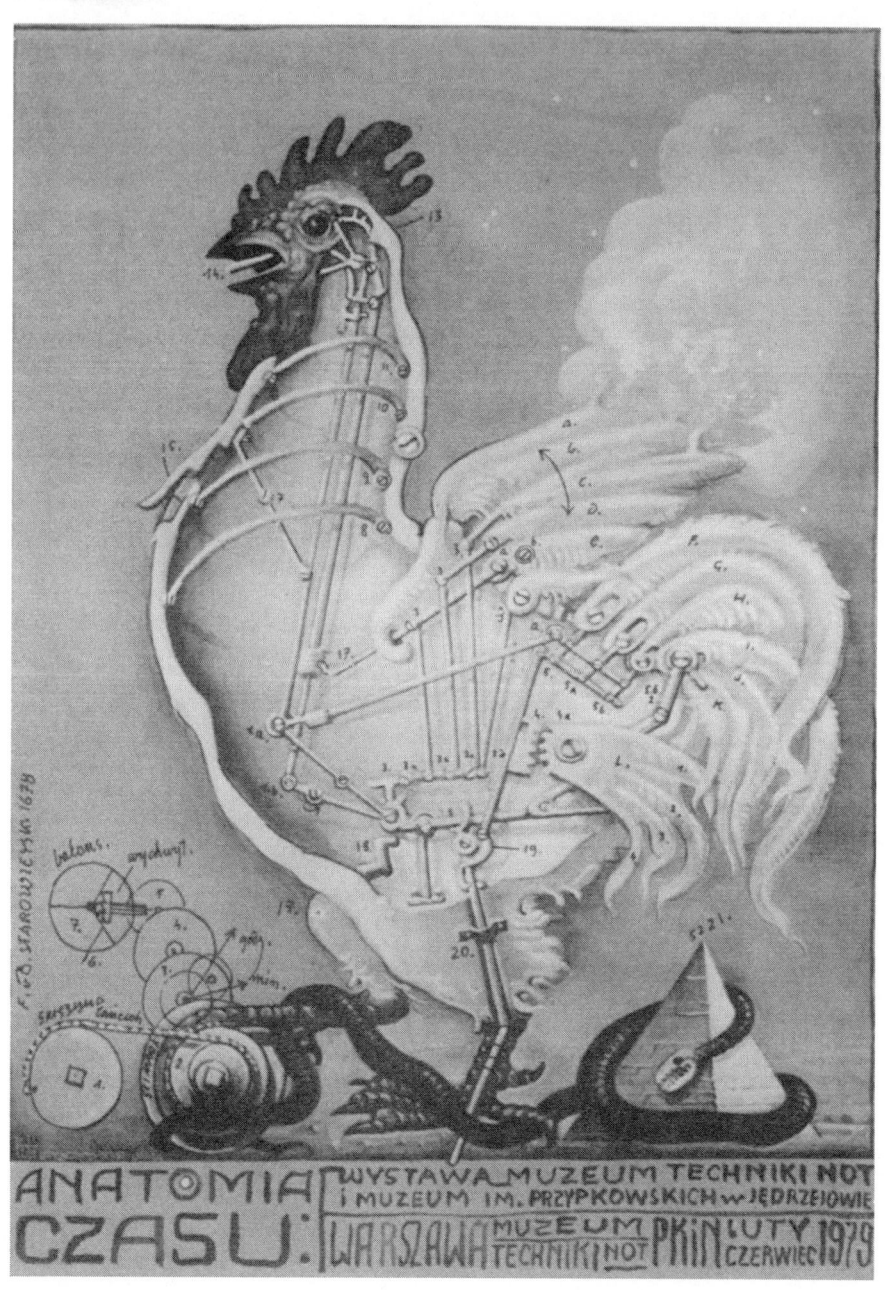

See explanation on page xxv.

The Study of Time IV

Papers from the Fourth Conference of the International Society for the Study of Time, Alpbach–Austria

Edited by

J.T. Fraser • N. Lawrence • D. Park

With 26 Figures

Springer-Verlag
New York   Heidelberg   Berlin

Dr. J.T. Fraser, Founder
International Society for the Study
of Time
P.O. Box 815
Westport, Connecticut 06881 U.S.A.

Professor Nathaniel Lawrence
Department of Philosophy
Williams College
Williamstown, Massachusetts 01267 U.S.A.

Professor David Park
Department of Physics and Astronomy
Williams College
Williamstown, Massachusetts 01267 U.S.A.

Library of Congress Cataloging in Publication Data (Revised)

International Society for the Study of Time.
    The study of time; proceedings of the first [fourth] conference of the International Society for the Study of Time.
    English or German.
    Vol. 2 edited by J.T. Fraser and N. Lawrence; v. 3. by J.T. Fraser, N. Lawrence, D. Park; v. 4 by J.T. Fraser, N. Lawrence, D. Park.

    First conference held in 1969 at Oberwolfach, Germany; 2d held in 1973 near Lake Yamanaka. Japan; 3d held in 1976 at Alpbach, Austria; 4th held in 1979 at Alpbach. Austria.

    Includes bibliographies.
    1. Time—Congresses. I. Fraser, Julius Thomas, 1923— ed. II. Haber, Francis C., ed. III. Müller, Gert Heinz, 1923— ed. IV. Lawrence, Nathaniel Morris, 1917— ed. V. Title.
QB209.155 1972   529   72-80472
ISBN-13:978-1-4612-5949-7

9 8 7 6 5 4 3 2 1

ISBN-13: 978-1-4612-5949-7   e-ISBN-13: 978-1-4612-5947-3
DOI: 10.1007/978-1-4612-5947-3

To the memory of

Georges Schaltenbrand

1897 – 1979

Opening Address*

Mr. Chairman, Ladies and Gentlemen:

Herewith I declare the Fourth Conference of the International Society for the Study of Time open.

Please accept my warmest welcome. I hope you may attain some feeling of timelessness in thinking about Time.

It is a pleasure for me to greet two past presidents who join us at our Conference, Professor Gerald J. Whitrow, Imperial College, London, and Professor David Park, Williams College, Williamstown, Massachusetts. Still another past president, Professor Satosi Watanabe, Hawaii, has sent a cable expressing his good wishes for the Conference; unfortunately, he was unable to come. Finally, let me mention the man who is directly responsible for our presence here. Without the untiring and energetic work of our Founder and Secretary, Dr. J.T. Fraser, neither the Society nor its Conferences and their published proceedings could have come into existence. It is not possible to express in words what his personal devotion to this task has meant; I may only mention the formulation of the general theme of this Conference, "Beginnings and Endings," the adaptation of it to the various areas of the sciences and the humanities here represented, and the daily care in handling the endless so-called "minor" questions of the organization. Indeed, these and a great many other burdens were taken over by both J.T. and Jane Fraser. We all owe them a great debt of gratitude.

May I add that during the three years since the last conference, Dr. Fraser and I could always rely on the continuous unselfish help and advice of Professors Nathaniel Lawrence and David Park (both of Williamstown, Massachusetts) as well as that of Professor Lewis Rowell (Bloomington, Indiana). I express to them my hearty thanks for their friendship and their readiness to build, and repair, bridges.

---

*Abbreviated version, partly revised by the Editors.

I have resolved to use this opening address to speak about the general character and the aims of our society rather than to report on work in my own area of research. That I choose to do so shows the strong difference between our society and other learned or scientific societies. Let us consider, for comparison, three typical examples of such societies: an astronomical society, a philosophical society, and the Royal Society, say at the beginning of the time of the Enlightenment. They may be compared in several important respects.

In an astronomical society the area of interest is given by rather well-defined objects and methods: it receives short specialized reports which contribute to the growth of knowledge. The common background knowledge is of physics, chemistry, and mathematics. Speculations are allowed and welcomed, but within fairly well-defined limits. Our society differs considerably from this picture. Speculations on the study of time have a much broader character.

In a philosophical society the area of interest is in principle much less restricted. Quite commonly the historically familiar themes, including those connected with Time together with problems of current society, predominate. Contributions are not necessarily measured with respect to new results, but rather with respect to widened insights. The common background knowledge is that of the historical culture, enriched by the vocabulary of our present society. Special interests tend, however, to develop special vocabularies. Speculations in the sense of grand philosophical views and pictures are normally avoided. When they are presented, they are often considered as crazy or as based on ignorance of facts in special fields.

Our society faces somewhat similar difficulties with respect to the need for common background knowledge, coupled with speculations of wide range. But it is by focusing our interest on one main category, Time, that we are able to restrict the vastness of our philosophical themes.

Here a comparison with the Royal Society or similar European academies two or two and a half centuries ago is useful. Here also the area of interest was restricted, as compared with philosophy in general. The emphasis was mostly on the exploration of Nature with minimal attention to religious and metaphysical ideas, although it was understood that in meetings of the Royal Society religious arguments were important. But speculation was the heart of the Royal Society's creative work. The principal difference between the Royal Society and our own is the amount of background knowledge assumed in our discussions. It is almost a biological fact that no single person today can know even the essentials of all branches of science and culture. This difference must be reflected in the work of our Society and must help to determine its aims.

Now, what are our aims? The decline of the belief in the success (or even the possibility) of one comprehensive philosophical system or of one system of metaphysics, coupled with what Dr. Fraser has called the "overwhelming amount of information reaching us through the sciences and a large variety of utterances coming to us through the humanities," call for reflection and action. Typically, the modern tendency is to "arrange" one's own life and work, and then to concentrate in small groups on certain shared ideas and interests.

But in some corners of the earth positive action is taken. Various interdisciplinary groups are surveying the ground for an effort to do what Aristotle did for his time, when he systematized an enormous amount of scattered information into a coherent picture of the world and man. He did this by discovering master categories applicable to all fields of the knowledge of his time and developing a logical structure of thought, both of them so broad that they were still usable twenty centuries later. The analogous challenge of tomorrow has, of course, quite different dimensions.

Our Society is one such group. Our category *Time* is rich enough to be endlessly challenging and yet not quite overwhelming. We are indeed an interdisciplinary society, but one which aims at achieving some "superdisciplinarity." Our long-term goal should be in the interest of a comprehensive and integrated understanding of time, in terms which will allow our conclusions to be adopted and synthesized without loss of meaning into still broader syntheses as they are undertaken.

Given this as an objective of our work, we have to take into consideration almost everything man has explored, thought about, done, felt, enjoyed, suffered. How to do this is the question. Of course, an encyclopaedia would not be of much help for us. Reading it would be tedious as well as useless.

One major problem is how to arrange our interdisciplinary conferences so that they will collectively lead to some progress without curbing the freedom we have customarily exercised in planning these conferences. The freedom we need is to find good speakers and researchers who can speak with professional competence on subjects central to our common interests. If there is to be collective progress, however, then we must also consider the long-term results of what we are doing.

We should think in terms of three classes of persons. I shall call them (1) Searchers, (2) Experts, and (3) Organizers.

(1) *Searchers.* These are scholars for whom a major consideration is a general understanding of time comprehensible enough to be adaptable to the special investigations of experts. Commonly such overarching studies arise among physicists and philosophers, but there is no reason why work of this sort should come only from these quarters.

(2) *Experts.* These scholars have as their primary function studies in the confines of a single or a very restricted number of fields. A physiologist might confine himself exclusively to the relation between brain synapses and recorded stimuli discriminations. A literary critic might study the changing concepts of time in literature during periods in the wake of scientific revolutions.

(3) *Organizers.* The function of these would be primarily to compose the program of our conferences with a view to mutual enrichment by the investigations of the various speakers. To some extent this has already been done, and each triennial meeting has had a special emphasis which has served as a focal point for that meeting. It is my opinion that we should consider doing more planning of this type, in the longer perspective of where we have done well and where we have done little, or not done it well enough.

It should be obvious that these three types of persons do not exclude one another. It would be better to describe three *functions*: (1) Generalizing, (2) Specializing, and (3) Organizing. Any of our membership should be, on principle, able at least to *assist* in performing any of these functions.

There are, of course, many problems that beset this sort of approach to our future. Those who organize a meeting will expect certain kinds of research and results from those who are functioning as experts; they may not be able—lacking expertise themselves—to anticipate the results. The experts in turn may find that what they have to offer must be considerably shaped by what is needed for the conference. The interplay between the two will require intelligent and considerate cooperation.

Expertise and generalization both require care in the speaker's choice of language. He must be able to make his argument clear to intelligent colleagues who do not know the vocabulary of his specialty. I call this "correctness on demand." In addition, the speaker must maintain a high level of professional responsibility without spending too much effort anticipating objections from other specialists. He is like an adventurer charged with the task of making his discoveries understandable to those at home. The compromise is a real challenge! Any speaker should understand, therefore, that to present a paper to a society such is ours is *more* difficult than to present the same results at a professional meeting.

The longer future, beyond any one particular conference of the society, requires that we think about harvesting the results of these conferences for further use. Since the appearance of G.J. Whitrow's well-known *The Natural Philosophy of Time* there has been, to my knowledge, only one consistent attempt to survey, and to use creatively, the tremendous amount of research that has accumulated in the study of Time. This is the work of J.T. Fraser, to which I shall return shortly. It should be the business of further conferences to encourage such overarching study, even though, when it is properly done, it must be on the scale of a book-length work.

Traditionally, such works are philosophical in nature, but they need not be. Indeed, since many fields of study have brought so much new light to the study of time, we would expect efforts at generalization to arise in many quarters, and indeed we should hope that this will be the case. This is especially important where special disciplines have themselves created new tools for analysis—and synthesis—that may be best used by those who have invented them.

Societies and conferences are not by themselves creative, but they are the medium through which we may reach a higher level of insight. Therefore, it seems to be in order to consider some types of conferences. Thus far we have held what seem to me *material-collecting* conferences. Certainly, at the outset, a society like ours might well begin in this fashion, but we should be able to distinguish these conferences by something other than the times at which they were held. We run the risk of a pleasant intellectual exercise, occurring every three or four years, without discernible step-wise development or coordinated coverage of major areas in the study of time.

What stands out most clearly in the last one hundred years of work in the natural sciences is that natural languages are not perfectly adapted to asking and answering the kinds of questions which must arise in any serious study of time. There has been great progress in the adaptation of mathematical methods to the comprehension of the physical universe, and we are not at all likely to ignore these results. We may expect that the natural sciences will continue to play a large part in the research of those who contribute to our meetings. No doubt this would put a greater strain on the problem of the humanities vs. the sciences. But we may hope that it would be a healthy strain, since there would be a primary challenge to the scientists to make themselves intelligible to the intellectual world at large, or to be suspected of being magicians or fakirs. We can avoid neither the sense of the "flow of time" nor its mathmatical counterpart as indicated in differential equations and stochastic mathematics.

In conclusion, I must point to the single fact that underlies the very existence of our society. Its founder and its operative soul, Dr. Fraser, has singlehandedly developed a remarkably universal view of the world and man, with the concept of Time placed in the central position.

In commenting upon this feat I want to make two points. (i) If, as in the early academies, we undertake a truly synthetic attempt at our understanding of man and world, our primary address must be toward making the humanistic and the scientific views of time responsible to one another. (ii) In this regard, I know of no other work comparable to that of Dr. Fraser. His devotion is admirable and lifelong. We can, of course, show precedents for some of his work, and find fault with other parts, but this is not a suitable approach.

What we should observe is that Dr. Fraser's work depends heavily on metaphysical and methodological principles, scientific and otherwise, which makes me think in terms of *principle-testing* conferences, wheree the main topic is the question of the values and functions of certain *procedures* of research in the study of time, contributed to and criticized by experts from various fields. I wait for an invitation to such a conference.

University of Heidelberg,   GERT B. MÜLLER
Federal Republic of   Fourth President
West Germany   International Society for the Study of Time

A Backward and a Forward Glance

*The Uses and Problems of the Study of Time**

The purpose of the International Society for the Study of Time (ISST) is to encourage the study of time along all lines of scientific and humanistic inquiry which its members see as appropriate. In accordance with the Society's Constitution, this object is achieved mainly by the organization of conferences where scholars and scientists may discuss their time-related work in the company of specialists from other fields.

With the appearance of this volume, the Society will have published over one-hundred selected and refereed papers from its four conferences, amounting to some 2200 printed pages. The cumulative index in this volume suggests the breadth of the members' interests.

This essay sketches the concerns which motivated and the principles that guided the Society's work during its fourteen years of existence. It describes some of the practical difficulties that were encountered and some of the ways in which solutions to those difficulties have been attempted. It concludes with recommendations about the direction of future development.

1. The Idea and the Project

The great variety of opinions about, and concerns with the nature of time that are voiced in these four volumes demonstrates the central significance of time in the intellectual and emotive affairs of Man. Indeed, temporal experience perhaps more

*I wish to express my appreciation to Professor Nathaniel Lawrence for his detailed critical review of this essay and to Professor David Park for his valuable comments both on content and form.

than any other aspect of existence is all-pervasive, intimate, and immediate. Life, death and time combine in a dialectical unity which may be hard to comprehend but which, nevertheless, is symbolically stated in all great religions. Time is an essential ingredient of all human knowledge, experience, and mode of expression; it is, somehow, intimately connected with the functions of the mind; it is the only dimension of our inner life and it appears to be a fundamental feature of the universe. No other aspect of reality seems to bear the same pertinence to the individual and collective concerns of man.

But a universally acceptable framework that could accommodate the multitude of views about time, one which could serve as a guide for critical studies, does not exist. It may never be possible to consider physical, biological, psychological, historical, literary, and philosophical notions of time under the same heading. Yet, a survey of the literature of time does not leave one with the impression of complete incoherence, but rather with a kind of feeling that researchers often have while examining unreduced data. Surely there is a design to be found; surely there are universal truths to be discovered; there must be a pattern hidden somewhere among the multiplicity of facts, utterances, and fiction.

Although a corporate opinion, such as the one I have just stated, was never formulated, the Society owes its 1966 inception to the labor of three men interested in and actively seeking a comprehensive understanding of time. The first edition of G. J. Whitrow's The Natural Philosophy of Time appeared five years earlier.[1] M. S. Watanabe began exploring the physical and philosophical aspects of time in a series of articles in the early 1930's; his first book of essays on the subject appeared in 1948.[2] The first edition of The Voices of Time, a cooperative exploration of the intellectual and empirical issues of a multidisciplinary study of time, was published in January, 1966.[3]

These three men and others who soon joined them—foremost among them, Professor Georges Schaltenbrand, to whose memory this volume is dedicated, Professor Gert H. Müller, fourth President of the Society, and William Gooddy, M.D., F.R.C.P.—held, so I believe, that the nature of time will remain hopelessly obscured until we learn to rely upon insights that satisfy man's rational understanding as well as his introspective knowledge of time, "as expressed by the sciences and the humanities," in the words of the subtitle of The Voices of Time.

There was no intent then, neither is there any now, to see the Society's work as leading to a collectively developed, universal theory of time, even could such a theory be formulated. Neither has there been a desire to seek a collective answer to the question, "What is time?" Creative work is better left to the individual, working alone. Also, "What is time?" may just be a shorthand for asking a multitude of questions, having the same form, concerning matter, life, man, and society. But it has been our hope that the volumes, as they accumulate, will supply a cumulative record of thought to which individuals, interested in various aspects of the study of time, or in identifying possible universals, will turn for material. There was and there is an intent, however, to use the concept of time as a means of building bridges between the many ways in which modern man encounters reality.

There was no serious precedent for organizing multidisciplinary conferences in the study of time having the scope envisaged for the Society. In a very real sense, the development of the Society, perhaps up to our own days, has been a research project. We did not intend to produce anthologies or an encyclopedia, or just to publish the records of symposia. We could not comment upon an established field of knowledge but had to discern, instead, the outlines of a possible new field of study. It was clear from the beginning that the contents of papers to be read at the meetings must be measured against the potential content of a not yet defined and delimited discipline.

If in those early days we had cared to list the purposes of the Society, it might have included the following:

- to encourage the search for new knowledge of time;
- to help decide which fields of knowledge may contribute to an increased understanding of time, and what they may contribute;
- to assist in the creation of an epistemology based on distinct views of time which were suspected to being inherent in various fields of knowing;
- to stimulate and maintain communication between the humanities and the sciences, using time as a common theme; and
- to learn about the nature of time by providing channels for a direct confrontation of multitudes of views.

With little effort, one could describe many more possible areas of inquiry, each worthy of a conference. However, we soon found that although it was possible to think of areas of inquiry which are interesting, practicality demanded that the Society proceed in the opposite fashion: select themes that will match the preparedness of those scholars and scientists who are qualified to address the issues *and* who can come to the meetings.

2. Departments, Dialects and Temperaments

Even before the founding of the ISST, while *The Voices of Time* was being written and edited, it became clear that interdisciplinary studies of time encounter certain difficulties created by the necessary compartmentalization of knowledge. These difficulties are not confined to methodology.

First, there is a language problem. Each field of knowing, be it clockmaking, genetics, or the comparative study of religions, has its own vocabulary which, in addition to words peculiar to the field, also gives particular meanings to otherwise common, garden-variety words. Each field uses certain stock phrases, often un-analyzed, and each has its preferred way of putting things.

Second, there are profound disagreements among professions regarding acceptable methods of reasoning. What by the standards of one profession is thought to be a correct and salutary argumentation may be regarded, according to the standards of another discipline, as useless or even reprehensible.

Third, there is a problem which may be called the personalities of knowledge. That personality traits are decisive in the choice of occupation is a truism. Individuals express their personalities in different ways by creating the principles and techniques of various branches of study. Knowledge so created bears the personality marks of its creators, including their views of time. The wisdom of the English language recognizes the differences of ambience that characterize groups of different species of animals. We speak of a flock of sheep, a pride of lions and a pod of seals. Likewise, a laboratory of physicists, a couch of psychoanalysts, a bevy of social scientists, and a museum of historians represent distinct preferences of temperamental ambience. It follows, and we have had many years of experience in confirming the truth of this claim, that the question, "What is time?" is not emotionally neutral. Individual judgments as to which domain of knowledge is best equipped to deal with this question tend to be dogmatic and non-negotiable.

Difficulties of language, reasoning, and temperament could be eliminated in one fell swoop if the ISST would organize conferences of specialists instead of multidisciplinary conferences. But this practice would negate the express purpose of the Society. Yet the pressure from specialist groups has been so strong that the committee, chaired by Professor David Park—third President of our Society—while drafting the Society's Constitution (adopted in 1974) felt it necessary to include in it the following provision:

> In preparing conference programs, the Conference Committee shall respect the Society's interdisciplinary character, and there shall be no conference of the Society devoted to discussions within a single intellectual discipline.

This provision of the Constitution has been implemented through the insistence of the organizers of the conference to hold only plenary sessions and eschew parallel sessions of restricted scope.

A glance at the roster of professional societies shows that the fragmentation of knowledge, demanded by specialization and amplified by increasing distinctness of language, mode of argument and temperament, is universal. The Royal Societies of the world, its National Academies, are all old foundations. The new ones have names like the American Academy of Microbiology. This fragmenting tendency persists at a time when an increasing number of people are calling for an end to it and demanding that more attention be paid to studies that seek to comprehend larger segments of human life and culture. Such studies are rare partly because of the parochialism that is inseparable from the necessary division of knowledge, but also because of the distrust of generalizations that are not controlled by an established discipline.

In this background of clearly contradictory demands, the study of time offered itself as a discipline broad enough in its concerns, yet subject to the control of recognized professional standards. Questions about time seem to provide the right degree of flexibility to relate diverse disciplines while respecting their integrities. Today, fourteen years after the Society was founded and six after its Constitution was adopted, we observe that members of the Society do find their own work, as well as their general understanding of its relation to human experience at large, advanced by exposure to people who study time from widely different points of view.

3. Working Assumptions

A usually unstated assumption beneath all collective intellectual work is that all who do it address the same or at least closely related issues. In the case of a multidisciplinary study of time, no such prior agreement necessarily exists. Whether or not such assumptions should be made and agreed upon has not been debated or even proposed at our four conferences.

It is clear that, since these conferences have been attended by over 400 scientists and scholars, there is an unspoken agreement on the validity of the proposition that the issues discussed do, in some fundamental way, relate to each other. The participants, as Lenin has once reputedly said, voted with their feet.

It is, nevertheless, appropriate to attempt to formulate such assumptions, if for no other reason than as a subject for debate. What may be regarded as necessary and probably sufficient assumption for the multidisciplinary study of time has been called the principle of the unity of time, and may be stated as follows:

1. When specialists speak of time, they speak of various aspects of the same entity.
2. This entity submits to study by the methods of the sciences; it can be made meaningful subject to contemplation by the reflective mind, and it can be used as proper material for intuitive interpretation by the creative artist.

One may object to the first assumption on the basis that time is not a well-defined notion but only an ordinary English word with certain semantic uses and specific forms, and it is used here only because our langauge is English. The reply is that one must begin deliberation somewhere, and it is assumed that all those coming to our conferences have some idea of what we mean by time.

The term "entity" does not imply a distinction between time as reality on the one hand and, on the other hand, a primitive category of existents called entities. It is used only to permit the description of relationships between the knower and the known—with details yet to be discovered.

The second assumption is a canon that authorizes the method of inquiry discussed earlier.

Reaching an ultimate understanding of the unity of existence has surely been one of the oldest dreams of the intellect. Such understanding has usually been sought through whatever mode of knowledge thinkers regarded as richest in meaning and broadest in scope. For the Greeks it was philosophy, for the Schoolmen theology, for us today it is science. Plato, Roger Bacon, and Giordano Bruno intended to account for the world in all its aspects, including that of time. But synthesizing approaches directed toward the creation of a unified view of time could hardly go back to periods which preceded the fragmentation of thought that accompanied the emergence of increasingly larger number of distinct fields of knowledge.

There are good reasons to believe that the radically increased interest in questions of time, as represented in the work of this Society and its members, is a reaction against that compartmentalization. The principle of the unity of time is an abstract way of seeking intellectual anchorage in a world of fragmented knowledge and incoherent values. In short, the collective work of the Society may be a response to a deep-seated social malaise.

4. Development and Experience

There is no need to record here the year-by-year travail of those who have directed this Society. But certain issues which characterize its first fourteen years, especially in connection with the organization of the conferences and the publication of the proceedings, should be identified.

From the very beginning the first and foremost organizational problem has been this: Who are the people to be invited to the conferences? How are the speakers to be selected?

Optimism among cloistered scholars and scientists is neither undesirable nor unknown. But it can easily lead to requests, impossible to honor. For instance, there has been a continuous demand placed upon the organizers of the conferences to invite "the very best people who will tell us all there is to know about time in. . . ." The bibliography appended to this volume beginning on page 234 may help the reader in completing the sentence. This request has been heard very many times. The usual and major difficulty in meeting such a reasonable demand is that people with the qualifications required, more often than not, do not exist.

It has been our intention, as already mentioned, that the proceedings of our meetings provide a record of continuing investigation, even if it lacks a common focus. Many people are interested in aspects of time and many have something clever to say, but contributions from people who have not worked with and studied the problems of time are unlikely to say anything that has not been said before, often much better, and sometimes a millennium or two ago. For this reason the Society has followed a rule, informal but strictly adhered to, that no one is invited to lecture who has not previously made contributions to his particular discipline, in a vein and on a theme which the organizers of the conference regard as qualitatively satisfactory and thematically relevant to the study of time.

Humanists as well as scientists, one would hope, follow their inner guiding lights, while remaining loyal to their particular fields of research. Naturally, they want to talk about their work; most people are not very willing to deviate substantially from their expertise. Nor would it be wise to ask them to do so. It has been our desire, therefore, to hear and publish arguments which can pass the criticism of the authors' peers. To the prior restrictions on whom to invite, we must then add this: Regardless of the qualifications or even the interest of a potential contributor, it would be ill-advised to solicit from him contributions on a theme with which he is not currently working.

Should an invited speaker be asked to clarify his terms for a multidisciplinary audience, when it lengthens his presentation and probably bores his colleagues? Should he be asked to do so in the written form of his paper?

Speakers must be permitted to use the kind of reasoning which they and their colleagues judge appropriate to their themes. Should everyone be asked, in addition to presenting his views, to defend his mode of reasoning? Will those participants who disapprove of the working methods of sociology withdraw disapprovingly while a sociologist holds forth about his concerns? Will potential readers of the proceedings who see in the contributions by physicists an exercise of interesting but useless abstractions underrate the volume if there are quite a few articles on physics?

Should the speakers be encouraged to speculate? If yes, to what extent? Are we interested in the most sensational, latest discoveries obtained by modern machinery and research or should we seek the lasting aspects of knowledge? If both, how are they to be apportioned?

By what argument can we face our critics who object that we have not had any mention of time and S-matrix theory? Of brain, music, and time? Time and the Aruba? Time, prophecy and extrasensory perception? Suppose they feel that the selection has been elitist leading to socially irrelevant programs? That too much stress has been placed upon social relevance and too little on the "factual" findings of the sciences? That there are too many papers or too few? That minorities are not represented? That we have had too few famous people?

Few of the participants can afford the travel expenses and room and board at the conferences. But to obtain support from foundations or universities, a participant usually has to show that he will be a speaker and is likely to publish a paper. This necessarily limits the number of people who can be invited, which, in its turn, might make our meetings too closed and supply the elitist-critic with reasonable fuel. For established professional societies such details are handled as matters of course, including the fact that very few of the papers read at conventions actually get published. Thus far, the ISST has chosen to publish the majority of the papers read at its meetings, but the continuity of this practice has in no way been assured.

We have stressed quality and not numbers. Membership, therefore, until after the Fourth Conference in 1979, was restricted to people who had been invited to read papers. In 1979 the Constitution was amended to allow the admission of Corresponding Members, including students.

Putting aside programming and membership, there are difficulties regarding the publication of papers. Since we are multidisciplinary, the editors of the proceedings had to cast about for many different pools of referees. It has been our practice to ask one senior reader to give us an opinion. When the opinion was negative, we have sought the views of a second reader. In case of a tie the decision had to rest with a third reader or with the editors themselves. Because of the many fields involved, the task has not been easy. The refereeing process has been further complicated by the fact that the arguments given earlier regarding the selection of speakers must also apply to the selection of referees. Again, for this and for some related reasons, Professor Park's Constitutional Committee judged it necessary to include in the Society's Constitution an Article which reads:

> Every member of the Society shall be ready to assist the Officers, Council, and the various subcommittees in discharging their duties to the Society.

On page xx there is a facsimile of the Instructions to Speakers for the preparation of their contributions to this volume. It speaks for itself.

INSTRUCTIONS TO SPEAKERS
for the preparation of their contributions to
The Study of Time IV

Contents. The contents should be intelligible to a readership of very diverse backgrounds. Specialized terms should be clearly defined and significant conclusions clearly stated. Whenever possible, mathematical development should be confined to appendices. Authors are encouraged to strike a balance between expertise and popular presentation. At the conference we talk to one another, but in many cases the author will wish to sharpen his paper for publication. The reader is to be envisaged as a sophisticated intellectual with publications of his own, but not ordinarily in the author's own special field of training.

The submission of the manuscript will be taken to imply that the material is original and has not been and will not be (unless first accepted for *The Study of Time IV*) submitted in equivalent form for publication elsewhere.

Length of manuscripts: 6000 to 7000 words. Publishing limitations make it necessary to hold the length of the papers, in general, to this figure.

Form and layout. Form, organization and referencing should conform to the professiona practices of leading journals in the author's field and must remain consistent within the paper References should be listed alphabetically at the end of the paper. Because of the multidisciplinary readership, journal titles should not be abbreviated. In the interest of accessibility to a wide range of readers, authors may wish to place technical arguments in brief appendices.

Every paper should be preceded by a summary not exceeding 300 words, intelligible without reference to the main text. It should not include abbreviations or references.

Manuscripts should be typewritten in double spacing throughout, including summary, footnotes and appendices. Footnotes and captions for illustrations should be typed separately at the end of the manuscript. A short running title of not more than 40 characters (including spaces) should be given if the full title is longer than that.

Illustrations, including graphs and tables, must be in camera-ready form. Each illustration, table or graph must be accompanied by a brief caption that explains its significance without reference to the main text.

Language. Papers must be received in acceptable literary English. Since the publishers employ no copy editors, this is entirely the responsibility of the author.

Tie-in with earlier proceedings. Volumes I, II, and III of *The Study of Time* contain almost 100 papers in all. Please consult them so as to prevent unnecessary duplication. Authors should refer to prior articles if they appear significant to their work, capitalizing on opportunities for continuity.

Refereeing. All papers will be refereed in accordance with the practices of professional journals. Delivery of the paper at the Conference does not automatically guarantee its publication. Accordingly, you are kindly requested to refer to your papers as "Submitted to *The Study of Time IV*" and not as "Forthcoming in *The Study of Time IV*" until you are advised of the acceptance of your paper.

Submission. Manuscripts *in triplicate* are due in the office of the Secretary not later than October 15, 1979. In case a manuscript is found unsuitable for the forthcoming volume, only one of the copies will be returned. Since we wish to make *The Study of Time IV* available without undue delay, this deadline will be strictly kept. Publication of manuscripts received after this date cannot be guaranteed.

5. Boundaries and Promises

Following the directives of the Council of the Society, a goal of our fifth conference will be that of introducing increased coherence into our discussions while maintaining the multidisciplinary character of our work. Specifically, we would like to articulate the concrete, cultural implications of distinct conceptions of time in our epoch.

To comply with this directive, the theme of the fifth conference has been selected as *Science, Time, and Society in China and the West*. We hope to hold this conference in cooperation with colleagues from the People's Republic of China.

According to preliminary plans, the conference will be comprised of three sets of paper. The first set will focus on the conception of time that can be abstracted from the study of the social and political institutions of the West. The second set will focus on the social and political strains that are manifest in countries which attempted to superpose, uncritically, Judeo-Christian concepts of time upon a different tradition. The third set of papers, and this will be the main portion of the conference, will focus entirely on Chinese concepts of time, and relate them to the tasks of the great modernization program of the People's Republic of China.

These plans, if carried out as conceived, will give the Society an opportunity to demonstrate one of its strengths: a unique and tested know-how for developing cross-disciplinary and cross-cultural discussions in an atmosphere of professional criteria and expertise.

Since the existence of the Society has become more widely known, there has been a continuous flow of letters addressed to it. They come from around the world, from people in all walks of life, belonging to different cultures. About one half of the mail relates to the Society's business as it is sketched in this introduction. The other half of the mail offers unusual interpretations of all kinds of phenomena, relating them to time; many amount to invitations to join a cause; others, and these are the most moving ones, offer simple insights into the loneliness of man.

Some letters arrive without the sender's name and address as if their writers were afraid of being ignored. A number of them contain technical suggestions for time machines. Some have drawings that suggest drug use or unbearable anxiety. Relativity Theory seems to be a favorite subject of many people lost in the moral relativism of our epoch. Some letters come from blind people reaching out for understanding and company. Some letters are in the writing of frightened old men, some in the hands of tense young women.

These strange particles demonstrate the power of the idea and experience of time in the human condition. And they illustrate the difficulty of giving a simple answer to the question, "What do we study when we study time?"

The suspicion had often been voiced that fascination with the study of time has its roots in the insight which it provides into man's destiny or, perhaps, into the absence of destiny. Therefore, even the most abstract scientific analysis of time harks back to fundamental issues relating to the existence of man.

Perhaps what a meaningful multidisciplinary study of time demands is an intellectual climate, rather than simply a new method of argument. Such a climate should permit creativity, common to all knowledge, to flourish, and aspects of reality previously separately understood, to produce their synthesis, by interacting through the idea of time. The task is difficult, but human aspirations seem to call for it. For, as in the quiet countryside of Robert Frost's New Hampshire so also in the countryside of the intellect,

> Some thing there is that doesn't love a wall,
> That sends the frozen-ground-swell under it,
> And spills the upper boulders in the sun.

J.T. FRASER,
Founder
International Society for the Study of Time

Notes

1. *The Natural Philosophy of Time,* second edition, Oxford: Clarendon Press, 1980.
2. Also, *The History of Time, seen through Physical Science,* Tokyo: Tosho, 1973 and *Time,* Tokyo: Kawade Shoboh Shinsha, 1974. (Both in Japanese.)
3. *The Voices of Time,* second edition, Amherst: University of Massachusetts Press, 1981. With a new introduction, "Toward an Integrated Understanding of Time."

Contents

Frontispiece

The poster (shown on page ii) is that of an exhibition, held at the Museum of Technology in Warsaw, Poland, between February 12 and June 20, 1979. Called *The Anatomy of Time*, it was organized by a small community of scientists and scholars devoted to the study of time.

The exhibit was attended by an estimated total of 100,000 visitors. A telegram of congratulations was received on opening day from the International Society for the Study of Time. It was received as a sign of world solidarity among those, anywhere, concerned with the role of time in the collective and individual affairs of man.

The intent of the exhibit was to encourage intellectual concern with time via the visual message of the displays. Captions were held to a minimum. Instead, the intricate beauty and the significance of the devices themselves were emphasized by the way they were displayed.

The newest items were instruments produced by the Polish precision industry of our own days. The oldest devices included an Arabic astrolabe from the 11th century, on loan from the Jagiellonian University in Cracow, and several sundials from the Przypkowski Museum in Jedrzejow, co-organizer of the exhibit.

The national press was generous in reporting in words and pictures the details of the exhibit. Talk shows on radio and television emphasized the social aspects of time: work, leisure, and the significance of related technology.

Simultaneously with the exhibit, we held lectures and discussions, and also convened a meeting of the Polish Chronosophical Society. It was its seventh meeting since its founding in 1974.

Perhaps the most popular item on exhibit was a real, live, rooster in a cage, near the entrance. It was lively, crowed seldom, but was not at all nervous. It lent authenticity to its mechanical counterpart on the poster and reminded people that there is much in the study of time about which even the most perfect clock can say nothing.

Warsaw   WALDEMAR VOISE

The Study of Time IV

My Time Is Your Time

N. Lawrence

This society was founded 13 years ago and was convened for the first time 10 years ago. This year we are meeting for the fourth time. I have been asked to speak from a background of a brief review of those meetings and to point my own remarks toward prospects for the future. I shall begin with general observations about those papers, indicating the areas in the study of time where we have made some progress, and giving a sampling of other typical areas where we have done little work.

I shall then turn to a problem of which we are all aware, how we might work toward a common understanding of time, while preserving the perspectives of our separate disciplines. But first let us turn to the past.

Our papers fall naturally into two classifications. I shall call them the "technology of time" and the "general theory of time." The technology of time is infinitely fascinating; without it any generalizations about time become comic or shallow. These papers have delighted me most. The general theory of time is more demanding. However, it is, after all, what we hope to contribute to, as members of the same society. We sense a common cause, even if modesty prevents us from going beyond our own domains. For myself, as a teacher of philosophy, the general theory of time makes the highest challenge. The pleasure of technical reading only partly satisfies that challenge. As to modesty, philosophy—as Winston Churchill said of Clement Atlee—has a great deal to be modest about; yet it must persist in the immodesty of its search, while being very modest about its achievements. Nothing I say today will alleviate that modesty.

The technology of time has flourished, though unevenly, in our hands. Past proceedings show how such things as social status and economic level make a vital difference in what we mean by "past," "present," "future"—those words of time which we like to think we share in a uniform way. Again, we find that escapement mechanisms in clocks lead inexorably to modes of industry and economy which they either control or express, as you will. Still further afield we find that least perceptible sensory data are not of the same temporal size, but vary with the sense organ in question. In addition, all of these are at great variance with the least expanse of conceptual time which is meaningful for, say, a physicist or a politician. These are but a few examples of a handful of conclusions to be found or surmised, by even a casual skimming of our proceedings.

Yet some areas of the technology of time are remarkably missing. We are all but barren in the field of law, for instance. In Anglo-American law I can become the new owner of your land or a portion of it by merely occupying it without complaint or action on your part for X number of years. A question arises: What determines the value of "X," i.e.; what is significant about 17 years rather than 16 years and 364 days? Here is a quantum jump in time not dreamed of in microphysics. There is an analogous problem in criminal law: some crimes, generally excluding capital ones, are said to "date." This means that one cannot be prosecuted for them after a certain period. Generally, the amount of time is a rough measure of the seriousness of the crime. Why this proportion and why these times, rather than others?

We are light also on economics. Economic laws are like evolutionary ones, reasonable after the fact but rather weak as to predictive power. Why? What is prediction anyhow? Once again, as with the law, the technology of time leads on to more general questions extending beyond the given field.

We have, in addition, had surprisingly little to say about the language of time. Some sort of general understanding of time is required to undergird its various technologies. But what language should be studied? Time is expressed in various Indo-European languages in much the same way, although the subtleties of any given verb system are rarely mastered by a non-native speaker. But suppose the Indo-European speaker studies Japanese. He finds to his surprise that the Japanese verbal, so called, inflects for time, but not in a way that meshes with western modes; it also inflects for attitude and courtesy. Furthermore, the Japanese adjectival as well as the verbal conjugates for time. Here is a language in which the cardinality and ordinality of time—so taken for granted in Anglo-European thought—are complemented by qualitative features of tense as well. We are accordingly warned, therefore, to search our own language for non-mathematical features of time. Furthermore, there arises a serious dilemma: In what language shall we assess the worth of one language's temporal modes as compared with those of another language? There may be no such language. We may have to turn to extralinguistic sources for our understanding of time, even though our explanations of, and communications about, time are largely dependent upon language. At the end of my remarks today, I shall propose that not all factors in our comprehension of time depend upon language for their demonstrability. In fact, they must not.

In summary of what we have done as a group so far, the proceedings of this society exhibit a wide variety of approaches to the subject of time, both in the technology of time and in the general theory. In the technology of time, we have several large unexplored areas. I have identified three: law, economics, and language. I have suggested that the nature of the language we use in the presentation of our thinking about time raises the question of how we may transcend—in some degree—the provincialities of language. For the rest of this paper I propose to consider a somewhat smaller problem, how we can conciliate conflicting claims about a general theory of time within a single language or family of languages.

Time is both unique and multifaceted. The uniqueness of time restricts what we say of it more than we commonly recognize. If time is a kind unto itself, then it is not a kind of anything else. There is no genus of things of which time is a species. Anything which we bring forward to illuminate the nature of time, therefore, is likely to require analogy, metaphor, etc. Any claim that time is literally a process, a thing, an idea, a relation, a motion, or—worse—that it is a fire, a river, a dream, simply won't do. With cautious charity we can allow that such talk serves to call up this, that, or the other aspect of time, but does not, in and of itself, cover enough, even figuratively. The uniqueness of time leads us to its multifaceted character.

We should not be dismayed by the multifaceted nature of time, save where some enthusiast claims universality for his approach. Consider a maple leaf, for example. The would-be minor poet tells you that it is a harbinger of spring, a symbol of the resurgence of life. The biologist tells you it's the most marvelous little starch machine we know of, and the nothing-but physicist insists it's really just all molecules, composed of one can hardly tell what kinds of uninvented or undiscovered baryons, leptons or quarks. It is only the exclusivity that gets in the way of understanding. We are willing to take the maple leaf as all of these. Wider relations are always there. There are other symbols in nature, richly explored in Zen paintings, for example. The leaf is not a starch machine by itself; it functions that way in a vast ecology of immediate and remote environments. The molecules, moreover, are just abstractions from an equally vast physical aspect of the world whose vocabulary of properties, such as solidity, color, shape, and sound, are not to be found in the language of microphysics at all, save as correlated with, but not explained by, the wave-particles which microphysics studies.

If such diversity is possible over a little maple leaf, we may expect a great deal of diversity about the nature of time. I shall choose two major types of thinking about time to illustrate this diversity. I choose them because they often appear in that tempting universal form that says or implies that other views correspond to a lesser, maybe even a misleading, "reality."

The first view of time is as somehow the creature of, or the inseparable companion of, self-hood, consciousness. This view can be traced back at least as far as Aristotle, who said a great many other things about time also. Aristotle puts it quite simply. He says we may "fairly" ask: if psyche did not exist, would time exist, for if there is no psyche "to count, there cannot be anything that is counted . . ."? But if nothing but psyche, or the rational part of psyche, "is qualified to count, there would not be time unless there were psyche [soul]."[1] It's possible that Aristotle left the way open for "psyche" to refer not just to individual psyches, but to an overarching world-psyche. The passage is liable to extensive debate, but what is important is the reasonableness of not being able to think of time save in connection with the self, the soul, the psyche. This notion was inflated to Brobdignagian size by Hegel, as we shall shortly see. He specifically generalizes it to a world-psyche.

The second view of time is about as far from the first as one can imagine. Instead of making it intimately connected with self and mind, this view treats time as utterly static, a mere mathematical dimension, remarkably like that of space. Extension in this view, has two forms, one is three-fold—namely space, the other unitary— namely time. This so-called "block universe" has no inherent properties of before and after. Counting is a subjective enumeration of what is timelessly simply *there*,

there is no change; things are not coming into existence and passing out of existence. One proponent says, "The objective world doesn't happen, it simply *is.*" We shall return to him in a moment. This view also has a principal root in the past, in Parmenides' denial of change. Parmenides rejects change on logical grounds, and with it time. I shall not comment on his reasons, save to say that for him, time without change didn't make sense, and thus, if change is illusory, so is time. The modern block universe saves time from banishment by changing it into a space-like dimension, much as Zeus might save a threatened wood-nymph by changing her into a statue.

Now, the interesting thing is that both of these views of time place some strain on common sense and, moreover, both can be found, in rough form, in the minds of mentally disturbed patients. Before proceeding to the rationales underlying the modern forms of these two views we shall look at our mental patients.

First, let us listen to a schizophrenic. He is haunted by the sense of changelessness. What might be, for a philosopher or a physicist an intellectual clarification, maybe even a triumph, is for him an emotional collapse. "Everything around me," he says, "is immobile. Things appear isolated, each one in itself, without suggesting anything. . . . There is an absolute fixity around me."[2] This case is from the well-known non-Freudian psychoanalyst, Eugene Minkowski. In his work on the psychopathology of time, he calls this syndrome "morbid geometrization" and "spatial thought."

Here is another schizophrenic, at least superficially less troubled. "I look for immobility," he says. "I tend toward repose and immobilization." On this man's view, "Everything in life, even sexual sensations, are [sic] reducible to mathematics." (LT 278, 279) A third patient says, "In reality there is no time. . . . I continue now to live in eternity; there are no more hours or days or nights. . . . Time is immobile." (LT 285, 286) Such temporal thought, says Franz Fischer, "appears more and more saturated with internal space," as the disease worsens. (LT, 275)

The transformation of time into space (or the immobilization of time) is not a feature of all sorts of dementia, however. Some paretic patients tend to temporalize space. Spatial questions get spatial answers but in terms of time. "Where have you been?" elicits the response, "Where I was before." "Where are you?" is answered with "I have been here since morning." "What are you doing?" a paretic was asked. "I am waiting for things to happen, and I am making plans." (*Ibid.*) Paretics—and this was in 1930—talk of automobiles that go 800 miles an hour and claim that their children age ten years on a day. (*Ibid.*) This is a caricature of Heraclitus, Parmenides' near contemporary and arch rival. All things change; everything is on fire. It is a pardonable exaggeration to say that Plato's metaphysics is an effort to conciliate the views of these two philosophers.

Parmenides' modern descendant in physics is another Minkowski, Herman Minkowski. Let us leave the "morbid geometrism" of the schizophrenic and turn to the mathematical geometrism of the physicist. H. Minkowski presents us with a block universe, a four-dimensional map, where the fourth dimension is time. As Minkowski says of what he calls the "postulate of the absolute world,"[3] it features the identical treatment of the four coordinates x, y, z, and t. A four-dimensional map, of course, cannot be truly produced in a visual image. Its true image is an algebraic one, rather than the somewhat misleading "light cones" which H. Minkowski uses as visual illustration. Time is taken in its measurable aspect only, stated in mathematical form. In this expression past, present, and future events have equal status. The spatio-temporal distance between any two events can be given in a four-variable expression, as can the extent of the events themselves. There is no need to superadd any passage of time or any account of change. What, then, are we to say of apparent change, apparent lapse of time? The classical answer to that question was given by the mathematician, Hermann Weyl. In an often quoted remark, cited above, Weyl says, "The objective world simply *is*, it does not happen. Only to the gaze of my consciousness, crawling upward along the life line of my body, does a section of this world come to life as a fleeting image in space which continually changes in time. . . ." "The latter world," Weyl explains, "is a subjective one."[4] Like many another author I have commented in print on this strange assertion. Its principal weakness is that it invokes a remarkably Cartesian split between subjective and objective worlds, without explaining in what sense they are the same world, and without identifying which world one is in, when he observes that there are two such worlds. The difficulty is a bit exacerbated by what is meant by "absolute." Minkowski speaks of the world as seen through the relativity postulate as the "absolute world." (op. cit., *ibid.*) Weyl, however, says of immediate experience with its data of change, that it is "subjective and absolute." (op. cit., *ibid.*)

It is true that H. Minkowski says that space and time by themselves are destined to fade away, (p. 75) but they fade away a lot more in the direction of space, which is not commonly regarded as mobile, rather than of time, which is usually associated with motion and change. As H. Minkowski himself says, "This geometrical treatment announces the following 'fundamental axiom': The substance of any world-point may always, with the appropriate determination of space and time, be looked upon as at rest." (p. 80)

In this comparison between the changeless time of relativity physics and that of the schizophrenic, I have emphasized the similarities: the fixity of time, the illusoriness of mobility, the suspicion that time in its most familiar form doesn't really exist* objectively, the likening of time to space, and the closely related restriction of it to mathematical measurability. Directly before one of his pendular swings, our schizophrenic says, "I seemed to myself to be a timeless being, perfectly clear and limpid as far as the relations of the soul are concerned, as if it could see its own depths. Like a mathematical formula." (p. 287) I suppress here the important differences until we have taken a look at the other extreme view of time, as primarily given in intimate association with self and consciousness. For this view I choose a monstrous quotation from Hegel, cited incidentally at the very first convening of this society.[5] "Time is the concept itself in its existence, as it presents itself to consciousness as an empty intuition; that is the reason the Geist appears of necessity in time and will appear in time as long as it has not conceived the pure concept of itself, i.e., as long as it has not abolished time." Time, says Hegel, is "the pure Self, seen from the outside but not yet conceived by the Self." The concept, by conceiving itself, "abolishes its time form." Hence, "time appears as the fate and necessity of the Geist that has not yet found its fulfillment in itself."

I shall undertake to explicate and unpack this wild-sounding passage.

Hegel, like the Stoics before him, and following a tradition as old as Plato, believed that, since human mentality can comprehend nature rationally, there is a fundamental rationale belonging both to the order of nature and to the human reason which understands that order. This "logos," as the Stoics called it, is immanent in nature and in man, but it is itself non-personal, pervasive, a spiritual presence giving life to all matter. It is divine, and human reason, sharing in it, shares a role in its existence. To live according to nature is thus a proper and intelligent act of worship. One must exercise self-discipline and a transcendent perspective to achieve the proper passivity before the totality of the Logos. There is also a proper *activity*, namely, the use of one's freedom, not futilely to oppose the divine order, but with wisdom to model one's microcosmic self on the macrocosmic Being, the Logos. For the stoics, the life of philosophy is an ethical life, founded on a metaphysical and religious persuasion.

For Hegel, however, metaphysics dominates. Pure matter, in his view, does not exist. Matter exhibits a rational order and is, insofar, invested with a kind of primitive mentality. But there is no such thing as an actual, pure, and separate mentality. Mentality requires materiality in order to be embodied. The simple rationale of nature, with its two necessary poles, mind and matter, is not the limit of reason, however. For Hegel, mind in mere material nature is only unconscious reason. Mind, however, seeks consciousness of nature and hence, ultimately, consciousness of itself. This is the primary quest of mind conceived in its role as Geist, spirit. The activity of spirit is not confined to the mere mechanical repetition of otherwise inert matter obeying physical law. Spirit is more than the rationale of the world discoverable through physical science. Its higher form is its embodiment in living matter. Here, rationality is more complex; it appears not only in physical reasonableness, but also in the end-directedness of living things. In living things the very parts of a creature cannot be understood as the outcome of mere physical law; they can only be understood for what they are in *functional terms*, their interdependency and inter-commitment to one another. This "for-ness" of parts for one another reaches its highest form in the human social commitment of person to person in society, through which consciousness-in-general comes to self-consciousness. The rule is simple: no society, no humanity; no humanity, no way for the spirit fully to become conscious of itself. The evolution of society is thus the evidence of a maturation of Geist seen as overarching consciousness—"*Bewusstsein überhaupt*". As with the Stoic Logos, the *Bewusstsein überhaupt* manifests itself in many individual consciousnesses, but outruns these personal embodiments, both in scope and in time. So time, seen massively from a transcendent perspective, is the concrete advance of the self-realization of Geist. It follows that for Hegel, real time is to be seen through the eyes of history, not physics. Physics is completely sterile as to the meaning of time. Ironically, physics can give no account of the very human rationality which treats the universe exclusively as a physical system.

So much, then, for Hegel. I've been coarse-grained in my highly personalized adaptation of his comments to our problem. I make no apologies; Hegel's thought flourishes at a high level of generalization. It requires cables to hold him to earth. He is important in another way, however, namely, in his influence on modern phenomenology and existentialism. I quote from Maurice Merleau-Ponty an assumption which is more tested by use than it is proved by argument. Merleau-Ponty, stressing the indissoluble mental pole to our experience, says, "We must not therefore wonder whether we really perceive a world, we must instead say: the world is what we perceive."[6] This prefatory remark, published in 1945, follows by exactly twenty-five years A. N. Whitehead's remark about 'apparent nature,' "We may drop the term 'apparent'; for there is but one nature, namely the nature which is before us in perceptual knowledge."[7] And it is astonishingly like a remark from Herman Minkowski's essay on the "block universe": "Not to have a yawning void anywhere, we will imagine everywhere and everywhen there is something perceptible." (op. cit, 76) The differences among the three men, of course, are in what is meant by "perception."

In Merleau-Ponty it is the dimensions of the perceiver which dominate, and the perceiver as a consciousness, a self. Late in a difficult book he says, "We hold time in its entirety, and we are present to ourselves because we are present to the world. . . . Time is the affecting of self by self. . . . It is of the essence of time to be not only actual time, or time which flows, but also time which is aware of itself . . . we are the upsurge of time." (424−426) He continues with, "I am not the creator of time . . . I am not the initiator of the process of temporalization; I did not choose to come into the world, yet once I am born, time flows through me, whatever I do. . . . we are the upsurge of time." (427, 428) He goes on to explain that we are nonetheless not merely passive before time, but active within it, both in being the authors of action in time, and in forming concepts of time which require us in some degree to distance ourselves from the mere present flux.

Even if Hegel and Merleau-Ponty seem equally obscure to you, you will see that Hegel's presentation of time as the condition of consciousness emphasizes consciousness generally; Merleau-Ponty's analysis stresses more clearly the role and status of individual consciousness. The touch of French individualism as to what constitutes a self is not surprising. Moreover, for Merleau-Ponty history does not create men; men create history. In this, too, certain aspects of time, however massive it is, must be traced back to individual consciousness. Here Merleau-Ponty meets common sense at least half-way. He directs us to our richest acquaintance with actual time, rather than abstract versions or reconstructions of time. Our lives are immersed in historical time; and we can hardly discuss history, even if we reduce it to a chronicle of events, without intimate and essential experience of consciousness.

I must confess that I am rather drawn to this central thesis. If I begin with time in its fullest complexity and in its most immediate presentation, consciousness and history are indissoluble aspects of it. We are still at liberty to abstract these troublesome aspects away, to see what is left of a cleaner, simpler, neater version of time conceived mathematically. But, should I start out with this version of time, as being the real, primary time, I can only get back to the richer temporality of consciousness and history by arbitrarily stuffing it with all those things, other than the physical order as such, which belong to lived time in the first place. Yet these things are not just contents of a skeletal framework supplied by physics. They have time elements of their own that the cardination or ordination of events can never explain. Most obvious among these is the astonishing phenomenon of memory, the foundation and abiding stockpile of things past and yet strangely present. I am unable to say consciousness has no vital connection with time, when I dwell on the operations of memory.

Yet for all this, what a strange world! For as you stare again, the physical world, in Merleau-Ponty's account, seems to have lost its autonomy. Put aside physics for the moment. What of prehistoric time, pre-human time, pre-living time? One has the uncomfortable feeling that they are taken as human constructs, ingenious, intricate, explicative, but ultimately rooted firmly within consciousness, single or collective. All this is good counterthesis for the block universe, but is it an adequate account by itself?

Now let's turn back to our clinically ill mental patients. The last schizophrenic we cited does not always live within the frozen time so reminiscent of Herman Minkowski. The very word "schizophrenia" suggests a patient who has two personalities. Hear him as he closes in on the other end of his pendulum swing: "I am stopped; I am projected from behind into the past. . . . There is no more present, there is only a going-backwards. . . . The past is so bothersome; it drowns me; it draws me backwards. . . . the past arose before me in a particularly vehement way. . . . Everyone finds himself—isn't it true?—in his own self." (287, 288) Here are the shades of Hegel and Merleau-Ponty; and then the patient, *in extremis*, obligingly adds the final touch, near-identification of self with time: "If there were no clock on the wall I would perish. Aren't I a clock myself . . . as I am a clock myself, everywhere in me things go pell-mell. I am all that myself. It is a flight, a way of getting away from yourself. I am ephemeral, and I am not here. . . . As I said, I am the living clock, I am a clock everywhere, it continually comes and goes." (288) The salient features of this outpouring are "Everyone finds himself in his own [past] self," the identification of the self with time as a "living clock," and finally the clock itself takes over: "it," says the patient (not *I*) "continually comes and goes."

Here we need, however, to notice the difference between the two kinds of theories—time as dimension and time as the field of consciousness—and their psychopathic counterparts. It is the *helplessness* of the psychopath, his *passivity*, his *inadequacy*. Whatever the block universe presents in a fixed universe, Minkowski gives us things to do, measurements to take, observations to make. Hermann Weyl is explicit about the subjective consciousness crawling up its own life line. This is a mixed metaphor impossible to conceive very clearly; nonetheless, there is a consciousness making its way through a world of change. Whatever the problems of the identification of time with consciousness, Hegel's individual person at least may participate in the advance of the World Spirit, much as the Stoic could share a rational life with the Logos. Merleau-Ponty's self can be active as well as passive in the temporality with which it is somewhat fused and in which it finds and develops itself, through both thought and action. The poor schizophrenic, however, is helpless in his world; he seems strapped to his chair while time is frozen into the immobility of the early relativity theoreticians or he is just as suddenly jammed into pell-mell chaos where all he can be is the crazy clock in a jumbled world whose tilt control has been broken.

We should observe one other compound difference of the schizophrenic from his scientific and philosophic counterparts. The first part of this difference is that the schizophrenic is unsteady. He may swing from one position to the other, and he never quite goes all the way: he says he "tends"' to immobility, and he has a "tendency" to immobilize life around him. He remembers that there used to be a future, etc. Only occasionally does he say things like "There is no time" or "I am a living clock."

There is a clear contrast between our two opposed theoreticians, on the one hand, and our schizophrenic, on the other: the mental patient is more fluctuating, more uncertain, less logically persistent. But above all, the patient cannot act. He is trapped; strictly he cannot *do*. Even the paretic who is "making plans" is not actually doing anything. The reason is clear. The mental patient's statements are not theories; they are cries for help, outbursts of desperation. He is trying to communicate and, at the same time, is preoccupied with his own existence. Moreover, his memory shows at least some recollection of what a more normal world looked like before it went away, or as it is seen from a great and melancholy distance—hence his anguish; hence his sense of distortion. His statements are reports, descriptions of his own emotive states, not primarily efforts to describe reality at all.

It is this contrast between action and thought that will give us some way of accommodating a diversity of approaches to time, without losing sight of its singularity and its unity. I shall borrow a distinction from Bergson, who finds in consciousness two closely connected aspects of awareness. Consciousness includes not only intellect, but something else he calls intuition. Intellect, says Bergson, has a natural affinity for matter. It looks for similarities, uniformities, likenesses, in order to group, to classify, and to form law-like accounts of the order of nature. All falling bodies obey the same law. All expanding gases follow another one. Every proton is a xerox of every other proton, as exchangeable as parts in a mechanical assembly line. On the other hand, intuition, which Bergson calls "disinterested instinct," is just the opposite. It clings to what is unique, particular, singular, once for all, never to be quite exactly duplicated again. The natural affinity of intuition is for life, obviously not life as watched, clipboard in hand, but as it is lived. We are reminded of the "morbid geometrism" of Eugene Minkowski, but even more of the geometric presentation of time in Herman Minkowski, when Bergson says that intellect triumphs in geometry, wherein "the kinship of logical thought with unorganized matter" is revealed. By "unorganized" he means matter without respect to organism. Living things are transient, particular, time structured. Non-living matter abides, unchanging. Conservation of matter and conservation of energy are natural ideals in physics.

There is a mistake in supposing, however, that this is the physicist's world and it's just another and valid way of looking at the world. Not so; the physicist needs passing time and the world of change as much as anyone else. Up on the magic mountaintop, where there are few theoreticians, we can indulge in a quieted and charted universe of no essential temporal movement. There are other good physicists, however, on the lower slopes of the mountain, in the experimental vineyard, preparing the wine for the dreamers. For them time has a preferred direction which doesn't depend on any formal deducibility from the Second Law at all. These physicists know all about the preferential direction and always have. They must treat time as an ordinal series, so that what works now has worked before and will do so later. What is said to be true—namely some physical law—without regard to time, means simply "It works—i.e., is true—*every* time." Theories that have no inherent connection, both for testability and for meaning, with the world of action, are uncomfortably close to the rootless world of the mental patient. Verifiability and falsifiability are essential parts of a scientific theory. But verification always leads us back to consciousness. It would be odd if the sequential establishment of scientific theory through changing time presupposed a false understanding of time, while the product of this striving turned out to be true.

The temporal act of confirmation has got to be as real as what is confirmed. The theory confirmed scientifically does not validate the experimenter, but only what he thought. The converse is true.. The experimenter validates the theory. Actually, the idealized time of either philosophical or physical theory is at the far end of a loop that begins with observations and acts, and returns to observations and acts. The vision at the far end of the loop shifts and changes, but the reality of how it was sought for and what it comes back to do not. In this sense lived time has priority, systematic priority. Every observation that leads us to temporality observed is filtered first through the lived time of an observer; and only then can it be laid down in an orderly way in the conserving intellect. But, on the other hand, the immersion in lived time—our chronic habit—gives us little perspective. Moreover, dwelling on the strong bond between consciousness and time may blind us to those autonomous features of the world which perhaps enter our lives only by permission of our consciousness, but come stamped with an authority beyond ourselves. It is risky to treat such externality under the strict metaphor of consciousness. How would we confirm that supposition? At the other extreme, in regard to the block-universe geometrism, it should be understood that because we strive for a fixed idea of time, it does not follow that time must be fixed. That would be to beg the question in favor of intellect, as if there were no communication about or understanding of what Bergson broadly calls "intuition."

When we ask, "Which is the real time, the unchanging one or the changing one?" we are left with that philosophically familiar and annoying answer, "Both." Our task, as I see it, is not to try to eliminate one in favor of the other. It is the direct opposite, to show their lines of connection, above all to remove from our discussion the use of vocabulary and syntax so theory-laden that our view of time is really contained implicitly in our premises. In previous remarks before this society I have supported the idea that language is rich with temporal metaphors and that we need a great plenty of such metaphors, taken together, to avoid the partiality of any one of them. It is not that time is merely a matter of language or even primarily so. However, the sprawling, only semi-orderly character of language permits the conservation of a wider variety of insights into time than what a highly schematic system can tolerate without scandal.

Even dealing with this problem in terms of time's movement or the lack of it, as I have done in this paper, is far too simple. The foregoing dissection has been undertaken only in a gross outline and would profit from refinement. However, I bring these remarks to a close with a summary and a suggestion.

I have urged that the time of action and the time of thought cannot get along without one another. For those who wish to identify time and consciousness too closely, there is a clear warning: Consciousness does not have its own way with time—not mine or yours, anyhow. So to treat the transindividual aspects of the world and its time under the rubric of consciousness is adventurous and maybe a little blinding, as well. On the other hand, for those who wish to chart the universe in four dimensions, of which time is one, there is a warning also: the foundations of your theory—which is only one of the recent ones—rest firmly in the time of action and experiment. This time is directional and dynamic, and—fascinatingly—less escapable than theories that emerge from it, we might say indeed from time to time.

Both types of thinking have their limits and the contest between them by ·no means establishes the boundaries of our interest in time. So I wish to take a half-step beyond that contest. First, they are both types of thought, as we have considered. Let us then try a fancy. Suppose that the universe's inner core is indeed revealed in some measure to thought, but that it is equally given to will. The thought of our mental patients showed a marked resemblance to various theories of time and self, as long as it was merely thought, but the will was crippled or withered in the case of the mental patients. The activities of observation and experiment undergird any scientific theory or any philosophical one, however, abstract or misty. Yet, if the universe is more like will, isn't that being pretty anthropomorphic? The answer is, no more so than when we make it like thought. Indeed, when is anthropos not anthropomorphic? Why not place the volitionality of nature alongside its rationality? Nature confounds our rationality in many ways; it confounds our volition in many ways. We are accustomed to think of nature's objectivity as nature existing apart from human perception. We are just as obdurately reminded of her objectivity, though, by the varying degrees of conformity or resistance to our *volition*. The many faces of nature should not be confined to some purely mental mode, whether of geometry or self-consciousness. We might do well to think of nature's time as sponsoring or generating many diverse metaphors, not as reconstructable from just one of them. Thinking of time under the metaphor of end-directed action has a marked advantage over thinking of it in terms of rationality or geometry or—as in Newton—the presupposition of mechanical law. Human action, at any rate, is intelligible only in the light of nature conceived as dynamic.

Time conceived in terms of action and activity does not ignore the theoretical side of time's nature, but it does show that the answer to the theoretical question, "What is time?" lies much more in the technology of time, namely, how it functions, than we commonly admit. Time is as time does.

Finally, and in this connection, I like the definition of time reported to me as coming from that nostalgic segment of the counter-culture called the "flower children": "Time," said one young woman, "is nature's way of keeping everything from happening at once." I like it because it sets me to thinking about the more arcane aspects of the Big Bang theory, for example. But I also like it because, since human acts are always undertaken for the sake of something, it suggests a kind of teleology in nature and, therefore, in time, which is not scientic, nor ethical nor aesthetic nor religious; yet which has something to say to each of these points of view—and perhaps to all points of view.

After all, is our interest in time just curiosity, or are we drawn to it by need?

Notes

1. I have quoted and paraphrased the translation edited by D. Ross, in the Students Oxford Aristotle, of the *Physics* 223a 21–28. The passage is troublesome; translations are not wholly in agreement as to its fine structure.

2. Eugene Minkowski, *Lived Time*, trans. Nancy Metzel (Evanston, Illinois: Northwestern University Press, 1970) pp. 276–277. Hereafter referred to as *LT*.

3. In "Space and Time," in Lorentz, Einstein, H. Minkowski, and H. Weyl, *The Principle of Relativity*, with notes by A. Sommerfeld, trans. Perrett and Jeffery (New York: Dover, 1923), p. 23.

4. In *Philosophy of Mathematics and Natural Science*, trans. Olaf Hilmer (New York: Atheneum, by arrangement with Princeton University Press, 1949), p. 116.

5. Eric Voegelin, "On Hegel—A Study in Sorcery," in *The Study of Time*, eds. J. T. Fraser, F. C. Haber and G. H. Muller (Berlin, Heidelberg, New York: Springer Verlag, 1972), p. 429.

6. In *The Phenomenology of Perception*, trans. Colin Smith (New York: Humanities Press, 1962), p. xvi.

7. *The Concept of Nature* (Cambridge: Cambridge University Press, 1920) p. 40.

Issues of Beginnings and Endings

Beginnings of Organic Evolution

A.G.CAIRNS-SMITH

Introduction

It is now generally believed that life arose on the Earth spontaneously, that the first systems capable of evolving indefinitely through natural selection were the outcome of normal physico-chemical processes. We may accept this as a reasonable premise without being committed to a more particular set of ideas embodied in the doctrine of chemical evolution. According to this doctrine the physico-chemical processes in question consisted of a preliminary build-up of our biochemicals (amino acids, sugars, and so on) in primordial waters followed by their polymerization and further organization into systems that could eventually reproduce and so become subject to Darwinian selection. This straight line view of the beginnings of organic evolution has been well discussed—for example by Oparin (1957), Calvin (1969), Miller and Orgel (1974) and Dickerson (1978). According to this view organic molecules are the main concern, and indeed those particular organic molecules that lie now at the basis of our biochemistry. Inorganic minerals insofar as they were involved served in a secondary role, in providing catalysts for the formation of small organic molecules or surfaces on which these might have congregated to make their polymerization more likely to occur. Bernal (1951) saw clays in this way—and this has been the general view since that time.

It has seemed to me that clays could have had a more direct significance, that they could have been the materials *out of which* the first organisms were largely made. I think that before the means of doing competent organic chemistry had evolved, such central control structures as genes, catalysts, and membranes, could not have been made, out of DNA, protein, and lipid. Clay minerals formed continuously in the early environment would have been far more suitable materials (Cairns-Smith, 1966, 1975, 1977; Cairns-Smith and Walker, 1974).

I doubt whether some probiotic soup was needed for the origin of life on the Earth. This idea is commonplace in elementary textbooks and has seeped into the general scientific imagination: a probiotic soup is widely thought to be predicted by astronomy, consistent with geology, supported by biology, confirmed by chemistry—and perhaps even demanded by logic. But this is not so.

I will consider these points first and then develop the idea that at the beginning of the evolution that was to lead eventually to life as we know it, there were organisms, starter organisms, of a quite different kind.

Background Considerations

Astronomy and Geology

As a first guess it was reasonable to suppose that the primordial atmosphere of the Earth would have been strongly reducing, like that of Jupiter, if the Earth formed from a cold cloud of gas and dust that was predominantly hydrogen. The original experiment of Miller (1953), demonstrating the easy formation of four of our most fundamental amino acids was designed on this basis. Most current theories of planetary evolution, while agreeing that the solar system had a cold start, take a more complex view of subsequent events. The first atmosphere was not simply part of the original gas cloud, nor was it necessarily strongly reducing. Walker (1976), for example, sees the early atmosphere as faintly reducing at most by the time conditions had settled down sufficiently for life to be possible: he says "a reducing phase, if it ever existed, had a short life and a violent end" and proposes a primordial atmosphere like our atmosphere now with the removal of all oxygen and the addition of at most a few percent of hydrogen.

Indeed the recent trend has been to see the early Earth as hotter and more dynamic than previously—and hence, running more rapidly towards its present condition (Siever, 1973). Moorbath (1977) suggests that a modified doctrine of uniformitarianism is still tenable as far back as 3.8 billion years—the age of the oldest known metasediments. These contain small amounts of graphite and plenty of carbonate (Allaart, 1976), suggestive of CO_2 rather than CH_4 in the atmosphere.

Clearly in such an uncertain field we should not be dogmatic: but in line with the scepticism of Abelson (1966) and Sillen (1965), the popular idea that life arose under a methane-laden atmosphere with the seas rich in organic molecules has little support from either astronomy or geology. The atmosphere may well have been neutral—mainly N_2 and CO_2. In that case organic molecules might still have formed locally, from CO_2 under the action of high energy ultraviolet light penetrating an anoxic atmosphere (we will return to this). But that same radiation would mitigate against the build-up of large "food supplies": it breaks as well as makes organic molecules (Rein et al. 1971).

Biology

At first glance molecular biology might seem to support a straight line hypothesis. All organisms now alive have a similar, quite small set of small organic molecules as fundamental units. These are often referred to as "the molecules of life" and some common source is strongly suggested. But really these amino acids, nucleotide bases and so on, are the molecules of *our* life. It is clear from the detailed similarity of all present organisms (in the genetic code, the amino acid set etc.) that the source was a common ancestor. The evident sophistication of this ancestor implies that it was quite high up the evolutionary tree. (See Figure 1a). We can say also that it had already settled on a *modus operandi* which subsequent evolutionary processes were unable to change.

This is not particularly surprising. Features in organisms tend to become fixed when they have been present for a long time—presumably because later changes presuppose them. But that is not to say that central biochemistry was always fixed; any more than that the present invariability of hair in mammals says that hair was always in our ancestry. (All we can say here is that hair was probably already frozen-in in the common ancestor of the mammals—a beast that had nevertheless non-hairy antecedents). Similarly it is too hasty to conclude that first life had all or any of the biochemicals that all life now shares (See Figures 1b and 1c).

Chemistry

There can be no doubt of the importance of Miller's famous experiment and of the subsequent work on the abiotic synthesis of biochemicals that it inspired (Lemmon, 1970; Stephen-Sherwood and Oro, 1973). The variety of conditions found to be fruitful, together with the discovery of biochemicals or their precursors in such remote places as meteorites (Kvenvolden, 1974) has made it plain that our biochemistry is based on a set of molecules that are easier to make than one might have supposed. But what exactly is the significance of this? That our biochemicals were formed in the early environment and then used as building blocks for first life? Not necessarily by any means. One might as well argue that because the bone material apatite is a mineral that would have been in the environment before there were bones, then the first animals with bones must have picked them up from the environment. Convergences of this kind may simply reflect the generality of the laws of chemistry. What happens in organisms may be similar to what happens elsewhere—but not because one occurrence causes the other. An evolved biochemistry might be expected to be based on a limited number of readily made stable components. That would be good engineering, an understandable outcome of natural selection. But it is not at all clear exactly when the selection of our particular set of biochemicals was made.

In any case, the success of the simulation experiments has been, at times, overstated. The products are invariably complex mixtures, usually largely polymeric tars. Furthermore some critical molecules of intermediate complexity, particularly nucleotides and lipids, have not been shown to be formed under simple uncontrolled conditions (Miller and Orgel, 1974, Dickerson, 1978).

Logic

There is a simple-minded logic that says that to make a machine you must start with the components. That is the way that we would make a machine. But evolution is different: all that we can say here is that evolution had to start with *some* machine

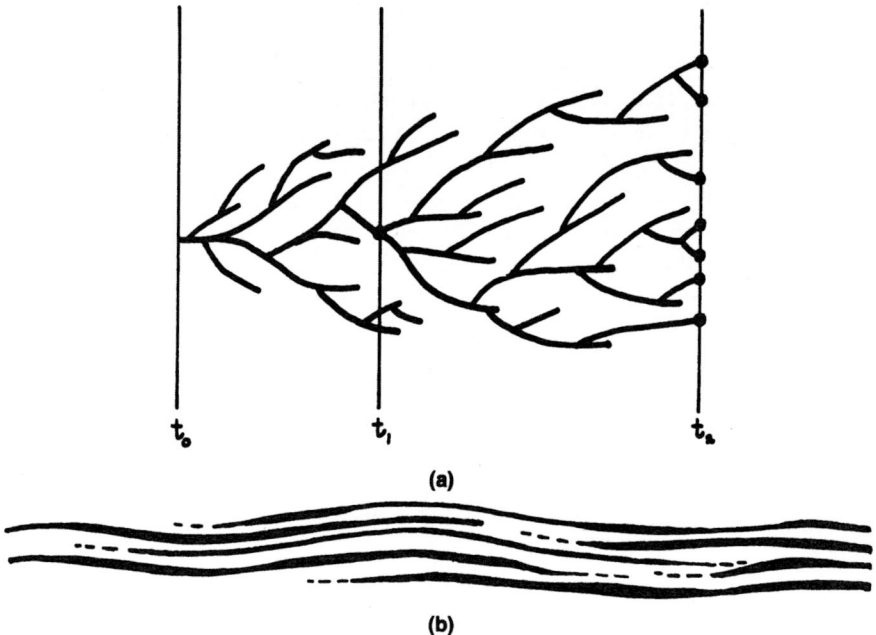

Figure 1. Models for three evolutionary effects that may be important for understanding very early evolution.

(a) *A Tree*. Darwin's tree of life has been heavily pruned—very few of the species that lived in the remote past have descendants now (at t_2). Going far enough back in time (here to t_1) there will only be one such species representing the last common ancestor of life at t_2. As t_2 increases, and species continue to become extinct, t_1 might move forward—but not backwards—separating it further from the beginning of evolution (at t_0), and the ultimate ancestor of life on Earth. That the distance between t_0 and t_1 is now very considerable is indicated by the great sophistication of the machinery that is common to all life.

(b) *A Rope*. The lines in (a) are really multi-stranded, because an organism consists of a collaborating set of subsystems. Thus while an extended line of organisms may be continuous—that is, with its members related through a sequence of minor modifications—the subsystems need not be: old subsystems (e.g. gills) may peter out while new ones (e.g. lungs) "peter in." The strands may be shorter than the rope, that is, with those at one end discontinuous with those at the other—they might indeed be made of different materials. Similarly, during early biochemical evolution a coming and going of subsystems could have disconnected the original from the final materials out of which organisms were to be made.

that was capable of evolving indefinitely through natural selection and that the components of *that* machine had to be supplied by the environment. The question is: Did the machine that could most easily start have to be similar to the machine that would be preferred eventually? It is a question of how radical very early evolution could have been.

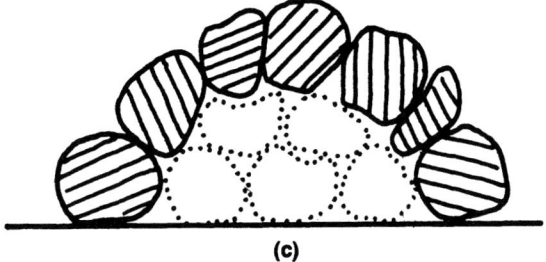

(c)

Figure 1. *(continued)*

(c) *An Arch*. The invariance of our central biochemistry—established by t_1 in (a)—is largely due to its subsystems leaning on each other: everything depends on everything, more or less. The whole system is much too complicated just to have happened; its present rigidity must be a consequence of an extended evolution. The problem here is similar to that of how an arch of stones might be built through a series of minor modifications—adding, subtracting or adjusting one stone at a time. This problem can only be solved with some kind of scaffolding of stones which are subsequently removed. The strong interdependence of the subsystems in our biochemistry is *prima facie* evidence for earlier subsystems now lost. There is no particular reason to suppose that the missing subsystems were made the same way as any of the remaining subsystems, and indeed reasons are given in the text for supposing otherwise.

Takeovers In Evolution

Was evolution at the biochemical level always comparable to the changes that took place when (say) the fins of fish transformed into the limbs of land animals: was there a continuous line of transformations, that is, with functions sometimes changing but structures remaining related? Or did early evolution sometimes proceed through takeovers, with functions sometimes persisting while structures changed discontinuously? This would be comparable to the change from gills in fish to lungs in land animals. Here a vital function was maintained through transformation that disconnected earlier and later means: first there were gills; then there were bimodal animals, with both gills and lungs (the latter derived from the alimentary canal); then, in many animals, there were only lungs. The whole process was gradual, but that is not to say that there was a succession of structures intermediate in form between gills and lungs. Evolution through takeovers requires no structural similarity between the first way of performing functions and later ways. Continuity between organisms in an evolving line does not imply a long term continuity of the subsystems that make them up. We might compare a line of organisms with a rope which may be continuous for a hundred metres although its individual fibres are only a few metres long (Figure 1b).

Clearly logic does not exclude the possibility of takeovers in evolution, since they happen. Furthermore it is not difficult to see why. We can think of an extended line of organisms as a kind of evolving technology in which new materials and manufacturing methods are being discovered from time to time. The discoveries are made gradually and in the first place in response to selection pressures operating on existing functions (Maynard Smith, 1975). Bone presumably evolved in the first place as a structural material. Once organisms had acquired the necessary manufacturing methods, bone was found to have other uses—for example for sound transmission in the mammalian ear. Keratin is a similar story: appearing in early land animals as a material for water retaining scales, it came to be used for claws, feathers, hair etc. (Fraser, 1969). A consequence of such discoveries is that sometimes a new material or a new technique makes an old one redundant.

Now consider the problem at a deeper level. Our last common ancestor already possessed the means of manufacturing protein. This is a piece of high technology analogous, perhaps, to our steel industry: protein is the central multifunctional material of modern organisms as, arguably, steel is the basic material of our present society. But before organisms had invented the means of putting together amino acids into particular genetically controlled sequences, would amino acids have been so generally useful? And what functions of poly-amino acids provided the selection pressure for the evolution of the equipment for protein synthesis? It is not to be assumed that this function was catalytic because enzymes now are so important. That would be like supposing that keratin was invented so that birds could fly. That is not the way evolution works: it does not see into the future. From the intimate dependence of an enzyme's function on the way its amino acid chains fold, and from the dependence of that on accurate sequence specification, it is more likely that some much less taxing function came first. Once accurate manufacturing machinery was there, for whatever reasons, this new stuff protein could demonstrate its versatility and in doing so, no doubt displaced other more pedestrian materials. What the immediate precursors of protein were is speculative, but that there were such precursors is I think much less so. Before there was protein there was surely other means to its now vital functions, as before there was steel there was bronze and before that a stone age (cf. Pirie, 1951).

Nor does logic demand that a genetic material be immune from takeovers. A first genetic material, let us say, can form spontaneously in non-biological surroundings. This is the basis of primary organisms which evolve a technology within which more sophisticated genetic materials become possible. One of these forms a second class of genes in the now genetically bimodal organisms. The secondary genes turn out to be more efficient and eventually come to control everything required for their own production. The first class disappear (see Figure 2).

The solution to the problem of the origin of life can only be impeded by insisting in the name of Ockam that only the pieces now visible on the chessboard are to be used. This ignores a well known and easily understood feature of evolution—it subtracts as well as adds. It is quite unlikely that the components most suitable for a first system would still be retained by an advanced system—given that a succession of additions and subtractions provides a means of radical change. Evolution had no prior knowledge of which molecules to choose initially to make the best machines in the future: in particular it was not to know that organic molecules—so difficult to control—would be ideal once means of control had evolved. Life would start, not with a predestined set of official biochemicals, but with whatever available units could most easily put themselves together into the kinds of hardware needed for a

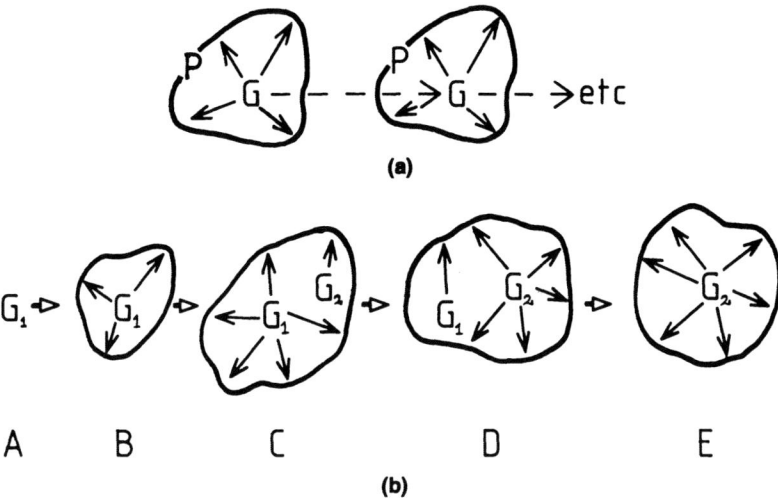

(a)

A   B   C   D   E

(b)

Figure 2

(a) An organism consists of a genotype and a phenotype. The genotype is a set of instructions—like a musical score. It is a set of particular patterns (genes) written on a special kind of paper—a genetic material. The genotype has two properties:

    (i)   It can be printed by a templating process (dashed arrows) and

    (ii) It specifies the production of a surrounding phenotype (solid arrows).

The phenotype is the expression of the genetic information (like a performance of a piece of music) and its function is to provide an environment that protects, and assists in printing and publishing the genes. (Publication day for the dandelion, for example, is when its seeds are carried off in the wind). The problem of the origin of life is largely that a very elaborate phenotype seems necessary to operate the present genetic material, DNA. (This score can hold a lot of information, but it needs sophisticated equipment to print it or to read it).

(b) A general solution to the above problem—genetic takeover. Here we suppose that a first kind of genetic material G_1 could operate without any phenotype (stage A). These primary organisms spread to fill increasingly difficult niches by evolving the means to specify a surrounding phenotype (stage B). A secondary genetic material, G_2 appeared later within such a phenotype on which it was wholly dependent (stage C). In subsequent genetically bimodal organisms the second genetic material proved to be more efficient and versatile, and eventually displaced the starter material (stages D and E).

system with evolutionary potential—genes, catalysts, membranes and such things. We seek the missing pieces of a stone age biochemistry (compare Figure 1c).

A Clay-Based Primary Life?

Millot (1979) has reviewed the salient features of clay structure and genesis. A typical clay crystallite consists of 2-dimensional polymers—unit layers—stacked on top of each other. The individual unit layers are made up of either three or four sheets of oxygen atoms (formally as O^{2-} and OH^- anions) with small cations, especially Si^{4+} and Al^{3+}, nested in (tetrahedral and octahedral) sites between these oxygens. In the very common kaolinite structure, for example, there are three sheets of oxygens in the unit layer with one sheet of silicons and one of aluminums between them. Here the arrangement of O^{2-}, OH^-, Si^{4+}, and Al^{3+} is such that the charges cancel exactly. A typical kaolinite crystal is a neat stack of neutral unit layers. In other clays, such as montmorillonite, there are 3 sheets of cations between 4 sheets of oxygens and hydroxyls. Here the charges do not balance so that the unit layers have an excess of negative charge that is compensated by larger cations, such as Na^+ and Ca^{2+}, between them. There is also commonly water between the layers of such clays. Unlike the cations within the layers, the inter-layer cations are quite mobile and can be exchanged by others—including organic cations. Water too can be partly or wholly replaced by organic molecules, and indeed most kinds of clays allow organic molecules of one sort or another between their layers (Theng, 1974).

Generally, montmorillonite layers are much less neatly stacked than the kaolinite type. Montmorillonite often has the appearance of crumpled linen under the electron microscope. The individual layers are flexible membranes. Individual kaolinite layers are flexible too, being swiss-rolled into tubes in the mineral halloysite.

The more recently discovered clay mineral imogolite consists of seamless tubes that have an internal pore of about 1 nm (Cradwick et al, 1972). The related allophane consists of tiny (3.5—5.5 nm) porous hollow spheres (Wada and Wada, 1977).

So if it were such things as membranes, tubes and vesicles that were needed for the first organisms, the clay minerals could most easily have provided them—the clays that we have been discussing form continuously in the environment from solutions derived from rock weathering (Millot, 1979). Presumably this would have also been so on the primitive Earth.

Of course a heap of tubes and membranes is not an organism. For that there must be organization under genetic control. There must be a long term genetic memory—presumably through a genetic material of some sort that could hold and replicate specific more or less complex patterns (Figure 2a). Are there clays that could do that? If so we might have a lead to the true beginners at evolution (G_1 in Figure 2b) which, unlike DNA, could work without preexisting machinery.

It is not difficult to see how clays might hold large amounts of information. For example there is considerable freedom in the ways in which the fixed cations inside clay layers can be arranged. Hence, rather as DNA holds information like a (1-dimensional) message tape, as a particular base sequence; so a clay layer could in principle hold a great deal of information as in a punched card—as a specific 2-dimensional array of features.

Two questions follow: (i) How could such 2-dimensional patterns be reproduced; and (ii) how could the patterns mean anything?

(i) The Replication Requirement

If during crystal growth a particular arbitrary feature in the arrangement of cations in a preexisting layer is copied in the next layer growing on top of it, then the resulting crystal should appear disordered to some extent *within* the layers—a particular form of the disorder representing the "information"—while at the same time ordered *between* the layers. One approach to the search for a clay genetic material, then, is through the search for inter-layer ordering in clay crystal structures. Although a difficult field of study for X-ray crystallography, there is evidence for such inter-layer ordering in several kinds of clays (Bailey, 1975).

Another approach to the search for clay genes is through synthesis. Of great interest here is the laboratory evidence for the replication of features of cation substitutions during clay synthesis that has been found by Weiss (1973). He found that a number of properties that depend on silicon atoms being replaced by aluminium atoms in synthetic clays can be inherited by successive seeding over several generations.

Here I would like to suggest a third approach—through the study of crystal morphology. Figure 3 illustrates what crystal genes might look like that held information in the form of a specific 2-dimensional pattern of some sort that was

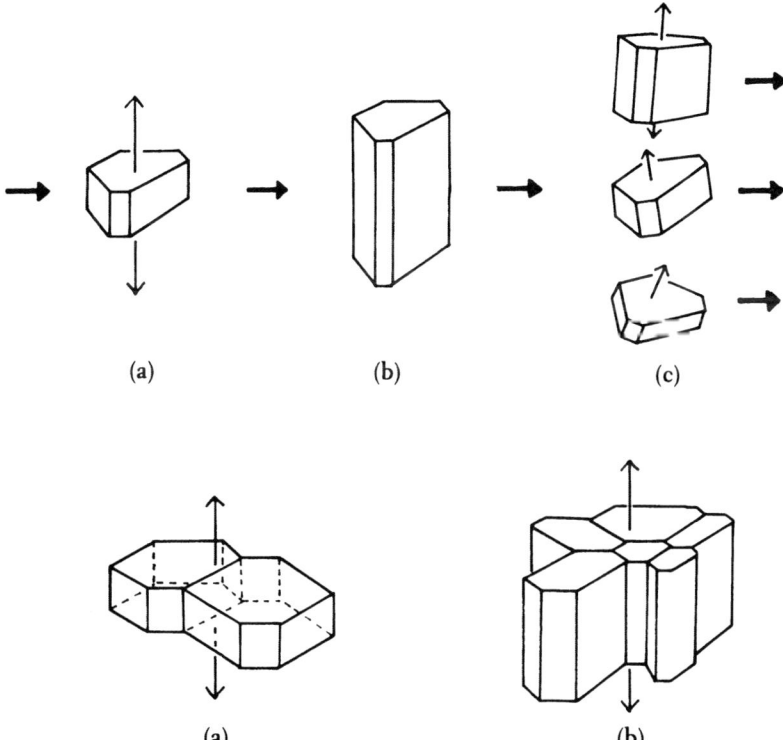

Figure 3. A crystal gene that replicates information written in 2-dimensions might be expected to have a tabular or columnar morphology. (a) shows a replication cycle consisting of (i) growth (exclusively) on two faces that are in the plane in which the information is held, and (ii) cleavage (exclusively) in this plane. The information might simply be the constant shape of the growing faces. This could be increased if the crystal was twinned and if the planes of intersection between its constituent domains remain parallel to the sides of the crystal (b and c). The information might then consist of a particular crazy-paving pattern of domains rather as the information in a punched card consists of a particular arrangement of holes.

replicated during growth in the stacking direction. A columnar or tabular morphology is to be expected since the sideways extent of new layers must not either grow or diminish as the crystal grows. This would either add "nonsense" information or lose information, respectively.

In its outer form at least, kaolinite often has the general sort of columnar appearance idealised in Figure 3c—when it occurs as "booklets," "concertinas" or "vermiforms." These crystals may resemble a tall stack of paper, in which each sheet has the same or a similar rather complex shape. As a result the columns are grooved or fluted. Many of these are shown in the electron micrographs of Keller, (1978) (and references therein). Long vermiforms, such as those in Figure 34 of Keller (1978) may give the appearance of having been extruded. But it seems more likely that the constant cross-section arises from growth having been exclusively in the stacking direction, as required for our putative genetic crystals. If this is so then "information," in the form at least of the constant shape and size of the layers, was replicated quite accurately as the crystals grew.

Fluting and grooving of crystals is an indication that the crystals are twinned—made up of domains as in Figures 3b and 3c. In that case more complex information might be reproduced during crystal growth—not just the shape of the individual layers but a particular crazy-paving patterning of domains written in each of the layers. (For our discussions here it is not necessary to know the underlying cause of the twinning, but it seems likely that it arises from there being domains with alternative arrangements of the aluminum cations within the layers—as in the pseudo-twinning discovered by Mansfield and Bailey (1972) in much larger kaolinite crystals).

At present we can go no further than to say that it is crystallographically plausible to suppose that during a layer-on-layer growth process specific defects in the layer below may be copied into the newly forming layer—and that there is some evidence for such effects. So let us move on to the next question. Even given such a spontaneous printing mechanism, could what was being printed have any significance—what could the information mean?

(ii) The Control Requirement

The meaning of genetic information it that it tends to preserve, replicate and propagate itself. Those messages are selected that can control their immediate environment to this end. For all life that we know of on the Earth today, the means to this end are immensely complex. Many intricate DNA messages must collaborate before any of them make sense—before they give rise together to an efficient vehicle, an organism, that ensures their propagation into the indefinite future. The messages in the first genetic material(s) must have had a much more direct meaning. Considering the kaolinite model, how could a message in the form of the particular shape and size of the replicating cross-section of a mass of columnar crystals tend to be preferred?

The answer to this question is: Quite easily. Many large-scale properties of materials are controlled by the shapes and sizes of their constituent particles. Soil chemists are familiar with the idea that rheological properties and the general porosity of soils are controlled by the shapes and sizes of clay and other particles. We might imagine clays forming in a sandstone from silica, alumina etc. dissolved in percolating waters. The mechanically protected pores in sandstone provide excellent reaction vessels—and indeed most sandstones contain clays (kaolinite and others) that have grown *in situ* (Wilson and Pitman, 1977). A clay will only continue to grow if it does not clog up the pores. A replicating clay, then, would be more effective, and the message it was replicating would become more prevalent, if that message consisted of a particle shape and size that gave rise to the appropriate porosity. Fraser (1978) has a more subtle idea here—that crystal shapes and sizes would have been selected that were (literally) in tune with vibrations in the environment). We would be justified in calling such a message genetic in the biological sense because it would be making things right for its own propagation. (Think again about DNA messages—that is what *they* do if now in a very indirect way.)

Where there were different kinds of replicating crystallites each would soon create a local zone containing millions and millions of copies. Because of effects of the replicating defect patterns peculiar to each zone, some zones would be more suitable for clay synthesis that others. Some zones would grow faster—like competing colonies of bacteria in a petri dish where the faster reproducers have made bigger patches by the time nutrient or space runs out. But evolution depends on such competitions continuing indefinitely. It depends on there being environments that are open systems (chemostats not petri dishes). Clays are very often formed in open systems—in sandstones for example—where endless supplies of nutrient weathering solutions percolate in, and where continuously or intermittently, clays are being carried away. Under such circumstances, according to the simple mathematics of the chemostat (Bull, 1974), a marginal difference in growth rate can lead to a faster replicating mutant displacing the other *entirely.*

On the medium term, then, one might imagine replicating clays evolving like this within a given initial environment that was generally favorable to clay synthesis. Through successive replacements, more efficiently replicating types would appear, that is to say, clay defect patterns would appear that contrived their surroundings better to favor the replication of clay defect patterns. On the larger term we could predict another trend. Clays that had evolved in a first easy environment would spread to adjacent, more difficult environments where clays could not previously have formed. In such places only evolved clays would be able to propagate. By that time you could hardly deny them the title of organism.

Sometimes kaolinite indeed seems to make things right for its own synthesis: booklet kaolinites have a particularly porous texture (Keller and Hanson, 1975). Not only should a porous texture favor clay synthesis by allowing an adequate through-put of nutrient solutions, but a high porosity seems to be favorable to kaolinite synthesis particularly (Johnson, 1970). Such positive feedback effects were suggested by Clemency (1976) in a particular case where adjacent kaolinite and montmorillonite each appeared to be maintaining conditions suitable to their own synthesis, by controlling the concentrations of ions in their surroundings. Such an effect might conceivably be a consequence of a natural selection of a very simple sort: from an initial population of crystallites those that tended to give rise to booklets were self-selecting because they continued to maintain suitable conditions for crystal growth while at the same time propagating their kind. If this were so then sets of booklets in a given micro-region that had been derived from common seeds should show detailed morphological similarities—of groove arrangements and cross-sectional area for example. Furthermore if they have been the products of a multistage mutation and selection process they might appear to be contrived in a more detailed way—for example the exact size and shape might be particularly suited to maintaining the conditions for kaolinite synthesis in a particular locality. Hence perhaps the very beginnings of evolutionary processes might be discoverable on Earth today.

I think, then, it is conceivable that kaolinite might be able to evolve under natural selection: superficially at least it seems to have an appropriate crystal structure, it can have an appropriate morphology and it can form under suitable conditions. Perhaps, it sometimes shows signs of having evolved. It is too soon to attempt to identify the primary genetic material(s), but we can say perhaps that kaolinite is a model for several of the features to look out for in clays.

The test question for any systems that may have been the products of natural selection is this: Do they reproduce and does their structure appear to be contrived so as to increase the efficiency of that reproduction? "Contrivance" is hard to define but may be easy enough to recognize if one is on the lookout for it. One can recognize human artifacts, like axe heads, because they belong to a small class of forms that turn up more often than you would expect on the basis of chance and normal physical processes—and because they can be seen to have a function.

Evolutionary Potential of Clays

We have seen, from our discussion of hypothetical evolving clays, that a genetic material might itself evolve insofar as information that it contains may affect its physical properties—and hence, in suitable circumstances the propagation of the information. Clearly this direct approach is limited: there is a limit to the things that (say) kaolinite could do directly, however intricate its defect structure might be. One might conclude, indeed, that the evolution of a clay could only be trivial. But that would be too hasty. In principle, indirect modes of control could give significance to specific intricate genetic patterns. One may see it indeed as a characteristic of the evolution of genetic control that it tends to become increasingly indirect. (The elephant is a very indirect way of propagating DNA sequences). Given the necessary reading equipment, the most abstract-looking messages can have a meaning. No, the intrinsic properties of a genetic material, the properties that it acquires immediately by virtue of information that it contains, although important to give it a toe-hold, are not the main factors in deciding whether in principle that genetic material could sustain a long-term evolutionary process. For this there are only two absolute requirements:

(i)   the genetic material must be able to hold and replicate accurately large amounts of information; and
(ii) it must come to control its environment indirectly.

Suppose that a clay unit layer with N unit cells consists of n domains. For reasonable values of N and n there can be an enormous number of possible crazy-paving domain patterns. Suppose for simplicity that the position of each domain is determined by one critical feature (for example the position of a central screw dislocation). For $N \gg n$ there would be $\sim N$ possible positions for each of the n features, that is N^n in all. A crystallite of 1μ across containing 10^6 unit cells per layer and 100 domains would thus be one of about 10^{600}. The crystallite could hold about as much information, that is, as a large protein molecule. To be retained from generation to generation the crystal growth processes would have to be very accurate. Assuming that, the next question is how the information might be expressed indirectly.

Indirect Techniques of Control

DNA replicates genetic information and protein does something about it. Although not essential, it is clearly a good idea to have specialist structures for what are very different activities—of replication and control. Being more complex and not being absolutely essential, such an arrangement is not to be expected initially: but being more efficient it is perhaps a predictable outcome of evolution.

For a crystal to be a good replicating machine it should probably be rather rigid with strong forces between the units so that few mistakes are made. Some typically well crystallized material like kaolinite, vermiculite or chlorite was perhaps the original counterpart of DNA. To be a versatile control structure, however, it is better to be flexible, like protein, so that information within it can make a variety of complex forms. To evolve very far, then, we might guess that an original clay genetic material would come to operate through more flexible intermediate structures analogous to protein. Organic polymers, perhaps even peptides, are a possibility given environmental monomer supplies. (Lahav et al. (1978) have shown that short chains of amino acids can be made on clays). But I am inclined to think that peptides would only have become a good idea much later: clays are not especially good at binding amino acids, nor at joining them together. To begin with there would have been easier ways of making secondary flexible control structures—by controlling the formation of other clays like halloysite or montmorillonite.

Suppose that under conditions A, crystal growth of a genetic material is favored, but that under conditions B—a different temperature or pH perhaps—another clay is favored that grows epitaxially on the genetic crystals. The way the second clay grows depends on the surface characteristics of the first and these in turn depend on the domain structure or other form of genetic information in the first clay. The second clay is then a kind of phenotype for the first—if the success of the first depends to some extent on the characteristics of the second. Perhaps, because of the secondary clays surrounding them, the first clays will grow faster when conditions A return, or will be protected against periodic inclement conditions C.

We are imagining now a situation where instead of a genetic clay making conditions right for itself directly, it makes conditions right for other clays, which in turn make conditions right for the genetic clay. This roundabout approach would have the advantage that the secondary clays would not have to be able to replicate the information that was partly responsible for their formation. Freed from this constraint there would be a wider range of possibilities open. The secondary clays need not be hard crystals; they could be flexible gel-forming membranes, or tubes or vesicles.

How Do You Do Organic Chemistry Without Enzymes?

If you ask an organic chemist this question he will mention apparatus. Particularly if a sequence of reactions is to be carried out automatically, he will talk about flasks, tubes, pumps, and so on. In the absence of the detailed microscopic control that the active centers of enzymes can provide there has to be more control on the larger scale. Solutions must be mixed, conditions adjusted, products separated, etc. Each operation may be simple in itself, but for a multi-step synthesis the level of contrivance required to create an appropriate sequence of operations is very considerable. Quite simply, specific organic molecules are usually difficult to synthesize. Yet before there were enzymes there must have been some means of doing organic chemistry: this must have become very effective to create consistent clean supplies of nucleotides and other components required before our protein synthesizing machinery could have even begun to evolve. I think that before there were enzymes, there was an altogether different kind of biochemistry that depended much more than our life does on genetically specified apparatus. Our life is protein life: biomolecular control depends absolutely on the unique socket-forming abilities of proteins, on their ability to recognize other molecules. Thus many reactions can take place in the same vessel without confusion. Once this ability had evolved, biochemistry would never have been the same again. But we must try to look back beyond that technological revolution. When you do not have the means of recognizing individual molecules, how do you keep control? I can see no alternative to having physically laid out production lines of some sort, with tubes, pumps, mixing chambers and so on. Then molecule A reacts with molecule X, not through some structure that recognizes both, but through some kind of glassware that localises both. Of course there is nothing probiotic about this: apparatus of any complexity could only be the result of contrivance, the product of evolution through natural selection.

As remarked earlier, in ordinary clays we have some likely-looking raw apparatus. (There is even a proton pump that works on a thermal gradient across finely divided $Al(OH)_3$, $Mg(OH)_2$ and $Ca(OH)_2$ (Freund and Wengeler, 1978)). And there are intricate interlayer compartments, catalytic effects at special sites—and so on. Could natural selection operating on the kinds of clay-based organisms that we have been imagining have put suitable apparatus together? Can we imagine at least the very beginnings of such construction work?

The Start of Clay-based Organic Biochemistry

CO_2 might have been a sufficient carbon source for the first essays in organic chemistry. CO_2 can be reduced in aqueous solution to small organic molecules (in particular formate, formaldehyde and oxalate) by the action of UV light (Getoff, 1962, 1963).

Now imagine an assemblage of genetic and secondary mineral crystallites exposed to the UV sunlight. Adsorbed CO_2 is partly converted to a mix of organic molecules. These further react. Often they are photolysed back to CO_2. The steady state concentrations that are arrived at will depend on complex factors including physical characteristics of the mass of crystallites—veriforms, membranes, tubes etc. This mass creates a richly heterogeneous set of microenvironments; some protected from the UV light, others tending to bind certain types of organic molecules, still others containing catalytic sites.

Ions of transition metals, like iron, might have been particularly important, both free and incorporated into clay layers. Such ions would be UV-shields but they would also themselves have photochemical effects. Radiation in the $200-300$ nm range can readily generate oxidising and reducing agents—Fe^{3+} from Fe^{2+}, for example, and radicals such as H (see Cairns-Smith, 1978). Where transition metal ions are incorporated asymmetrically in a clay membrane one might expect asymmetric effects—protons, hydrogen atoms or solvated electrons being released preferentially on one side of a clay membrane. Thus chemical potentials could be generated, further complicating the situation. There would be microenvironments of different pH, pE etc., and inevitably flows generated between them. All these, at first, quite fortuitous effects would influence the yields and locations of different organic species. Let us concentrate attention on one—oxalate.

Oxalate is one of the species recognized by Getoff as a product of inorganic UV photosynthesis. It is also easily destroyed by UV in the presence of Fe^{3+}. Oxalate concentrations might thus be expected to be particularly sensitive to the kinds of factors that we have just discussed. Oxalate binds aluminium and is also a catalyst for kaolinite synthesis (Wey and Siffert, 1962).

Let us suppose, then, that kaolinite could have been a genetic material as already discussed. We suppose that domain patterns in a population of replicating kaolinite crystals control the morphology of the crystals themselves and, through epitaxy, the shapes, sizes and charge distributions of secondary clays. The genetic information thus exerts a remote and partial control on the curling, folding and interparticle interactions of the secondary clays and hence on the architecture of the maze of pores and interconnected compartments that any clay mass inevitably makes. If the net effect of all this is to increase slightly the oxalate concentration surrounding one particular set of replicating crystals with one particular domain pattern, then that pattern could become more prevalent with time. In line with the earlier discussion on the advantages of a suitable porosity, if a genetic pattern in any way contrives its immediate environment to make or accumulate organic molecules that are favorable to clay synthesis then that pattern is fitter in the biological sense: it has more offspring more quickly. A chance assemblage of pores, compartments and so on might bring about some marginal effect. In that case, apparatus could already be under construction by natural selection with further improvements to be expected. The ultimate limit to such improvements would be set by the amount of genetic information being accurately printed. As we have seen, that might have been very considerable.

We might predict that metal binding agents generally would have been important components of early phenotypes that were creating conditions to encourage the replication of an inorganic crystal gene—whether kaolinite or not. Krebs acids and amino acids are metal binding agents. Perhaps these were present in primary phenotypes for that reason and hence became incorporated into the system that was to take over from it.

Bringing in Nitrogen

I think that however the competence was evolved to make amino acids and then nucleotides, the primary organic biochemistry would have started from CO_2 and first made mainly C-H-O compounds. These are still at the heart of our biochemical pathways—nitrogen almost looks like an afterthought (Lipmann, 1965; Cairns-Smith, 1975). With a clay controlled organic biochemistry, nitrogen *could* have come in late. In any case, the reactivity of amino groups would have made organized organic chemistry difficult until suitable means of handling them had evolved. (As Abelson (1966) points out, amino acids and sugars, for example, tend to react with each other to give complex brown products). The simplest way of imagining the eventual introduction of nitrogen would be through the evolutionary discovery of N_2 fixation. Schrauzer and Guth (1976) have found that $Fe(OH)_2$ can convert N_2 to ammonia and hydrazine, the reaction being promoted by UV light. Nitrogen fixation should thus not have been beyond the devices of wholly inorganic control machinery.

There need never have been external soup, then, to provide "nutrients"—nitrogenous or otherwise—for first life. I suggest that different kinds of organic molecules were made locally in the phenotypes of early clay-based life—when they could confer a selective advantage and when their synthesis became possible. Perhaps, the ground plan of our biochemical pathways was gradually sketched with the aid of, eventually, some very exotic clay machinery. (Perhaps some signs of this machinery might be found in the most ancient sediments). Only much later, I think, within the phenotypes of some line of evolved primary organisms were the first nucleic acid-like molecules invented to set the stage for the beginnings of a new kind of genetic control machinery (Figure 2). But that is another story: the true beginnings of organic evolution were, I think, in a different part of chemistry.

Conclusion

From its remote origins organic evolution has proceeded through continuous successions of organisms; but that is not to say that the subsystems in organisms now are related continuously to initial subsystems. Through takeovers, functions may be maintained while the structures carrying out these functions change (the rope analogy, Figure 1b). Thus there is no overriding restriction imposed by current biochemistry on the material nature of the first means of carrying out functions such as the transmission of hereditary information.

Indeed we would not expect that the materials most suitable for first life would have continued to be so as techniques of fabrication evolved. To begin with selection would favor materials that happened to organize themselves into appropriate forms: later, materials would be favored that could *be* organized. In essence this is why I think organic evolution (organic in the biological sense) would have started with inorganic materials (inorganic in the chemist's sense). Many inorganic polymerizations—for example the formation of clays from dilute solutions—are simultaneously crystallizations. (This is not so for organic polymerizations). Local thermodynamic control can thus correct errors: when crystals are growing slowly enough a wrong ion adding will usually, sooner or later, come away again. The process is slow, but it can lead to a highly organized polymer. Fortuitously such polymers are often membranes, tubes etc.—and they might have had an early biochemical use. Even that most difficult and critical of functions—the printing of genetic information—may well be a fortuitous effect in certain growing clay crystallites. So perhaps that is the way that life could have started. But it is not the way it would have gone on—given that organic molecules became involved.

Organic materials are potentially more able to *be* organized. Once put together correctly, two carbon atoms may stay together indefinitely. Just because of this, though, there is much less scope for automatic error correction: mistakes stay put too. Generally it takes a skilled organic chemist, or an enzyme, to avoid a tarry chaos. Thus, while it is often said that organic molecules are ideal for building biomolecular machinery, this is only partly true. Organic molecules are ideal for life in the kind of way that a violin is ideal for making music—in the hands of an expert who can control the richness of possibilities. Organic molecules are ideal for evolved life, but first life was not to know this. I think that, like so many beginnings, the beginnings of organic evolution were special.

Notes

Abelson, P. H. 1966. Chemical events on the primitive earth. *Proceedings of the U.S. National Acadmey of Science* 55: 1365–1372.

Allaart, J. H. 1976. The pre-3760 m.y. old supracrustal rocks of the Isua area, central west Greenland, and the associated occurrence of quartz-banded ironstone. In *The early history of the Earth*, Windley, B. F., ed. London: Wiley, pp. 177–189.

Bailey, S. W. 1975. Cation ordering and pseudosymmetry in layer silicates. American Mineralogist 60: 175–187.

Bernal, J. D. (1951). *The physical basis of life*. London: Routledge and Kegan Paul.

Bull, A. T. 1974. Microbial growth. In *Companion to Biochemistry*, Bull, A.T. et al., eds., London: Longman. pp. 415–442.

Cairns-Smith, A. G. 1966. The origin of life and the nature of the primitive gene. Journal of Theoretical Biology 10: 53–88.

———1975. A case for an alien ancestry. Proceedings of the Royal Society B189: 249–274.

———1977. Takeover mechanisms and early biochemical evolution. BioSystems 9: 105–109.

———1978. Precambrian solution photochemistry, inverse segregation and banded iron formations. Nature 276: 807–808.

———in press. *Genetic Takeover*. Cambridge: Cambridge University Press.

Cairns-Smith, A. G. and Walker, G. L. 1974. Primitive metabolism. BioSystem 5: 173–186.

Calvin, M. 1969. *Chemical evolution: molecular evolution towards the origin of living systems on the earth and elsewhere*. Oxford: Clarendon Press.

Clemency, V. 1976. Simultaneous weathering of a granite gneiss and an intrusive amphibolite dike near Sao Paulo, Brazil, and the origin of clay minerals. *Proceedings of the International Clay Conference*. Mexico, p15.

Cradwick, P. D. G., Farmer, V. C., Russell, J. D., Masson, C. R., Wada, K., and Yoshinaga, N. 1972. Imogolite, a hydrated aluminium silicate of tubular structure. *Nature (Physical Science) 240*: 187−189.

Dickerson, R. E. 1978. Chemical evolution and the origin of life. Scientific American 239, 3: 62−78.

Fraser, J. T. 1978. *Time as conflict*. Birkhauser, Basel. p73.

Fraser, R. D. B. 1969. Keratins. Scientific American 221 2: 86−96.

Freund, F. and Wengler, H. 1978. Proton conductivity in hydroxides as related to transport phenomena in soils. Paper read at the 6th international clay conference, Oxford.

Getoff, N. 1962. Reduction of carbon dioxide in aqueous solution under ultraviolet light. *Zeitschrift Naturforschung* 17B: 87−90.

Getoff, N. 1963. Effect of ferrous ion on formation of oxalic acid from aqueous carbonic acid by ultraviolet irradiation. *Zeitschrift Naturforschung* 18B: 169−170.

Johnson, L. T. 1970. Clay minerals in Pennsylvania soils in relation to lithology of the parent rock and other factors—I. *Clays and Clay Minerals* 18: 247−260.

Keller, W. D. 1978. Classification of kaolins exemplified by their texture in scan electron micrographs. *Clays and Clay Minerals* 26: 1−20.

Keller, W. D. and Hanson, R. F. 1975. Dissimilar fabrics by scan electron microscopy of sedimentary versus hydrothermal kaolins in Mexico. *Clays and Clay Minerals* 23: 201−204.

Kvenvolden, K. A. 1974. Natural evidence for chemical and early biological evolution. *Origins of life* 5: 71−86.

Lahav, N., White, D. and Chang S. 1978. Peptide formation in the prebiotic era: Thermal condensation of glycine in fluctuating clay environments. *Science* 201: 67−69.

Lemmon, R. M. 1970. Chemical evolution. *Chemical Reviews* 70: 95−109.

Lipmann, F. 1965. Projecting backwards from the present state of evolution of biosynthesis. In *The origins of prebiological systems*; ed. Fox, S. W. New York: Academic Press, pp 259−280.

Mansfield, C. F. and Bailey, S. W. 1972. Twin and pseudotwin intergrowths in kaolinite. *American Mineralogist* 57: 411−425.

Maynard Smith, J. 1975. *The theory of evolution*. 3rd ed. Harmondsworth: Penguin Books: 287.

Miller, S. L. 1953. A production of amino acids under possible primitive earth conditions. *Science* 117: 528−529.

Miller, S. L. and Orgel, L. W. (1974). *The origins of life on the earth*. New Jersey: Prentice-Hall.

Millot, G. 1979. Clay. *Scientific American* 240, 4: 76−84.

Moorbath, S. (1977). The oldest rocks and the growth of continents. *Scientific American* 236, 3: 92−104.

Oparin, A. I. 1957. The origin of life on the earth. Edinburgh: Oliver and Boyd.

Pirie, N. W. 1951. In Bernal *The Physical Basis of Life* 1951. 71.

Rein, R., Nir, S. and Stamatiadou, M. N. 1971. Photochemical survival principal in molecular evolution. *Journal of Theoretical Biology* 33: 309−318.

Schrauzer, G. N. and Guth, T. D. 1976. Hydrogen evolving systems. 1. The formation of H_2 from aqueous suspensions of $Fe(OH)_2$ and reactions with reducible substrates including molecular nitrogen. *Journal of the American Chemical Society* 98: 3508.

Siever, R. 1973. In *Origins of life: chemistry and radiochemistry*, ed. Margulis, L. Berlin: Springer-Verlag. p. 170−180.

Sillen, L. G. 1965. Oxidation state of the Earth's ocean and atmosphere I. A model calculation on earlier states. The myth of the "prebiotic soup." *Arkiv Kemi* 24: 431.

Stephen-Sherwood, E. and Oro, J. 1973. Chemical evolution. Recent syntheses of bioorganic molecules. *Space Life Sciences* 4: 5–31.

Theng, B. K. G. 1974. *The Chemistry of Clay-organic Reactions*. Bristol: Hilger.

Wada, S. and Wada K. 1977. Density and structure of allophane. *Clay Minerals* 12: 289–298.

Walker, J. C. G. 1976. Implications for atmospheric evolution of the inhomogeneous accretion model of the origin of the Earth. In *The early history of the Earth, ed. Windley, B. F. London: Wiley.* pp. 537–546.

Weiss, A. 1973. In paper presented at the fourth international conference on the origin of life, Barcelona.

Wey, R. and Siffert, B. 1962. Reactions de la silice monomoleculaire en solution avec les ions Al^{3+} et Mg^{2+}. In *Genese et synthese des argilles*. Eds. du C.N.R.S. Paris, pp. 11–23.

Wilson, M. D. and Pittman, E. D. 1977. Authigenic clays in sandstones: Recognition and influence on reservoir properties and paleoenvironmental analysis. *Journal of Sedimentology and Petrology* 47: 3–31.

The Beginning of the Beginning in Western Thought*

D. CORISH

From Homer on, the common Greek word for 'beginning' is ἀρχή. The verb is ἄρχω, and it can mean 'to rule' as well as 'to begin,' though the latter seems to be the original meaning. According to Buck's *Dictionary of Selected Synonyms in the Principle Indo-European Languages*, "words for 'begin, beginning' are most commonly based upon notions like 'seize upon' or 'enter upon,' but there are also other and diverse sources." It is possible, according to Buck, that ἄρχω 'begin' and ἀρχή 'beginning' come from an old aorist form of ἔρχομαι to come,' the development being 'came to,' 'started,' 'began'.[1]

Instead of starting with the earliest uses of the word ἀρχή and working on chronologically, I shall speak first about a later use—that of Aristotle and his followers—for we are often dependent upon later reports of what earlier thinkers said, and we have to be on our guard not to let such late reports simply dictate to us what we should think of the earlier thought. We have to read between the lines to some extent, and avail ourselves of the insights of modern scholarship in order to achieve some reasonably critical view.

For Aristotle, the word ἀρχή takes on the technical meaning of 'principle', which though it may be identical to 'beginning' in some senses, is not necessarily a temporal meaning at all. As St. Thomas Aquinas later says: hoc nomen, principium, nihil aliud significat, quam id, a quo aliquid procedit—the word 'principle' signifies merely that from which anything proceeds.[2] In the Aristotelian system the four causes are principles: the material cause or stuff from which anything is made; the formal cause or type of the made thing; the final cause or purpose of the made thing; and the efficient cause or actual maker of the thing.[3] None of these ἀρχαί, principles, is *prima facie* a temporal beginning, though analysis might reveal in any or all of them some dependence upon temporal concepts.

Aristotle himself in Book Delta of the *Metaphysics* lists six meanings of ἀρχή: (1)

---

*See "Acknowledgment" at the end of this paper.

a physical or mathematical starting-point, such, he says, as that of a line or a road; (2) the most appropriate starting point, such as, he says, "the point from which we can learn most easily"; (3) a foundational part from which something arises, such as the keel of a ship, or, as some say, he reports, the heart (others say the brain) of an animal; (4) an external origin, such as the father of a child or the hard words that start a fight; (5) the originative will—and so, he says, appealing to the 'rule'—meaning of ἀρχή, the magistrates are called ἀρχαί; (6) the beginning of scientific knowledge, for example, the hypotheses of a demonstration. "It is common," he says, "to all ἀρχαί to be the first point from which a thing either is or comes to be or is known; but of these [ἀρχαί] some are immanent in the thing and others are outside. Therefore the nature of a thing is an ἀρχή, and so are the elements of a thing and thought and will, and essence, and the final cause."[4]

Again it is clear that none of these meanings of ἀρχή is *prima facie* simply 'temporal beginning.' We notice that the first example he gives is that of the beginning of a line or road, an example suited to space and its mathematics rather than to time. The closest we come to simply 'a beginning in time' is perhaps meaning (4): an external origin, such as the father of a child or the angry words which cause a fight—but even here the emphasis is on the externality of the ἀρχή to the thing caused, rather than on the temporal priority of the ἀρχή.

It is very typical of Aristotle not to appeal here to what we might have supposed to be the obvious temporal meaning of the word ἀρχή, but to seek instead more general meanings—as he supposes them to be. And he has more than once thought spatial terms to be more fundamental than temporal. So, for example, he says in Book iv of the *Physics* that the words πρότερον and ὕστερον, which we may very accurately translate 'prior' and 'posterior', primarily indicate spatial positions[5]—though in fact it costs us some effort to think of spatial positions as exhibiting priority and posteriority, whereas we naturally think of temporal positions as having these qualities. Aristotle goes on in this place to attempt to derive temporal priority and posteriority from the spatial by means of these same relationships in motion—prior and posterior positions occur in space, and so they occur also in motion through space, and consequently in the time which the motion takes[6]—but, as I have pointed out in another paper, in making this derivation he begs the question, inherently treating motion as temporal in the first place.[7] Again, as I have tried to show in a second paper, Aristotle is still bedevilled by this attempt to derive temporal order from spatial when he comes to deal with the specific temporal meanings of 'prior' and 'posterior.' Thinking of these terms, as he does, as having spatial meaning primarily, he is unable to see his way to a straightforward prior-posterior temporal order—that is, as we should say, an earlier-later temporal order—but, as a guarantee that the prior and posterior are really temporal, as opposed to spatial, he has to envisage a link to time in terms of the *now*. This gives him not a straightforward prior-posterior order of time, but a hybrid prior-now-posterior order, which, as I have argued in my paper, is deficient in the essential mathematical characteristic known as connectedness.[8] Interestingly enough, the now in this account is for Aristotle an ἀρχή—but not a simple temporal beginning. It is rather a principle by reference to which the temporal 'prior' and 'posterior' are defined as properly temporal rather than spatial. I shall speak of this again.

It must be said, of course, that for Aristotle the notion of the beginning has no cosmic significance, as it does for some earlier Greek thinkers. For Aristotle believes that the world has no beginning and is eternal.[9] It is noteworthy also that in that discussion Aristotle uses the term ἀρχή in an obvious temporal way. He says, for example, that the circular motion which he envisages the cosmos to have is without beginning (ἀρχήν) or end and goes on without ceasing for infinite time.[10] So we must not imagine that Aristotle is incapable of assigning a temporal meaning to ἀρχή. It is simply that when it gets down to what he thinks to be fundamentals, temporal meanings do not figure in his account. Time for him, as his attempt to derive temporal from spatial properties shows, is somewhat less easy to grasp than space as a fundamental physical dimension.

It is, as I indicated, rather typical of Aristotle to miss or omit simple temporal meanings of terms in favor of what he considers to be more fundamental, but what we may well consider to be much more dubious, meanings. I am inclined to think, however, that it is not with Aristotle simply a matter of choice that space should be considered more fundamental, somehow, than time. There is for him, it appears, a real difficulty in envisaging time as an entity independent of space. I am at the moment engaged in research which is tending, I think, to show that a great deal of Greek tradition, up to Aristotle at least, has difficulty with the concept of time, as opposed to the concepts of motion and space. I will not argue that position here, as it would take too long, takes us too much into the minutiae of the subject and into the quarrels of scholars about it, and take us too far away from our central concern with the concept of the beginning; I will merely mention that I think a good argument can be made for Plato's having, in the *Timaeus*, confused time with motion[11]—as Aristotle, it seems, accused him of doing[12]—and of earlier thinkers, Gorgias and Melissus in particular, not having adequately distinguished between time and space itself. I shall refer to this again later. It cannot be stressed too much that the most systematically developed Greek Science is geometry, the science of space, and that dynamics, the science which sees time playing something like an equal part with space in an account of the world, is not a Greek, but a later development.[13]

But we should not of course entertain the idea that the Greeks knew nothing about time. They knew a great deal about it, and began the first laborious attempts to separate out time as a scientific subject in its own right—though the history of that development has yet to be written. What complicates the issue is that although we can trace in Aristotle and his predecessors a certain tendency not to distinguish time enough from other factors such as motion or space, and can trace a growing improvement in this respect, a growing awareness of time, a making less gross confussions, as one thinker succeeds another—nevertheless there exists in even the earliest Greeks a perfectly sophisticated sense of time, as evidenced in the use of tensed language, temporal adverbs, prepositions, and so on. Aristotle too, when he is not at pains to explain time in accordance with his paradigm of space, can give a perfectly straightforward account of time, such as Newton, for example, might use. In the *Categories*,[14] for instance, Aristotle mentions time as an example of a quantum in its own right—but in the mature and more elaborated *Physics* and *Metaphysics* time is a quantum only because space and consequently motion through space are.[15] One is forcefully reminded of St. Augustine's famous complaint: "Quid est ergo tempus? Si nemo ex me quaerat, scio; si quaerenti explicare velim, nescio—What then is time? If nobody asks me, I know. But if I try to explain it to someone who asks, I do not know."[16] This is very much the situation of the Greeks up to, and including Aristotle.

But it is worth remarking here, as a comment on the development of temporal theory among the Greeks, that it is not really surprising that people should know things implicitly long before they develop the explicit theory. The history of scientific thought is in a sense precisely the history of a develpoment in terms of a growing explicit. New theories are not plucked out of the air. A need for a particular theory grows in terms of questions that arise, and a theory is ultimately accepted as satisfactory because it answers those preconceived questions. The theory is, as it were, ultimately *recognized* because of what had already been in people's minds. What had been implicit becomes explicit—and from the retrospective vantage-point of the historian the stages of the rendering explicit can be traced. The danger, for even very good historians, lies in the temptation to give credit to a favorite thinker for knowing explicitly what in all historical probability he knew only implicitly.

A case in point is the historian's refusal to believe that when Plato speaks in the *Timaeus* of a situation before time existed he is merely contradicting himself. Plato must be deliberately using the fanciful language of myth here, so the thinking goes, because "no sane man"[17] could believe that anything was before time. Yet there may be ample evidence here that Plato did simply make a mistake, lulled, it may be by the mythical trappings of his account, and perfectly used to temporal language but not yet quite used to the full logical demands of temporal theory—which is another matter entirely from the skilled use of ordinary temporal language. Another case in point is the question of Aristotle's appreciation of a simple, straightforward earlier-later order of time—such an order as is to be found in the Newtonian treatment of time. Aristotle speaks in several places as though he envisaged such an order. For example, he speaks in the *Physics* of the now as the "link" of time, in the sense that it joins two successive entities, the past and the future,[18] the successive being, as Aristotle says elsewhere, "something posterior"—and he gives the perfectly temporal example of the succession of the second day of the month to the first.[19] Yet when it comes to giving an explicit account of the temporal prior and posterior Aristotle is unable to be that straightforward. Because of his belief that 'prior' and 'posterior' are primarily *spatial* terms, he cannot be sure that he is dealing with a purely temporal 'prior' and 'posterior' without appealing to the 'now' of time. And an appeal to the now or present, as we know, is quite at variance with a prior-posterior, that is, with an earlier-later, account of time, which requires no reference whatever to any present, or any tensed language. It appears that Aristotle is aware of the earlier-later order of time, but he is aware of it only implicitly, and when it comes to explicit theoretical statement he is seen to be somewhat lacking.[20]

So we need not be surprised to find, in Aristotle and earlier Greeks, a perfectly straightforward appreciation of time in their ordinary use of the Greek language, whereas there seems to be lacking (from our modern vantage-point of course) a fully developed capacity to treat of time as an independent subject or to use it as such in a scientific account of the world. It is possibly such a lack of theoretical control of time (more positively, it is an indication of a not yet fully developed theoretical control of time) which explains the non-appearance of 'the beginning,' the ἀρχή, as a simple

temporal beginning in Aristotle's account of the ἀρχή in *Metaphysics* Delta, Chap.
1. Again it explains the use of ἀρχή in a temporal setting but with a non-temporal,
or not primarily temporal, meaning in ch. 11 of the same book. There the now is
considered an ἀρχή, a principle of some kind, by reference to which (and not in a
simple earlier-later way) temporal priority is determined, things in the past, Aristotle
says, being prior when they are further from the now, but things in the future being
prior when they are closer to the now.[21]

Similarly in *Physics*, iv. 14 Aristotle says:

> What is *before* is in time; for we say 'before' and 'after' with reference to the distance
> from the 'now', and the 'now' is the boundary of the past and the future; so that since
> 'nows' are in time, the before and after will be in time too; for in that in which the 'now'
> is, the distance from the 'now' will also be.[22]

There, as I have mentioned before, the now serves to guarantee that the before and
after, defined, awkwardly and for the nonce, as 'distances from the now', are in time,
for they are originally for Aristotle spatial terms. But as 'before' and 'after' cannot in
the first place be taken as temporal by Aristotle, so the ἀρχή which the now is cannot
be taken as purely temporal either. For the now as providing a point of reference in
terms of which temporal 'before' and temporal 'after' may be defined is clearly not a
simple temporal starting point, since befores and afters may occur on either side of it,
on the past side as well as on the future. The ἀρχή here is not a temporal beginning
but simply a principle of reference—something which, being in time itself, puts the
before and after defined in terms of it in time also.

All of this is much more complex, and much more confused, than a simple
earlier-later temporal order, like the Newtonian, and it argues a comparative
non-development in the temporal thought of Aristotle; an inability as yet to envisage
in theory time as a dimension of the physical world not derived from space. The
development of dynamics had to wait for that conception of time as independent,
whatever may be said for the subsequent merging of time with space again (but not,
as in Aristotle, the derivation of time from space) in the modern space-time theory of
relativity.

But if the Greeks up to Aristotle were engaged in the not quite successful struggle
to separate time theoretically from space, there is no doubt that they also handled
temporal concepts with unconscious ease when they were untrammelled with
theoretical demands. That is why the original meaning of ἀρχή is very probably as
the lexicon gives it: 'temporal beginning' primarily, though other meanings signifying
'origin,' such as, for example, 'cause,' might quickly and easily accrue ot it. So in
Homer ἐξ ἀρχῆς means 'from of old,' and Menelaus in the *Iliad* blames Paris for the
troubles: ἕινεκ᾽ ἐμῆς ἔριδος καὶ ᾽Αγεξάνδρομ ἔνεκ᾽ ἀρχῆς. Many evil things have
fallen out, he says, because of my rage and because of its beginning by Paris.[23] Here
ἀρχή could be translated 'cause'—but the temporal reference is clear.

We must, I think, concede the general notion of time and the notion of beginning
as primitive. It is very difficult to see where they could have come from more directly
than from the simple experience of time itself. It is easier, perhaps, to see such a
notion as 'cause' or as 'principle' arising out of the notion of 'beginning' in its pure

temporal meaning. 'Post hoc, ergo propter hoc' may be a fallacy to more sophisticated minds, but it is also very likely the route of causal discovery for more primitive ones. But, at any rate for the Greeks, all the temporal concepts, tenses of verbs, temporal adjectives such as πρότερος, 'former', and temporal adverbs and conjunctions such as πρίν, 'before' are there from Homer. And it might be wise to suppose that for Aristotle and others the difficulty with temporal theory was not any lack in their sense of time or in their command of temporal language so much as the temptation to make time conform to their more mathematically manageable sense of space.

Now, granted that Greek writings as early as Homer contain the notion of a temporal beginning, we may next ask whether some kind of specific beginning is envisaged by the early Greeks, and if so, what it is. Here we might list the ἀρχαί noted by Aristotle and others: water for Thales, the unlimited for Anaximander, air for Anaximenes, fire/change for Heraclitus, τὸ ὄν, that which is, for Parmenides, the elements, earth, air, fire, and water for Empedocles, the unlimited number of all kinds of things for Anaxagoras, atoms and void for the atomists, and so on—but these are all ἀρχαί in Aristotle's technical sense of the term: principles of origin rather than, necessarily, temporal beginnings, though possibly there is not much, if any, distinction in the Presocratics between principles and temporal beginnings—but it will suit our present concern to concentrate on those beginnings which we can label explicitly temporal.

Here it may be noted that we do not find among the early Greeks any such initial phrase as ἐν ἀρχῇ, 'in the beginning,' as we find it in the Septuagint Book of Genesis and in the Gospel of St. John. Even Hesiod, who starts out to give an historical account, as it were, of the first beings, begins: πρώτιστα Χάος γένετ'— "First of all Chaos came into being."[24] As M. L. West remarks: "He tells us simply that Chaos came into being first; he does not tell us how."[25] Just before Hesiod begins this account, in his preliminary invocation to the Muses he uses the phrase ἐξ ἀρχῆς just as Homer does. He says that the Muses hold the mansions of Olympus ἐξ ἀρχῆς, from of old. We can hardly translate that 'from the beginning,' I suppose, since the beginning is, strictly, when Chaos came to be, which is long before the Muses, daughters of Zeus and Mnemosyne, appeared—but then again this is the kind of accuracy which we probably should not foist upon so early a writer.

What we do have among the early Greeks are statements to the effect that such and such existed forever. One of the earliest of these is the statement of Pherecydes of Syros (date uncertain but not later than the middle of the sixth century, B.C.) that Zas, or Zeus, Chronos, and Chthonie existed always, ἦσαν ἀεί. These are Zeus, the sky-deity, Chthonie, the earth-deity, and, most interestingly for us, Time. Pherecydes also evidently identifies this latter God with Kronos, the father of Zeus in what may be called the standard myth, the God who swallows his own children. The great classical scholar Wilamowitz thought that Time was too abstract a God to be found in the 6th century, and therefore this god must be identified as Kronos, not Chronos, but many scholars now agree with Werner Jaeger that the identification of Chronos, Time, with Kronos, swallower of his own children, is a "patent bit of etymologizing."[26] (One is reminded of Shakespeare's "And time that gave doth now his gift confound.")[27] As G. S. Kirk puts it in *The Presocratic Philosophers*, Χρόνος, which is widely supported in the sources, is almost certainly correct, the other two figures [Zas and Chthonie] are etymologizing variants of well-known theogonical figures, and we naturally anticipate a similar case with the third figure."[28] The abstraction objected to by Wilamowitz need not be of a very high level. As Kirk suggests: "Pherecydes probably took the Kronos of legend, asked himself what the etymology was, and arrived at the obvious answer, Chronos or Time—a familiar and simple concept."[29] As he points out, there is other evidence also to show that Pherecydes liked etymologies.[30]

The phrase 'always existed' is of some significance. Homer had called his gods 'always existing', αἰὲν ἐόντες, even though they had origins—but they are immortal. ἐόντες however is the present participle of the verb 'to be'; Pherecydes on the other hand uses a past tense, ἦσαν ἀεί, 'always existed'. We shall see that this kind of phrase in later writers comes to mean 'had no beginning'.

Time figures also as an originative deity in the so-called Orphic cosmogonies, ranging perhaps from the sixth century to the beginning of the Christian era. As Comford, writing in the *Cambridge Ancient History* says:

> At the best [the Orphic doctrines] interpenetrate the whole mystical tradition of Greek philosophy, Pythagoreanism, Platonism, Stocisim, Neoplatonism, and Christianity. At every stage the influence may have been reciprocal, so that the content of Orphic belief was perpetually modified. No history of its development can now be traced.[31]

In this cosmogony, Cornford says:

> Ageless Time is figured as a winged serpent with the face of a god flanked by the heads of a bull and lion. The mature tendency of the Greek imagination was towards the expurgation of the more grotesque elements, but this was counteracted by syncretistic influences from oriental quarters; and between the two it is impossible now to reconstruct the vision of the sixth-century Orphic.[32]

W.K.C. Guthrie accepts, with caution, the belief that the Orphic views were imported from the East, and, with even greater caution, that the tradition is sixth or fifth century.[33] As Guthrie points out, the Orphic tradition of the fashioning of the primal egg in the *aither* by Time is "a conception foreign to the Hesiodic cosmogony."[34] Kirk sees the identification of the abstraction 'Time' with a multi-headed winged snake as suspect if taken to be pre-Hellenistic Greek, but points out that others see no difficulty in an early ascription of the Orphic doctrines.[35]

But let us turn from these theogonies back to more definitely dateable philosophical tenets. The Time of Pherecydes, we saw, "always existed." "An analagous declaration is seen, some two generations later, in Heraclitus' world-order, which no god or man made but always was, and is, and shall be."[36] There is also a fragment—which Kirk accepts, but it is possibly spurious—from Epicharmus of Syracuse, some thirty years later, say, than Heraclitus, to the effect that the gods were always there, and that Chaos cannot, as in Hesiod, be the first god to come into being, because it is impossible for a first thing to appear from something into something. Plato was accused by some of plagiarizing from Epicharmus, and it is possible that some of the fragments are forged to support that accusation.[37]

But we soon come to the express concept 'without beginning,' in Parmenides (early 5th century), who says that which is is "without beginning, without end," ἔστιν ἄναρχον ἄπαυστον.[38] And Parmenides' disciple Melissus says: "Since then it did not come into being, it is and always was and always will be, and has no beginning or end, but is unlimited."[39]

There is some reason to believe that Melissus did not adequately distinguish between beginning in time and beginning in space. "Nothing," he says, that has a beginning and an end is either everlasting or infinite,"[40] and again, of that which is: "As it is always, so also its μέγεθος (its spatial extension) must be infinite."[41] And later the Sophist Gorgias, evidently influenced by Melissus, produced a curious argument to the effect that what is everlasting has no beginning and is therefore boundless and therefore has no spatial position; for if it had spatial position it would be in something, and so not boundless.[42] So, it turns out, what is everlasting has no spatial position, for it is boundless. This gross confusion between beginning in time and beginning in space is not baldly made by Melissus, but it seems likely enough, given his other statements and the general climate of spatio-temporal confusion that prevails until the time of Aristotle at least, that he did not come to the point of distinguishing adequately between the two, so that confusions were still possible.

But at any rate the concept of 'the beginning' is being separated out, on its way to becoming a term of scientific interest in its own right. Melissus picks out the concepts of beginning and ending even more. He says: "For if it came into being it would have a beginning (for it must have begun, coming into being at some time). And an end (for it must have ended, coming into being at some time). But since it neither began nor ended, but always was and will always be, it has neither beginning nor end."[43] The language is somewhat uncertain of itself and repetitious, but the definition of 'begin' as 'to come to be in time' is incipiently there.

With Empedocles (middle of the fifth century) we come to the word 'beginning' used absolutely. "Come then," he says, "and I will tell you first of the equal-aged and the beginning, ἡλικά τ᾽ ἀρχήν."[44] The beginning in this case seems to be the elements, earth, air, fire, and water, which are also equal-aged with each other, none of them being senior to the others. The word 'beginning' here is probably losing some of its temporal significance and becoming what Aristotle explicitly recognizes it as: a principle, not necessarily temporal—although it still has strong temporal overtones for Empedocles, because he says in the same fragment: "From these elements *came into being* all that we *now* see clearly." That is, he thinks of the cosmogonical process in its historical, temporal, tensed aspect, and not, more abstractly, in terms simply of its elements and their combination.

Similarly Anaxagoras, a somewhat later contemporary of Empedocles, thinks strictly in terms of temporally arranged cosmogonical processes, putting his account in such tensed language as: "At first it began to revolve from a small area, but afterwards it revolves [present tense] more widely, and afterwards again it will revolve more widely still."[45] Note that things *began* to revolve. There is, or rather was, some beginning. For Anaxagors Mind started the cosmic revolution ἀρχήν at the beginning.[46] And everything comes from this primal revolution, started by Mind.

A somewhat different beginning is given by the contemporary Pythagorean Philolaus. He is reported as calling the one the beginning of everything.[47] This is in accordance of course with the Pythagorean doctrine that everything is of the nature of number, and we may suspect that 'beginning' here does not bear a strictly temporal connotation. The authenticity, however, of the fragment is disputed. But the Atomists, a little later, are speaking of the real nature of things as being atom and void, and in a way that lends itself to present-tense, or indeed tenseless, expression—the real nature of things *is* atoms and void.[48]

Plato, in his cosmogonical account, the *Timaeus*, reverts to talk of a temporal beginning. But Plato is not content with a beginning *in* time; he speaks in a way that clearly indicates a beginning *of* time. The Demiurge, the creator, introduces time into the cosmos as a quality of order which will make it resemble yet more its eternal pattern.[49] As Aristotle points out, Plato is the only one of the Greek philosophers thus to conceive of a beginning of time itself.[50] Otherwise it is traditional among the Greeks to think of time itself as without beginning and everlasting. For Anaxagoras, who, as we say, envisaged a beginning of the ordered cosmos in a vortex or revolution introduced by Mind, nevertheless there was a time before that first revolution started, for he says that *before* the separation off of things occurred, with the revolving motion, everything was all together.[51] For Parmenides indeed there is no beginning only because there is no change at all, no time, but for Melissus, the somewhat unorthodox disciple of Parmenides, the world is like that of Heraclitus: a world which is, always was, and always will be. And this is in agreement with the Chronos of Pherecydes, who existed always. Aristotle himself reverts to this tradition of everlasting time, and for him too, we saw, as for Heraclitus and Melissus, the world itself, and not just time, is everlasting. Plato, in assigning a beginning to time, does not manage to be quite consistent, for, as I mentioned before, he speaks of a condition *before* time came to be, and it is unlikely that the general mythical form of his narration can save him here.

For Plato the word ἀρχή is, as Gregory Vlastos, says, a 'weasel-word.' It may mean any, or all of (i) beginning, (ii) source, (iii) cause, (iv) ruling principle, (v) ruling power.[52] So ἀρχή is already taking on for Plato some of the wealth of meaning it will have for Aristotle, and losing its more purely temporal meaning. In the *Phaedrus*, Plato says that the ἀρχή does not become from anything, for what becomes from something else is clearly not itself the ἀρχή.[53] Here ἀρχή might be translated either 'first principle' or 'beginning'—but 'temporal beginning' would probably be too strong a translation in view of the fact that Plato, in the *Timaeus*, as we have seen, *does* think a temporal beginning, indeed a beginning of all time, comes to be. ἀρχή is losing its temporal significance, and Aristotle will, by omission at least, as we have seen, get rid of it entirely.

We may bring this account of the Greek concept of 'the beginning' to a close by wondering why the Greeks, who tended to dislike the infinite in space, were inclined to accept it as regards time. The universes of both Aristotle and Parmenides are bounded in space but not in time—but Melissus, again the heretical disciple, makes his cosmos infinite in both space and time. Again it must be said for Parmenides that he is not really for limitless time but against all change, and so can accept no beginning and no end. Oddly enough though, he does see his universe limited in space, for it is decreed, he says, that that which is should not be without limit.[54] It is the same feeling for space that makes the cosmos of the Pythagoreans and the cosmos of Aristotle limited. It is perhaps something about the immediate experience of space and the immediate experience of time which makes the envisaging of one with boundaries and of the other without boundaries natural. At any rate this is the way the Greeks, with some exceptions, tended to see them, and in spite of two rather odd facts. One of these facts is that Euclidean geometry, the most Greek of sciences, ultimately calls for infinite space, one of the most un-Greek conceptions perhaps; the other odd fact is that, as I have reported before, there was a certain tendency among the Greeks, up to Aristotle at any rate, to confuse time and space together. Perhaps the recurrent insistence in the Greek mind that time was without beginning was one natural reaction to that unnatural merging.

Acknowledgment

I am grateful to Nathaniel Lawrence for some excellent suggestions which I have tried to incorporate into this article, Professor Lawrence (and many will agree with him) thinks it may be a kind of modern provincialism which makes us detect certain logical errors in ancient temporal theory, and makes us look for a development in the direction of our own scientific views. "The risk," he says, "is that oblique usage by Plato and Aristotle, interpreted as error or confusion, saves us from having to dig out lost meanings that might well illuminate our own." That risk does exist, but, I should like to point out, there is a temptation on the other side too (and, to my mind, the fall here is much more common): that of thinking that a favorite thinker, be it Aristotle, Plato, or whoever, always knew full well what he was about, so that when he seems merely confused he must be supposed instead to be teaching us a subtle lesson—or, more generally, we may find ourselves attributing meanings or intentions which have little textual support. Of the two views, with their attendant risks, it makes more sense to me to think that there was, generally speaking, and not necessarily continuously, a development, rather than that there was not. The *prima facie* evidence, which further research reinforces, is that some of the earlier Greek thinkers did not sufficiently distinguish time, as a subject of scientific investigation, from space; that Plato, as so early an authority as Aristotle charges, did not sufficiently distinguish it from motion; and that Aristotle himself, though not making such confusions, tried to derive temporal properties from those of motion and space. These are definite signs of a development of temporal thinking among the Greeks, and, until we have indisputable evidence to the contrary, the notion of development remains the most powerful working hypothesis, giving clear direction and order to research—granted that any hypothesis, like any falsifiable scientific theory, has to run certain risks.

Notes

1. Buck, C. *Dictionary of Selected Synonyms in the Principle Indo-European Languages* (Chicago: University of Chicago Press, 1949).
2. St. Thomas Aquinas, *Summa Theologica*, Part 1, Question 33, Article 1.
3. Aristotle, *Physics*, Book 2, Chap. 3, 194b, 16–195a, §§ 4.

4. Aristotle, *Metaphysics*, Book Delta, 1012b, 34—1013a, §§ 21.

5. Aristotle, *Physics*, Book 4. Chapter 2, 219a, §§§ 14—16.

6. *Ibid.*, 219a, §§§ 16—19.

7. Denis Corish, "Aristotle's Attempted Derivation of Temporal Order from That of Movement and Space," *Phronesis* 21 (1976).

8. Denis Corish, "Aristotle on Temporal Order: 'Now,' 'Before,' and 'After,' *Isis* 69 (1978).

9. *De Caelo*, Book I, Chap. 10, 279b, §§ 4.

10. *De Caelo*, Book II, Chap. 1, 284a, §§ 8—10.

11. See especially Plato, *Timaeus*, 37D, 5—38A, §§ 8. Some scholars such as A.E. Taylor, *A Commentary on Plato's Timeaeus*, London: Oxford University Press, 1928), pp. 190—191 tend to see Plato as identifying time with motion; others, such as Harold Cherniss, *Aristotle's Criticism of Plato and the Academy* (Baltimore: The Johns Hopkins Press, 1944), p. 418 tend to see him as saying only that time measures change.

12. Aristotle said that "some" identified time with the motion of the universe (*Physics*. 4, 10, 218a 33—b1) and the tradition, from Eudemus and Theophrastus, is that he meant Plato.

13. "The Greek sciences of astronomy, optics, and statics had all been elaborated geometrically, but there was no such representation of the phenomena of change. Archimedes' famous work in statics was not paralleled by any equivalent kinematic system which admitted of representation in the form of mathematical propositions." Carl B. Boyer, *The History of the Calculus and Its Conceptual Development* (New York: Dover Publications Inc., 1959), p. 71.

14. Aristotle, *Categories*, Chap. 6, 4b, pp. 22—25: 5a, p. 6. For a comment see Corish, D., "Aristotle's Attempted Derivation of Temporal Order from That of Movement and Space," p. 247.

15. Corish, D., "Aristotle's Attempted Derivation of Temporal Order from That of Movement and Space," pp. 246—247.

16. St. Augustine, *Confessions*, Book 11, 14.

17. "No sane man could be meant to be understood literally in maintaining at once that time and the world began together [*Timaeus* 38b 6], and also that there was a state of things which he proceeds to describe, *before* there was any world." Taylor, *Commentary*, p. 69.

18. *Physics*, Book 4, Chap. 13, 222a, §§§ 10—12

19. *Physics*, Book 5, Chap. 3, 227a, §§ 4—6.

20. Corish, D., "Aristotle on Temporal Order: 'Now,' 'Before,' and 'After,' " p. 73.

21. *Metaphysics*, Book Delta, Chap. 2, §§§ 1018b, §§§ 14—19. See also 1018b, §§§ 9—16.

22. Ibid., 223a, §§ 4—8.

23. Homer, *Iliad*, Book 3, Line 100.

24. Hesiod, *Theogony*, Line 116.

25. M.L. West, *Hesiod Theogony* (London: Oxford University Press, 1966), p. 192.

26. Werner Jaeger, *The Theology of the Early Greek Philosophers* (Oxford University Press, 1947), pp. 68—69.

27. William Shakespeare, Sonnet 60.

28. G.S. Kirk and G.E. Raven, *The Presocratic Philosophers* (Cambridge: Cambridge University Press, 1962), p. 56.

29. Ibid., n. 1., pp. 56—57.

30. Ibid., p. 56.

31. F.M. Cornford, *The Cambridge Ancient History*, vol. 4, Chapter 15, p. 533.

32. Ibid., p. 536.

33. W.K.C. Guthrie, *Orpheus and Greek Religion*, 2nd ed. (London: Methuen, 1952), p. 91.

34. W.K.C. Guthrie, *The Greeks and Their Gods*, (Boston: Beacon Press, 1954), pp. 318—19.

35. Kirk and Raven, *The Presocratic Philosophers*, p. 43. For early ascriptions and theories of oriental influence see M.L. West, *Early Greek Philosophy and the Orient* (London: Oxford University Press, 1971).

36. Kirk and Raven, *The Presocratic Philosophers*, p. 55.

37. Ibid., p. 55. See also Kathleen Freeman, *The Presocratic Philosophers* (Cambridge: Harvard University Press, 1946), p. 134.

38. Hermann Diels, *Die Fragmente der Vorsokratiker*, 10th ed. ed. with additions by Walther Kranz, (Berlin: Weidmannsche Verslags-buchhandlung, 1960−61). Parmenides, Fragments 8 and 27.

39. Diels-Kranz, Melissus, Fragment 2.

40. Ibid., Fragment 4.

41. Ibid., Fragment 3.

42. Diels-Kranz, Gorgias, Fragment 3.

43. Diels-Kranz, Melissus, Fragment 2.

44. Diels-Kranz, Empedocles, Fragment 38.

45. Diels-Kranz, Anaxagoras, Fragment 12.

46. *Ibid.*

47. The report is from Iamblichus, a neo-Platonist of the fourth century A.D. See Freeman, pp. 221−222.

48. See, for example, Democritus, Fragment. 9.

49. Plato, *Timaeus*, 37C-D7, 38B, §§ 6−7.

50. Aristotle, *Physics*, Book 8, Chap. 1, 251b, §§ 17.

51. Diels-Kranz, Anaxagoras, Fragment 4.

52. Gregory Vlastos, "The Disorderly Motion in the *Timaeus*," in R.E. Allen ed., *Studies in Plato's Metaphysics*, §§, n.3 (New York: The Humanities Press, 1965), pp. 397−98.

53. Plato, *Phaedrus*, 245D 1−2.

54. Diels-Kranz, *Parmenides*, Fragment 8.

Death, Literature, and Its Consolations

G. H. FORD

> "The artist carries death within him like a good priest his breviary."
>
> —Jakov Lind

An early scene in James Joyce's novel, A *Portrait of the Artist as a Young Man*, shows the protagonist, Stephen Dedalus, as a child attending a boarding school outside of Dublin. One day he opens up his geography textbook and notes how he has written his name on the title page:

Stephen Dedalus
Class of Elements
Clongowes Wood College
Sallins
County Kildare
Ireland
Europe
The World
The Universe

The little boy, as Joyce says, tries reading the page from bottom to top until he comes to his own name:

> That was he: and he read down the page again. What was after the universe? Nothing. But was there anything round the universe to show where it stopped before the nothing place began? . . . He could think only of God. God was God's name just as his name was Stephen. . . . It made him very tired to think that way. It made him feel his head very big.

In Joyce's delightful passage surely most of us will recognize a shared experience of our growing up, although the emphases in the questions may differ in accordance with individual interests. The incipient student of science would work from the top of the chart downwards to the big questions of: What is the universe? What is matter? And the incipient student of literature would work instead from bottom to top in order to ask the nature of his or her identity: Who am I? he asks. What will become of me?

Both sets of questions relate, of course, to time, but the literary-style ones emphasize the realization that the *individual* is yoked to time; it is the "I" or "me" who is threatened by time's changes. In Joyce's novel young Stephen is haunted by a human skull which he sees ornamenting the desk of the Jesuit rector of his school, although finally, inspired by a vision of what he calls "the angel of mortal youth and beauty, an envoy from the fair courts of life," he flies away from deathly confinements and embraces life: "I go to encounter for the millionth time the reality of experience and to forge in the smithy of my soul the uncreated conscience of my race."

These words are followed on the page by two words: "The End." Final pages of this kind, like final curtains in a drama, correspond to endings in music, and they raise similar critical problems. These, however, are not the kinds of endings to be treated in the following discussion. My topic is death and dying, as they are portrayed in literature, and although sometimes the death of an individual character—his or her *ending*—may coincide with a final curtain, there is obviously no requirement that this must be so.

This topic of death in literature offers a vast quantity of examples on which to draw. When the topic was first proposed to me, a year ago, I began browsing through the two volumes of *The Norton Anthology of English Literature* (of which I happen to be one of the editors), and I was astonished to discover that on almost every other page, of its five thousand pages, is some reference to death. And in virtually every one of the thousands of novels and plays in print there are death scenes. As Touchstone observes in *As You Like It*:

And so from hour to hour we ripe and ripe,
And then from hour to hour we rot and rot;
And thereby hangs a tale.

The tale that hangs thereby is of extraordinary length, and one result of this abundance has been my decision to restrict illustrations to works written in English. It would be tempting to expand the discussion by including such a highly relevant text as Tolstoy's story of the death of Ivan Ilyitch, but inasmuch as we are swamped with materials without crossing the Channel, I hope the insularity may be excused.

The only other apology that may be needed concerns the topic itself, which, in some quarters, is considered offensive. In the 19th century, the taboo subject was, of course, sex. In our century, the taboo subject (at least in my country, the United States) has been death. In his book, *Western Attitudes Towards Death*, Philippe Ariès cites Geoffrey Gorer's observation that 19th century children were told that they had been brought by the stork or found under a cabbage-leaf in the garden, yet they watched their grandfather die. Today's children study sex techniques in grade school, but when grandfather is no longer in evidence the children are told that he is "resting in a garden of flowers."[1] More recently, however, the taboo seems to be losing its force in America, at least among adults. Indeed, one of the contributors to a recent collection of papers, *Perspectives on Death and Dying* (1979), after citing a saying of the poet, Yeats ("Man has discovered death"), protests that the rediscovery of death may have become almost a fad in America today.[2]

One doctor who played a prominent role in this change of attitudes was Elisabeth Kübler-Ross, in her early work such as *On Death and Dying*(1968). Dr. Ross reports that when she began her investigations in the 1960s it was very difficult, at first, to get either doctors or patients to talk about death.[3] These inhibitions, as we have already indicated, seem never to have affected our writers. Let us begin with a line that has appealed to several writers. It was reputed to have been devised on the occasion of the execution of Mary Queen of Scots in 1587: "In my end is my beginning." Apparently what Queen Mary's motto meant was that her death would be viewed as a martyrdom and would hence incline later generations to become Roman Catholics.[4] In his time-conscious *Four Quartets* T. S. Eliot repeats Mary's line but also, and more significantly, opens his poem by turning the expression around. "In my beginning," he writes, "is my end."[5] The death process starts with birth, that is, and beginning and ending are, in a sense, one.[6] Eliot's paradox is a more sophisticated play on the idea expressed in Robert Herrick's little anthology-piece in which the poet addresses a garden of short-lived flowers and comments on a basic life-death cycle shared by flower and man alike:

> We have short time to stay, as you,
>   We have as short a Spring;
> As quick a growth to meet Decay,
>   As you, or any thing.
>   We die
>   As your hours do * * *

Noteworthy here is how much the poet has packed into one line: "As quick a *growth* to meet Decay"—whereby a platitude has been made memorable through language.[7]

But this issue of the role of language is one to which we can return shortly. Our concern first is with Queen Mary's endings and beginnings, and what becomes of time. Here, as often happens, a short passage from Shakespeare supplies us with a useful text. It is from *Henry IV*, the scene in which the young leader, Hotspur, is dying. Just before the end Hotspur comments memorably and rather mysteriously:

> But thought's the slave of life, and life time's fool,
> And time, that takes survey of all the world,
> Must have a stop.

This passage was used by Aldous Huxley as the title for his novel about death and dying: *Time Must Have a Stop*. If I understand the passage, Hotspur means that death, for the individual, is the end of time. What is involved in that stopping has, of course, been variously represented. For millions of Christians, throughout the centuries, eternity begins with death, and writers sharing such a faith variously portray life eternal. It may be a life of punishments as implied in the hymn, "Days of Moments Quickly Flying," by Edward Caswall:

> As the man dies,
> Such must he be,
> All through the days
> Of Eternity.[8]

Or it may be bliss in a heavenly city. Typical of such literature is John Bunyan's *Pilgrim's Progress*, a book that until our day was one of the most widely-read works in our language. As a brief sample of such literature, here is Bunyan's account of what happens to his hero, Christian, and to his companion, Hopeful, after they cross the river of death:

> Now I saw in my dream that these two men went in at the gate; and lo, as they entered, they were transfigured, and they had raiment put on that shone like gold. There was also that met them with harps and crowns, and gave to them: the harps to praise withal, and the crowns in token of honor. Then I heard in my dream that all the bells in the city rang again for joy, and that it was said unto them, "ENTER YE INTO THE JOY OF OUR LORD" (Matthew xxv. 21). I also heard the men themselves, that they sang with a loud voice, saying, "BLESSING AND HONOR GLORY AND POWER, BE TO HIM THAT SITTETH UPON THE THRONE, AND TO THE LAMB FOREVER AND EVER" (Revelation v. 13).

Although other Christian writers may be more sophisticated than Bunyan, they nevertheless share a similar sense of eternity. The sermons of John Donne may be cited and also his poems such as his sonnet: "Death Be Not Proud" which concludes with his paradox:

> "One short sleep past, we wake eternally
> And death shall be no more; Death, thou shalt die."[9]

For the 19th century we might follow Donne with another devout Anglican poet, Christina Rossetti, who, like Donne, writes of death as a release ("When I am dead, my dearest, / Sing no sad songs for me."). And in the 20th century a similar view of death and an afterlife persists in several writers, as for example, in Evelyn Waugh's novel, *Brideshead Revisited*, where the deathbed conversion of Lord Marchmain is central.

In most writers of our century, however, and in many writers since the Renaissance, Bunyan's vision is no longer in focus. As stated by the authors of a recent textbook, *Understanding Death and Dying*: "Unlike the medieval, whose guiding concern was his preparation for afterlife, the contemporary has, in the words of Nathan A. Scott, 'relocated the ultimate problem of human existence from the dimension of Eternity to the dimension of time.' The interest in death no longer centers on what follows . . ."[10] And not only for these recent writers but also for many earlier writers since the Renaissance has there been exhibited this shift of attitude towards an "afterlife." For such writers there is little certainty that when we have shuffled off this mortal coil there is a heavenly city to be found on the other side (so to speak) in that "undiscovered country, from whose bourn / No traveller returns." For them, if we are to exit from this world with dignity, it must be without the special consolations of a heavenly dispensation.

What are these alternative consolations, or supplementary consolations, let us say, in place of the strictly Christian ones? The obvious place to look for a demonstration will be in the plays of Shakespeare.[11] When the protagonist in a Shakespearian tragedy arrives at what Othello calls "my journey's end . . . my butt, / And very seamark of my utmost sail," he or she confronts his doom with a courage and self-assurance that promise for the character an exit with dignity. But what is truly consolatory for us as we watch these painful scenes is not just the courageous stance of the protagonist but the language in which the event is couched for us, the poetry, that is. To choose one illustration among so many memorable finales is difficult, but let us settle for Queen Cleopatra despite the more obvious attractions of King Lear. Cleopatra, like Lear, has (to put it mildly) been misguided. At the end of the play her lover, Mark Antony, is reported to be dead, after his defeat by Caesar, and her misguided vacillations have contributed to the downfall of her lover and of her country. But as she resolves to die, she reassumes heroic stature, and her past errors are obliterated by the glorious blank verse with which she greets death. To her maids she says:

> "Give me my robe, put on my crown. I have
> Immortal longings in me. Now no more
> The juice of Egypt's grape shall moist this lip.
> Yare, yare, good Iras; quick. Methinks I hear
> Antony call. . . . Husband, I come!
> Now to that name my courage prove my title!
>   ** ** ** Peace, peace!
> Dost thou not see my baby at my breast,
> That sucks the nurse asleep? . . .
> As sweet as balm, as soft as air, as gentle—
> O Antony . . ."

And her maid, as the Queen dies, says farewell to her:

> "Now boast thee, death, in thy possession lies
> A lass unparallel'd. Downy windows, close;
> And golden Phoebus never be beheld
> Of eyes again so royal! Your crown's awry.
> I'll mend it and then play."

What is the effect on an audience of lines such as these? We may affirm (I think) that although our world may be a "tough" world as Kent calls it in *King Lear*, a "rack" on which the protagonist has been stretched, yet because it is a world that can produce poetry as beautiful as these lines from Cleopatra's dying speeches, we are consoled and are reconciled to the human dispensation. Our emotional catharsis hence derives from the wizardy of the poetry.

The consolatory effects of poetic language can be illustrated by a more recent example if we look into the ending of the protagonist of the great novel by Malcolm Lowry, *Under the Volcano*, first published in 1947. The protagonist of this novel, Geoffrey Firmin, is a brilliant scholar but also a drunkard and is as misguided in his actions as Lear or Othello or Cleopatra. At the end of the novel Firmin is shot by a fascist police-chief in Mexico, and his body is thrown into a ravine followed by the body of a dead pariah dog. Firmin's dying words are prosaic enough; he sobers up to remark, wryly: "Christ, this is a dingy way to die." We may be reminded here of the gruesome prosiness, in Kafka's *The Trial*, of the protagonist's last words as he is being stabbed to death: "Like a dog!" he says. But Lowry, one of the most poetic of our 20th century novelists, doesn't leave the death-scene on this grimly naturalistic level. A friend of the dead man, reading a volume of Elizabethan plays formerly owned by Firmin, encounters a passage in Marlowe's play, *Dr. Faustus*, which he applies to Firmin's death. It is the comment by the Chorus on the death of Faustus:

> "Cut is the branch that might have grown full straight,
> And burnéd is Apollo's laurel bough
> That sometime grew within this learned man.
> Faustus is gone: regard his hellish fall."

Lowry knew where to go for his poetry, and Marlowe's lines on Faustus operate in the novel almost in the way of Horatio's lines after Hamlet's death:

> "Goodnight sweet prince,
> And flights of angels sing thee to thy rest!"

This death-scene in a 20th century novel seems to me an impressive example of how the Shakespearean style may help to reconcile us to the fact of death. My point can be reinforced by an essay on Lowry's novel by the American writer, William Gass (published two years ago), who said of Firmin's death-scene: "Despite the fact that the scene is excessively operatic . . . there is no death in recent literature with more significance."[12] The overall effect of this scene is strikingly different from the death of Private Roth in Norman Mailer's novel, *The Naked and the Dead*, in which a character also falls to his death down a rocky ravine. But Roth's exit is a pointless one; his is a representively naturalistic death-scene, in which consolations are altogether absent.

One other death-scene involving Shakespeare is from biography rather than from literature, the scene of the death of Alfred Tennyson on a moonlit night in 1892. According to our tastes in the 1980s, this scene may seem much too elaborately staged, for as the 83-year-old poet lay on his death-bed, surrounded by his family, the moonlight poured in through the oriel windows of his room and bathed the bearded face with soft shadows. As his doctor reported afterwards about the Lauriate's noble exit:

> On the bed a figure of breathing marble . . . his hand clasping the Shakespeare which he had asked for . . . and which he had kept by him to the end; the moonlight, the majestic figure as he lay there, 'drawing thicker breath,' irresistibly brought to our minds his own [poem] the 'Passing of Arthur.'[13]

For our purposes the fascinating touch here in this scene is the volume of Shakespeare clasped consolingly in the dying poet's hands. We know that as death approached, Tennyson asked his son to read to him from three of his favorite plays, including *Cymbeline*,[14] with its famous lyric, "Fear no more," which is addressed to the supposedly dead body of the heroine of the play, and which makes highly appropriate music for a death-scene:

> Fear no more the heat o'. the sun,
>   Nor the furious winter's rages;
> Thou thy worldly task hast done,
>   Home art gone, and ta'en thy wages.
> Golden lads and girls all must,
>   Like chimney-sweepers, come to dust.

These same consolatory lines were later to be used by Virginia Woolf as a theme-song or recurring motif in her novel about time and death, *Mrs. Dalloway.*

The examples so far provided may give the misleading impression that only the Elizabethans were equipped with the resources of a poetic language adequate to confront the fact of death, but obviously such later poets as Keats and Tennyson are of the same mould. Keats, who was to die young and who knew he would die young, wrote his deathless lines on the nightingale's song:

> Now more than ever seems it rich to die,
> To cease upon the midnight with no pain
> While thou art pouring forth thy soul abroad
>   In such an ecstasy![15]

And Tennyson, who was to die in old age, had been early haunted by a sense of man's mortality, and in his "Tithonus" wrote of it in language as rich as Shakespeare's. His is the story of a lover doomed to live forever and who yearns instead to be part of the normal life-death cycle: "Of happy men that have the power to die."[16]

Our list of writings which rely upon poetic language to cushion us to death and dying could be vastly extended beyond Shakespeare and Keats and Tennyson. Instead I propose that we turn now to more prosaic works of literature in order to discover whether comparable consolations may still be found. George Eliot's novels make an excellent test case, it seems to me, for her style of writing is certainly prosaic, and there are no heavenly cities in her account of the human dispensation.

To illustrate the point here, I propose to examine a long paragraph from Eliot's novel of 1870, *Middlemarch,* but before citing his passage let me fill in the context of the scene a little. Mr. Casaubon is a clergyman-scholar, in his fifties, who, for 42 chapters of the novel has had little claim on the affections of readers. It seems inconceivable that Mr. Casaubon could ever arouse our sympathies because George Eliot, up to this point, has unflinchingly shown him to be dryly pedantic, stilted in speech, extremely self-centered and selfish, physically unattractive and also not in good health.

In this scene that I am going to cite, Mr. Casaubon is out walking in his garden where he has been talking about his ill-health with his doctor, a vigorous young man named Dr. Lydgate. The doctor tries to feel sorry for his patient, of course, but he also feels some amusement at the elder man. As Eliot remarks: "He [Dr. Lydgate] was at present too ill acquainted with disaster to enter into the pathos of a lot where everything is below the level of tragedy except the passionate egoism of the sufferer."—Under Casaubon's persistent questioning about his health, Doctor Lydgate finally agrees to tell the older man that he has a serious heart condition and will probably not have long to live. Here is Eliot's paragraph about Casaubon's response to his death-warrant:

> Lydgate, certain that his patient wished to be alone, soon left him; and the black figure with hands behind and head bent forward continued to pace the walk where the dark yew-trees gave him a mute companionship in melancholy, and the little shadows of bird or leaf that fleeted across the isles of sunlight, stole along in silence as in the presence of a sorrow. Here was a man who now for the first time found himself looking into the eyes of death—who was passing through one of those rare moments of experience when we feel the truth of a commonplace, which is as different from what we call knowing it, as the vision of waters upon the earth is different from the delirious vision of the water which cannot be had to cool the burning tongue. When the commonplace "We must all die" transforms itself suddenly into the acute consciousness "I must die—and soon," then death grapples us, and his fingers are cruel; afterwards, he may come to fold us in his arms as our mother did, and our last moment of dim earthly discerning may be like the first. To Mr. Casaubon now, it was as if he suddenly found himself on the dark river-brink and heard the plash of the oncoming oar, not discerning the forms, but expecting the summons. In such an hour the mind does not change its lifelong bias, but carries it onward in imagination to the other side of death, gazing backward—perhaps with the divine calm of beneficence, perhaps with the petty anxieties of self-assertion. What was Mr. Casaubon's bias his acts will give us a clue to. He held himself to be, with some private scholarly reservations, a believing Christian, as to estimates of the present and hopes of the future. But what we strive to gratify, though we may call it a distant hope, is an immediate desire: the future estate for which men drudge up city alleys exists already in their imagination and love. And Mr. Casaubon's immediate desire was not for divine communion and light divested of earthly conditions; his passionate longings, poor man, clung low and mistlike in very shady places.

We are a long way here from the rich poetic language and rhythms of Shakespeare's dying Cleopatra. Nevertheless, George Eliot's paragraph is, in its own way, also charged with feelings. What this paragraph conveys is a sense of pity for our human lot, for the conditions of our mortality (we all must die), and even compassion for a character's limitations and pettiness of spirit which will lead Mr. Casaubon to ignominious actions just before his death. What Wilfred Owen said about the poems he wrote in World War One might apply nicely to Eliot's novel: "The poetry," Owen said, "the poetry is in the pity."[17]

Her capacity for pity and imaginative understanding is a quality of George Eliot's writing that caused the French critic, Ferdinand Brunetière, to argue that Eliot was a greater writer than Flaubert because she is a greater realist—for, Brunetière, insisted, pity is a part of life, and because Flaubert tried to exclude pity, his picture of Emma Bovary's death is a flawed and incomplete picture.

On this issue of pity some discrimination is called for, however, as the problem of some of the death-scenes in Dickens' novels illustrates. Early in his career Dickens discovered how to tap the springs of pity among his contemporaries by his open-stopped scenes of the deaths of children: Little Nell, Paul Dombey, Tiny Tim, and, later, Jo in *Bleak House*. Even the most sophisticated early Victorian readers wept copiously with pity over these death-scenes, but by the end of the century tastes had changed, and the scenes had been generally discredited. By the 1890s, Oscar Wilde could remark that a reader must have a heart of stone to read the death of Little Nell without laughing.[18]

Dickens' blunders in some of his death-scenes (at least in the eyes of posterity) seem to have been a special case, for in the best of his immediate successors in novel-writing, pity is a powerfully effective element, as it had been in *Middlemarch*. Thomas Hardy, for example, in his *Tess of the D'Urbervilles*, softens the harshness of the final scenes of Tess's capture at Stonehenge, and her death by hanging, through a note of tenderness summed up in the epigraph from Shakespeare which appears on the novel's title page: "Poor wounded name. My bosom as a bed shall lodge thee." One also encounters something similar in the curious novel by Joseph Conrad, *The Secret Agent* (1907), in which the incessant irony and the depressing effect of the murky London scene are undercut by a sense of pity which becomes evident in the central scene of the novel. This scene, in chapter VIII, features an elderly woman who is taking her last cab-ride across the River Thames (likened by Conrad to the River Styx). Across the river she will move into an old folk's home to die. During the course of this last ride her half-witted son, Stevie, tries to take pity on the whole world—which "ain't an easy world" as the cabman says. Even the decrepit and worn-out cab-horse Stevie wishes he could take to his bed for comfort and pity.

Our final extended illustration comes from another 20th century novel, *Sons and Lovers*, by D. H. Lawrence, from which we'll extract the final page for inspection. One of the reasons for selecting this final page is that it may provide some supplementary materials for Marianna Torgovnick's paper dealing with closures in novels.

At the end of Lawrence's novel, the protagonist, Paul Morel, has suffered the death of his mother from cancer, a mother to whom he was passionately attached. He has broken off relations with his mistress and also with his former fiancée, after which he takes a tramway-ride which transports him from the city out into the country where he will confront the night sky and will reflect on his feelings about life and death. Here is the passage:

> When he turned away he felt the last hold for him had gone. The town, as he sat upon the car, stretched away over the bay of railway . . . Beyond the town the country, little smouldering spots for more towns—the sea—the night—on and on! And he had no place in it! Whatever spot he stood on, there he stood alone. From his breast, from his mouth, sprang the endless space, and it was there behind him, everywhere. The people hurrying along the streets offered no obstruction to the void in which he found himself. They were small shadows whose footsteps and voices could be heard, but in each of them the same night, the same silence. He got off the car. In the country all was dead still. Little stars shone high up; little stars spread far away in the flood-waters, a firmament below. Everywhere the vastness and terror of the immense night which is roused and stirred for a brief while by the day, but which returns, and will remain at last eternal, holding everything in its silence and its living gloom. There was no Time, only Space. Who could say his mother had lived and did not live? She had been in one

place, and was in another; that was all. And his soul could not leave her, wherever she was. Now she was gone abroad into the night, and he was with her still. They were together. But yet there was his body, his chest, that leaned against the stile, his hands on the wooden bar. They seemed something. Where was he?—one tiny speck of flesh, less than an ear of wheat lost in the field. He could not bear it. On every side the immense dark silence seemed pressing him, so tiny a spark, into extinction, and yet, almost nothing, he could not be extinct. Night in which everything was lost, went reaching out, beyond stars and sun. Stars and sun, a few bright grains, went spinning round for terror, and holding each other in embrace, there in a darkness that outpassed them all, and left them tiny and daunted. So much, and himself, infinitesimal, at the core a nothingness, and yet nothing.

"Mother!" he whimpered—"mother!"

She was the only thing that held him up, himself, amid all this. And she was gone, intermingled herself. He wanted her to touch him, have him alongside her.

But no, he would not give in. Turning sharply, he walked towards the city's gold phosphorescence. His fists were shut, his mouth set fast. He would not take that direction, to the darkness, to follow her. He walked towards the faintly humming, glowing town, quickly.

This passage seems to me to be different from George Eliot's or Conrad's, for the quality of the pity is somehow different. What is offered, and quite grippingly I think, is an insight into what growing up may mean. We began our discussion with James Joyce's *Portrait* and the child, Stephen Dedalus, asking questions about his identity and the limits of the Universe. Here, in Lawrence, a child has grown up and is now confronting the experience of time and change and an awareness of his own isolation and future death. The death of a parent leads Paul Morel, like Hamlet, to consider suicide. And, like Pascal, whom he echoes, Paul finds that the silence of eternal spaces terrifies him. The final sentences seem to me ultimately affirmative, but in Lawrence, as in Keats, there is no shirking the darkness, whose presence is made memorably palpable by the passage.

Sons and Lovers was published in 1913. Lawrence himself was to die seventeen years later. In his last years, as death approached, he returned in his writing to the topic of dying, this time in a poem he called *The Ship of Death*. The perspective has changed now:

> We are dying, we are dying, so all we can do
> is now to be willing to die, and to build the ship
> of death to carry the soul on the longest journey.

Lawrence here, in his late years, is drawing on one of the oldest images of death and dying, the mysterious final voyage, present in Norse myths of course, but also extensively used by writers in English, for example by his predecessor, Walt Whitman whose "Voyager" must bid a final farewell to the shore. As Whitman addresses the voyager:

> "To port and hawser's tie no more returning
> Depart upon thy endless cruise old Sailor."

The final voyage is a recurrent figure in Tennyson's poetry also, as in his lyric, "Crossing the Bar", his monologue, "Ulysses", and his narrative, "Morte d'Arthur." In all of these poems the tone conveys the sadness of parting but also an affirmative and consolatory note.

Literature representing the final voyage prompts us to speculate. Which of the four elements, earth, air, fire, and water, supplies images of death and dying most commonly? Our answer would probably be *earth*, and there would be Walt Whitman, with his *Leaves of Grass*, to bear us out: ("And as to you Corpse I think you are good manure, but that does not offend me, . . . I reach to the leafy lips, I reach to the polished breasts of melons.") Nevertheless, even without sustaining evidence from a computer, I'd venture to suggest that water imagery predominates in our literature, whether it is the sea-voyage or simply the movement of water itself, by which a sense of time passing is conveyed to us. Virginia Woolf wrote a whole novel, *The Waves*, in which the beginning and endings of the lives of individual characters are likened to the motion of ocean waves. Earlier, Robert Browning, in a remarkable passage, used the same image in a monologue in his long Victorian murder story, *The Ring and the Book*. The speaker is the murderer, Count Guido Franceschini, on his last night alive before he is executed. "You never know what life means until you die," Guido remarks, and adds: "Even throughout life, 'tis death that makes life live." He clings to life of course, but at times finds his special kind of consolation in reflecting that we are all waves, and all doomed eventually to crash upon the rocks on shore. Addressing the clerics who are visiting him in his prison-cell he warns them:

> I see you all reel to the rock, you waves—
> . . . all bound whither the main-current sets,
> Rockward, an end in foam for all of you!
> What if I be o'ertaken, pushed to the front
> By you crowding smoother souls behind,
> And reach, a minute sooner than was meant,
> The boundary whereon I break to mist?
> Go to! the smoothest safest of you all,
> Most perfect and compact waves in my train . . .
> Will rock vertiginously in turn, and reel,
> And emulative, rush to death like me."[19]

Guido's final vision of moving waters takes place in his condemned cell. "Well," as Walter Pater observed (quoting Victor Hugo) "we are all under sentence of death but with a sort of indefinite reprieve—les hommes sont tous condamné à mort avec des sursis indéfinis." Before our wave crashes onto the shore, that is, there is what Pater calls an "interval" when we may console ourselves with what he calls "art and song."[20]

As a brief finale, let us consider an especially complex treatment of death and of the soul's mysterious voyage and of "art and song" too. It is the striking poem of meditation by Lawrence's contemporary, William Butler Yeats: "Sailing to Byzantium." Yeats writes:

> And therefore I have sailed the seas and come
> To the holy city of Byzantium.

But for Yeats the "holy city" is no longer Bunyan's Christian heaven, with which our discussion began. It is a city not of saints but of "sages" whom Yeats addresses:

> Consume my heart away; sick with desire
> And fastened to a dying animal
> It knows not what it is; and gathers me
> Into the artifice of eternity.

The voyage here is still a mysterious one and the destination mysterious too, for unlike *Pilgrim's Progress*, let us say, the fresh beginning is not specified. Nevertheless these late poems by Yeats on age and death function restoratively for most readers who can say Queen Mary's words again, but without knowing just what it is that may be in store for us: "In my end is my beginning."

Notes

1. Geoffrey Gorer, cited by Philippe Ariès, *Western Attitudes Towards Death* (Baltimore: 1974), pp. 92–93.

2. Robert Jay Lifton, "The American Experience of Death," *Perspectives on Death and Dying* (Middletown, Conn: The Connecticut Scholar, 1979), pp. 64–65. The epigraph for my essay is also taken from this article, p. 77.

3. Elisabeth Kübler-Ross, "What Is It Like to Be Dying?" in *Understanding Death and Dying*, eds. Sandra G. Wilcox and Marilyn Sutton (Port Washington, New York: 1977), pp. 99–111.

4. On the general topic of martyrs and their expectations of perpetuation, see Eugene Weiner's paper presented at the Alpbach conference

5. An earlier example of revising Mary's motto occurs in D. H. Lawrence's novel, *The Rainbow*. See George H. Ford's essay in *The Study of Time*, 3 ed. J. T. Fraser et al., (New York, 1978), p. 552.

6. J. T. Fraser, *Of Time, Passion, and Knowledge* (New York: 1975) p. 413, summarizes the argument by R. J. Quinones, in *The Renaissance Discovery of Time* (Cambridge, Mass.: 1972) that in Renaissance tragedies "the most significant moment of life, that of death, is identified with the first moment; the final and first hour become one." Quinones also noted that according to tradition, Shakespeare was born on April 23 and died on April 23 (Fraser, p. 504n.).

7. Cf. a statement by J. T. Fraser referring to "the unresolvable conflict between growth and decay." *Time as Conflict* (Basel, 1978), p. 92.

8. Geoffrey Rowell, *Hell and the Victorians* (Oxford: 1974), p. 82n.

9. The conceit of the death of death is also used in Shakespeare's sonnet number 146 ("Poor soul, the center of my sinful earth") which concludes:

> So shall thou feed on death, that feeds on men,
> And death once dead, there's no more dying then.

The overall image of Time as devourer is offset by the idea of love devouring time which culminates in Andrew Marvell's poem "To his Coy Mistress" where the speaker proposes to his lover that they should "devour" time rather "than languish in his slow-chapt power."

10. Sandra Wilcox and Marilyn Sutton eds. *Understanding Death and Dying*, p. 5. See also Nathan A. Scott, Jr., ed., *The Modern Vision of Death* (Richmond, Virginia: 1967).

11. On Shakespeare's final scenes in his tragedies, see Walter C. Foreman, Jr., *The Music of the Close* (Lexington, Kentucky: 1978). Foreman's final chapter deals with *Antony and Cleopatra*. See also Brent Stirling's *Unity in Shakespearean Tragedy* (1956).

12. William Gass, *The World Within the Word* (New York: 1978), p. 23.

13. *Alfred Lord Tennyson: A Memoir by his Son* (London: 1897), pp. 428–29.

14. *Ibid.*, p. 425. See also the informative chapter, "Deathbeds," in John Reed's *Victorian Conventions* (Athens, Ohio: 1975), pp. 156–71.

15. It is of interest that Keats was the favorite writer of another death-conscious 20th-century novelist, like Virginia Woolf. This was F. Scott Fitzgerald, whose novel, *Tender is the Night*, takes its title from Keats's ode.

16. Tennyson's "Tithonus" is another instance of a poem that provided a novelist with a title. Aldous Huxley drew from the poem for the title of his novel about a California businessman who sought to live forever: *After Many A Summer Dies The Swan* (1939).

17. *The Poems of Wilfred Owen* (London: 1933), p. 40.

18. See the chapter on Little Nell in George H. Ford's *Dickens and His Readers* (New York: 1965), pp. 55–74. For other and more effective aspects of Dickens' treatment of death, see Garrett Stewart, "The New Mortality of Bleak House." *English Literary History* (Summer, 1978), pp. 443–87.

19. Robert Browning, *The Ring and the Book*, 11, lines 2351 ff.

20. Pater's "Conclusion" to *The Renaissance*.

On the Beginnings and Endings of Time in Medieval Judaism and Islam

S. L. GOLDMAN

In discussing Medieval Judaic and Islamic conceptions of the beginning and of the ending of time I will first describe different conceptions of the World's being-in-time and then ask, with respect to these different conceptions, three questions:

1. Did time exist before the World did?
2. Will time exist after the World ceases to exist?
3. Are all World events in time?

I. On the Being-in-Time of the World

In medieval Judaism and Islam, it is safe to say that everyone is, in one way or another, a creationist holding to the view that the World was created by God. But having said this, one has said almost nothing of substance because of the wildly divergent cosmosgonies that one can find represented in the literature. I will arrange these in three broad categories.

(1) At one extreme, one finds those thinkers for whom the act of creation is logical and atemporal, being an expression of the intimacy of the connection that is perceived to exist between the nature of the Creator and the resultant creation. Thus, in al-Farabi and later in ibn Sina, creation is a bridge between an entity necessary in itself (God) and another entity, contingent in itself but necessary through another.[1] This latter entity is the World, manifestly contingent at the level of human experience, but revealing to the philosopher an underlying necessity that can only be accounted for in a contingent World, by anchoring this contingency in some other, wholly necessary, entity: God. Al-Farabi and ibn Sina want us to agree that there cannot be possibility unless there is also a necessary subject of which possibility can

be predicated. This argument was considered by Maimonides to be one of the strongest arguments mounted by those who defended the eternity of the World. In the *Guide for the Perplexed*, he called it a "forceful argument," but rejected it on two grounds.[2] It is, first of all, based on an equivocation. When the Aristoteleans argue that the "actual production of a thing [by some other thing] is preceded in time by its possibility and that this possibility cannot be self-subsistent but must itself possess a non-contingent substratum, they cannot conclude that God is the perpetual [because necessary] ground of the actually existing, contingent, World, which thus necessarily exists forever, only logically [but not temporally] preceded by the existence of God." This is because the word "action" is used homonymously here. It is predicated first of a corporeal *or* incorporeal-but-substantial subject on the one hand, and then of an incorporeal *and* non-substantial subject on the other.

Secondly, provoked by the force of this Farabian/Avicennan argument, Maimonides proceeded to construct what he tells his readers is a "high rampart erected around the Law and able to resist all missiles directed against it.[3] "We admit the existence of these properties, but hold that they are by no means the same as those [properties] which the things possessed in the moment of their production; and we hold that these properties themselves have come into being from absolute non-existence."[4] The arguments of the Aristoteleans, he concludes, thus "only have demonstrative force against those who hold that the nature of things as at present in existence proves the creation. But this is not our opinion."[5] Maimonides' own goal was to prove not the temporal creation of the World, but only the *possibility* of that creation having been in time "and this possibility is not refuted by arguments based on the nature of the present universe, which we do not dispute."[6]

In the century after Maimonides, Levi ben Gerson argued the vacuity of proclaiming a contingent yet eternal world, if only because, given infinite time, every possibility predicable of the World would have to be realized, including its non-existence, which would contradict the necessary ground supposed for its (acknowledgedly) contingent existence.[7] Furthermore, to claim the actual infinity of the World in time is to fall into numerous absurdities because the eternity of the World is an actual infinity and not a potential one, like infinitely divisible line segments or infinitely countable (but not counted) numbers. Here, in the case of time, the infinite *thing* is actually before us in its infinity (leading to well-known paradoxes involving increases in size of actually infinite quantities); there, in the case of lines and numbers, we speak of potentially infinite *acts* that have not in fact been accomplished.

(2) The preceding represents one extreme, cosmogonies in which the World is conceived to be created and yet eternal, fully temporal but without a beginning in time, so that time also is eternal. At the other extreme lie cosmogonies in which the existence of the World is radically contingent and explicitly temporary. Two are of special interest *vis-a-vis* the being-in-time of the World, namely, the views of the Kalam and of the radical rationalist physician Razi.

The atomism of the Kalam is remarkable at the very least for the remorseless consistency of the observance by the Mutakalimun of their twin principles that accidents are incapable of temporal endurance and that substance cannot exist without accidents.[8] The resultant atomic theory of "time," however, is atemporal. Time for the Kalam is an epiphenomenon and a peculiarly illusory one at that. It is not at all a *locus* of action, of happening. When unfolded and applied to the analysis of natural phenomena, this Kalamic theory of the creation of accidents requires the denial, as in al-Ghazali's *Destruction of Philosophy*, of all causal connection in Nature.[9] The Kalamic World is thus as drastically contingent in its existence as it possibly could be. In fact, it has no *continued* existence at all. The World exists for an instant, an instant which bears not the faintest implication for its further existence in a subsequent instant. This is precisely the opposite pole, both in fact and in spirit, from the superficially contingent but actually necessary existence of the World in al-Farabi and ibn Sina, a necessity entailing the eternity of the same World that exists only for an atemporal instant for the Mutakalimun.

Less drastically, and far more dramatically, the physician Razi also proclaimed the contingency of the World that God had created. In this case, the World was brought into existence by God at a certain point in time, for a certain period of time, after which it will cease to exist. Razi taught that five principles of Being co-exist eternally: the Creator God, the World Soul, Original Matter, Absolute Space, and Absolute Time. The World Soul possesses a life of its own, but it lacks knowledge, in particular, it lacks knowledge of its own principle of being. As a result of this ignorance, the World Soul dwells upon the multitude of forms with which it senses that Original Matter is pregnant, tormenting itself with a desire to materialize these forms, and unite with them, a desire that it is incapable on its own of realizing.

The Creator God, witness to this misconceived longing of the World Soul for a mode of existence inappropriate to its proper mode of being, chooses to educate the World Soul to what this mode of its being is, as a means of eradicating the World Soul's desire for communion with Original Matter. To this end, the Creator fashions the material world and plunges the Soul into it, so that in the course of coming to a consciousness of itself as an entity whose existence is sharply distinct from the existence of the material frame within which it finds itself, it will recognize its own nature. At that time, the material world will cease to exist and the Sphere of Being will again reduce to the five eternal principles listed above.

(3) In between the two extremes of a necessary and hence eternal World and a radically contingent World, there lie many other possible accounts of the Sphere of Being and of the beginning of the World. These can broadly be grouped into naturalistic, mystical, and ideal accounts of the beginning of the World.

The major naturalistic Judaic accounts of the World's beginning are those of Maimonides and Levi ben Gerson.[10]

Maimonides began his philosophical analysis of the being-in-time of the World by establishing that Aristotle had never demonstrated the eternity of the World nor had he ever *claimed* to have demonstrated its eternity. Aristotle had only concluded that the World's existence was *possibly* eternal and had established to his own satisfaction the likelihood that it *was* eternal.[11] For his part, Maimonides resolved to establish the parallel possibility of the createdness of the World and then to allow the Biblical and Prophetic traditions to decide a matter that cannot be decided philosophically. Indeed, he wrote, it was precisely to reveal the limited power of philosophical analysis, and the ultimate superiority of spiritual revelation, that the Torah openly proclaimed what philosophy cannot decide between these two views.[12] Maimonides is thus an Aristotelean nature philosopher seeking a necessary and naturalistic account of phenomena *as they now* appear in human experience, while altogether sheltering from philosophical speculation the question of Nature's coming into being.

About Gersonides' account of the being-in-time of the World, I will say only this for now: the World as it now exists, is a necessary consequence of the motion resulting from the union of form and matter that was the onset of creation and marked the first instant of time.[13] That the World is a necessary consequence of this union implies that no other World than this one could exist. This raises interesting metaphysical questions concerning the true source of the being of this form and matter, and of matter's power to exert this degree of control over form. The crucial point of novelty in Gersonides' otherwise Aristotelean, rationalistic, and naturalistic cosmology, is his insistence that the World *had* a first instant of being in time. This, in turn, implies that every temporal interval must have an onset and that the "now" is not merely a boundary *between* two temporal intervals, but is also a limit of a temporal interval. This was, and is, widely perceived to be an anti-Aristotlean position. It is not at all clear, however, that Aristotle would have disagreed with Gersonides' definition of time as "the measure of motion·as a whole according to the instants which form the boundaries of the motion, but not according to the instants that only distinguish the prior from the posterior."[14] Furthermore, this says nothing of the possible endlessness of future time, a subject to which I will return shortly.

A second variant of cosmogonies lying between the eternal creationism of ibn Sina and the radical contingency of the Kalamic World in particular and of Islamic occasionalism more generally, are the mystic cosmogonies. Of these, two are especially interesting.

In Sura 28 of the *Koran* we read: "Everything goes to destruction except His Face." To many Islamic theologians and philosophers, this text was the basis for proclaiming the transience of the creation and the absolute ontological gulf separating Creator and creation. But for the 13th century Sufi ibn al-Arabi the text plainly entailed quite the reverse, namely, that there *is* nothing else but the Creator and His Face![15] The creation is nothing else but God's Face, that is, is nothing else but an externalization of the Divine nature. As such, it is in a sense timeless, in not being subject to the destruction connoted by the passage of time as this is made much of by Aristotle in his *Physics*. The existence of the World is temporal in that it is an expression of an act of will on God's part, a choice to look at Himself so to speak and it is this unfolding of His covert nature by holding it at arm's length [metaphors are manifestly inescapable here] that defines the being of the World. To the extent that God does not change His mind, the World will not be destroyed in the way that Sura 28 might have suggested would be the destiny of that which *was not* God. This makes it possible, by the way, for the World to be eternal in the forward direction. With some stretching, this view may also be compatible with the World's absolute eternity, but that compromises again, as in eternal creationism, the integrity of God's *choice* to externalize His nature. Nevertheless, what is striking about ibn al-Arabi's conception is his abolition of the distinction between God and the World and his having been able to survive the inevitable charges of pantheism that had resulted in the persecution, and even execution, of Sufi thinkers in the eleventh and twelfth centuries. Almost as striking as this are the similarities between ibn al-Arabi's account of the being of the World and the accounts of the slightly earlier and contemporary Spanish Kabbalists.

The basic texts of Spanish Kabbalism are the *Zohar*, a collection of mystical texts which surfaced in the 13th century, one thousand years of Midrashim, that is, collections of Rabbinic Biblical commentary, and the *Sefer Yetsirah*, or *Book of Formation*, which must be dated no later than the seventh century and quite possibly as early as the third or fourth centuries.

One of the components of the *Zohar* is the *Sifra DiZniutha*, or *Book of Concealment*, which explicitly speaks of God's act of creation as a matter of arranging the Divine Face.[16] More specifically, it speaks of God's decision to create a World that in some sense was not God and of God's failing to succeed in doing so. These primitive World's quickly fell into disarray. At last God first "arranged His Face" and *then* created a World that was an expression, not of the Face perhaps, but at least of the "logic" of the arrangement of the Face. This is very plausibly based on a Midrash, indeed an often-repeated Midrash, that speaks of the Torah as having been created before the World and of God looking into the Torah in order to create a "viable" World.[17] One such Midrash even says that God created Worlds before this one, but they all fell apart until He first created the Torah and then looked into it, *et cetera*.[18] This entails interpreting the Torah as God's Face, in line with the above, and that is a very congenial interpretation for the Kabbalists. For the author of the *Sifra DiZniutha* this process of arranging the Divine Face in a way that resulted in a stable creation was the true content of the fourth sentence of *Genesis I*: "And it was evening and it was morning, one day." That is, the text does not say "the first day," as with all the other days, but "one day" implying a single, unique process that was a precondition for the following days and not merely the first of them. This notion in the *Sifra DiZniutha* seems to me isomorphic with ibn al-Arabi's notion of the World as an externalization of the Divine nature, an occasion for God to see Himself, a *necessity* if God is to see Himself. And this fits very well into Zoharic notions that the World was created in order for God's complex nature to achieve completeness,[19] to achieve a full realization, perhaps analogous to the need to physically separate the sexes so that in deliberately reuniting physically, the occasion should be created for a profound unification that was impossible when the first Adam was apparently self-contained and alone.

In both the Midrashic and Zoharic literature we repeatedly encounter the notion that God created the World in order to externalize His nature so that the Divine Name could be unified.[20] The symbol of this externalization is the going forth from God of a point of radiance, a spark of light. Initially, God "tried" to articulate His nature within Himself and failed. He then "turned away" from His self-reflection, let there be some thing that was not God (here the Kabbalistic notion of the contraction of the Godhead, in order to *create* a region of the primordial Sphere of Being that was *not* God comes in) and God was able through this means to achieve what self-reflection could not. Thus in the Midrashic collection called the *Pirke de Rabbi Eliezer*, or the *Chapters/Lessons of Rabbi Eliezer*, we read, commenting on the opening sentence of *Genesis*, that God created the Heavens from light and the Earth from the snow under His Throne of Glory. This Midrash was a painful one for Maimonides,[21] who argued vigorously against any attempt whatsoever to claim that God created the World from a preexistent stuff, as in Platonic creation theories, in order to preserve the absolute distinction between God's necessary being and the ultimately contingent being of nature. Clearly, Rabbi Eliezer's comment is to be taken metaphorically and it may be that the correct route to take in deciphering the metaphor is the Kabbalistic one: that Rabbi Eliezer here alludes to the primal stages of the Divine decision to externalize God's Self in order to complete it.

The third prominent line of Medieval Judaic cosmogony is an idealistic one that is somewhere between rationalistic naturalism and Zoharic mysticism. It is the cosmogony of Philo of Alexandria (admittedly too early for the period), of the *Sefer Yetsirah*, of Solomon ibn Gabriol, and of Abraham bar Hiyya, the great Spanish naturalist, mathematician, and astronomer.

The account of the coming into being of the World given in the *Sefer Yetsirah* involves time in a more fundamental way than in any cosmogony yet considered. That is to say, in the *Sefer Yetsirah*,[22] time emerges as an explicitly postulated, dynamically significant dimension of the Sphere of Being. Time is proclaimed to be the "order of the macrocosm" as Man is the order of the microcosm, to paraphrase Judah HaLevi's formulation in his *Book of the Kuzari* of the teaching of the *Sefer Yetsirah*.[23] There are three "true witnesses" to the creation of the World constituting three enduring aspects of the World through which the original act of creation is permanently refracted and through which the intrinsic "logic" of this act, the architecture of the Sphere of Being, becomes accessible to us. These "true witnesses" are *olam*, the creation taken as a coexistent whole; *shana*, time (literally, year); and *nefesh*, spirit. These three dimensions of Being are "witnesses" to the common rationality and unitary source of all beings. The elementary constitution of the World, the primordial air, water, and fire (not yet the material elements of the Greeks), together with the "automatic" dynamism of the Divine nature as it accomplishes its decision to create, still only define an abstract framework for the Sphere of Being. This achieves concrete articulation only when these first six dimensions are each and variously permuted with the three "witnesses": world, time and spirit. The resulting universe is then founded on a common wisdom whose abstraction by us bears "true witness" to the unity of the World and its rootedness in the Godhead.

In his essay "On the Creation of the World,"[24] Philo divided *Genesis I* into two parts and set up *Genesis II*, not as a repetition drawing attention to a specific facet of *Genesis I*, but as its continuation. The first sentence of the Torah is completely self-contained. It recounts God's abstract decision that there should be a World, and this decision then unfolds in the course of the rest of *Genesis I*. The decision to create leads to the *idea* of light and of darkness, of nothingness and of something. These lead to the incorporeal substance of water and of an immanent force called the Spirit of the Lord, and these two lead to Light. Light and the Spirit of the Lord constitute an "image" of God's creative *logos* and they together effect the explication of God's creation-idea, which is realized at the level of *yetsirah* or fashioning (as oppoed to *briah*, creating) in *Genesis II*. The being of time shares all of the features of being generally, on this account. Time has an amorphous character at first, as a real though latent and as yet undefined existent, until the creation idea next takes on abstract clarification (on the Fourth Day) and then substantial existence on its own, when the World takes on *its* material existence. This conception of the notion of time coming into existence, of time then having an ill-defined existence followed by a sharply defined one keyed to the material existence of the World, appears again in Joseph Albo in the 15th century and will be referred to below.[25]

Solomon ibn Gabirol was an 11th century Neoplatonic philosopher-poet of considerable originality and considerable influence on Medieval Latin and Renaissance thought.[26] The pivot of ibn Gabirol's cosmogony is God's will, as opposed to God's thought or knowledge as in the Medieval Aristotelean tradition. God chose to create a World and this choice is intrinsically incomprehensible, indeed it defines the bounds of comprehensibility. We are challenged to unravel the logic of God's creation in order to awaken in us a sympathy for the microcosmic structure of our own soul in the course of which awakening we will understand our place in the Sphere of Being. Ibn Gabirol's focus on an incomprehensible act of will as the starting point of the existence of the Sphere of Being is characteristically Judaic. As the Hebrew Prophets, in particular Isaiah, often insist, God's ways are not comprehensible to Man (this seems as well to be the burden of Moses' vision of God's nature in Exodus 33). We see an intimate affinity here between the Neoplatonist ibn Gabirol, such Spanish-Jewish Aristoteleans as Abraham ibn Daud (who otherwise excoriated ibn Gabirol as a sophomoric philosopher[27]) and Maimonides (in the "high ramparts" he erected around the Torah's teaching of creation), and the Zoharic Kabbalists, among whom there developed a notion of a four-level hierarchy to creation, the topmost level of which was the level of God's speech in Genesis I, a level that was beyond rational or even broader intellectual reach.[28]

Finally, and somewhat curiously, the Aristotelean nature philosopher Abraham bar Hiyya, an important Spanish-Jewish mathematician and astronomer of the 11th century, held to what I have here called an idealistic conception of the creation bearing some similarities to Philo of Alexandria's view. Bar Hiyya held, as did almost everyone else, that form and a generic matter were the ultimate constituents of the World.[29] Of the two, form was superior in that it could exist without matter, although only imperceptibly, while matter could not exist without form. Both of these in some real sense existed, with all of their characteristic properties, inside God, in His "secret place," until He saw fit to bring them forth. With this act of bringing forth and of uniting form and matter, God launched the existence of the World and simultaneously the existence of time. All talk of time prior to this act is only a 'common way of speaking' and does not refer to a temporal phenomenon. Thus, as in Philo, the full content of the World is already developed within God, but only latently. It is worth a passing mention that bar Hiyya implicity, like Gersonides explicitly some two-hundred years later, seems to hold that the World as it currently exists is the necessary consequence of the union of form and matter, that any attempt to unite these primordial elements would result in just this World. This is thus the *only* possible World that God could have made out of form and matter insofar as these undergird existence.

II. On the Being of Time in These Speculations Concerning the Beginnings and Endings of Being

With respect to the preceding cosmogonic speculations, it is possible to ask three questions that give us more specific insight into the conceptions of time that they embody.

1. Is the existence of time coextensive with the existence of the World?

This question entails two secondary ones: (a) Did time exist before the World existed? (b) Will time exist after the World ceases to exist?

(a) Did time exist before the World existed?

Philo and Sa'adia[30], Abraham bar Hiyya and al-Ghazali, Maimonides and Gersonides, Joseph Albo and Isaac Arama[31], a selection of thinkers spanning the entire chronology and ideology of Medieval Judaic thought, all responded to this question in the negative. For all of them, time was a property of the physical world and could have no existence independent of the motions of physical bodies. Maimonides was especially stern about firmly denying the existence of time before the existence of the World. Time, for him, was almost not real at all, being only a property of a property (motion) of substance. It was extremely abstract and therefore a difficult subject for contemplation even by the conceptually sophisticated. Furthermore, ordinary linguistic usages made it so common to speak of time in fantastic contexts, in contrast to careful philosophical usages, that people came to believe that time existed in the ways that it could be talked about. In such cases, he wrote, "we do not mean time in its true sense . . . [but] only use the term to signify something analogous or similar to time [i.e., a before-and-after ordering relation] . . . [thus] it cannot be said that God produced the Universe "in the beginning" as the opening passage of *Genesis* is commonly translated. "If you admit the existence of time before the creation, you will be compelled to accept the theory of the eternity of the Universe, for time is an accident and requires a substratum".[32]

On the other side, various Midrashic texts, Razi, and the great late 14th century Rabbinic philosopher Hasdai Crescas defended the view that time *did* exist before the creation of the World. Razi, of course, maintained this position because he had made time an eternal substance, co-existing with the Creator God. Crescas is, in a sense, much more interesting, because he mounted a detailed and deliberate defense of the view that time pre-existed the physical world, in explicit opposition to Aristotle's teaching on this point. Crescas' primary philosophical work is embedded in a tract entitled *The Light of the Lord*. The philosophical section of this work, resurrected and subjected to a monumental critical apparatus by Harry Austryn Wolfson,[33] consists of a critique of the 25 Aristotelean propositions with which Maimonides had opened the second section of his *Guide for the Perplexed*, identifying them as demonstrated philosophical truths. Crescas aimed at showing that they were in fact *not* demonstrated and that their opposites, the existence of a vacuum, of infinite space, of other worlds, of time independent of physical bodies, were philosophical, and also physical, possibilities. Crescas defined time as "the measure of the continuity/duration of motion or of rest between two instants."[34] Wolfson makes much of the choice of "duration" rather than "continuity" in this definition,[35] but for our purposes it is enough to note Crescas' further argument that making time the measure of motion roots its actual existence, tied to physical bodies, in the conceptual presupposition of measurability, an act that must be rooted in a pre-existent soul. Thus time existed *as* time, as a distinct measurability of before and after, in the soul of God prior to its physical existence as a property of the motion of bodies.

Finally, there are a number of Midrashim that speak of time as existing before the creation of the World, most notably in order to locate the creation of the Torah *before* the creation of the World. In a text that seems closely related to the *Sefer Yetsirah* (although in which direction the influence flowed cannot be determined),[36] we read that God created six abstract principles before creating the World and that these six—"water," "wind" and "fire," darkness, wisdom, and light—determined the six "dimensions," the architecture I called it before, of the World. Note that in the context of *Genesis*, the number 6 has temporal overtones, while (perhaps not independently) in Pythagorean numerology it has metaphysical ones and Philo makes much of the perfection of six as related to its appearance in the creation epic (perfection here means that it is the sum of all of its factors: this defines a Pythagorean perfect number).

(b) Will time exist after the World ceases to exist?

For al-Farabi and ibn Sina and their Aristotelean followers, the necessary character of the being of the World entails the co-eternity of the World's existence with that of God, who is the necessary ground of its existence. For very different reasons, Sa'adia Gaon, Nachmanides, and Maimonides also believed in the forward eternity of the World and therefore of time.

The 10th century Sa'adia of Fayum, Gaon or chief religious-law authority for world Jewry in his time, taught that the Messianic World-to-Come (Olam Haba) would be a physical world, different from the present one but still physical, and would be eternal.[37] Moses ben Nachman (Nachmanides), 13th century leader of the Spanish Kabbalists, Rabbinic authority, Bible commentator, and physician held that the doctrine of the Messianic redemption of the dead referred to a temporary rebirth to be followed by a subsequent rebirth to bodily immortality in a new and eternal corporeal world.[38] For Maimonides, the World *could* cease to exist, as its beginning was contingent upon God's continued willing it to exist; *were* it to cease to exist, time too would cease to exist because time is merely an accident of the motion of the World's matter; but in fact it *will not* cease to exist, and therefore neither will time, there being nothing in Scripture entailing an end to the World.[39]

For Gersonides, Joseph Albo, and al-Ghazali, although for slightly different reasons, the answer must be: *were* the world to cease to exist, time too would cease to exist. Albo, who was a student of Crescas', did not accept his teacher's argument that the necessary conceptual pre-existence of time *vis-a-vis* its physical existence as the measure of the motion of bodies, defined a real existence for time. For Albo, the existence of time in the soul (of God, say) before the creation was abstract only and not truly temporal. It was only the idea of a before-and-after ordering relation. Furthermore, in order to protect the distinction between God and the World, Albo held that the World *had* to come to an end, rejecting Maimonides' view that nothing in Scripture entailed an end to the World, and that time *would* then cease to exist.[40]

On the other side, Crescas, of course, held that were the World to cease to exist time would revert to its former real existence in the soul of God, while Razi had made time coeternal with the Creator so that there was no connection at all for Razi between the existence of time and the existence of the World.

2. Is time a thing at all?

For all of the Aristoteleans, al-Farabi and ibn Sina as much as bar Hiyya, Maimonides, and Gersonides, time is the measure of motion and is therefore an inseparable property of motion and at second remove of bodies. There can be no talk of its substantiality on this view.

For Razi, time is explicitly proclaimed to be a substance and an eternal one (defining and determining eternity?), and time is also an *ens reale* for the Medieval Neoplatonists, from Isaac Israeli in the 9th century[41] to the Islamic Brethren of Purity[42] and ibn Gabirol in the eleventh to Crescas in the 14th and 15th (insofar as Crescas shows some Neoplatonic characteristics).

The great Islamic naturalist al-Biruni,[43] writing in the 11th century, wrote of Razi that he had juxtaposed one Arabic term for time, *mudda*, to two other terms: *zaman*, used to refer to time as an accident of bodily motion; and *dahr*, used to refer to time as the duration of the soul, apparently in the Plotinian sense of time being the life of the soul. Thus Razi adds the notion of time as substance to the more common notions of time as a property of motion and the more "solid" (but not yet substantial) notion of time as a property of the substance soul. To these three distinct conceptions of time we can readily add several more that recur in Medieval Judaic and Islamic metaphysics.

(a) In the *Sefer Yetsirah* we encounter a truly novel conception of time as a dynamic principle of nature, ordering the Sphere of Being in a characteristic and significant way. As expressed by Judah HaLevi in the *Kuzari*, time is the *locus* of dynamism in the *Sefer Yetsirah* account of the structure of Being. On the one hand, there is a naturally destructive aspect of time, as Aristotle had commented in the *Physics*. On the other hand, the time that is a principle of Being in the *Sefer Yetsirah* is there enriched by Providence having "instituted the masculine and feminine principles in order to preserve the species in spite of the decay of the individual."[44] HaLevi here speaks specifically of animals, but it is generically true in the *Sefer Yitsirah* that everything has "masculine" and "feminine" aspects, these being the names of the constructively dynamic interactions a thing undergoes between its own nature and the natures outside of itself.

(b) In the *Midrash* and in the *Zohar* we encounter strongly evolutionary, or at least developmental, conceptions of time. In *Genesis Rabbah*, for example, there is recounted a dispute between Rabbi Yehuda and Rabbi Nehemia, both prominent authorities, over the interpretation of the six days of creation.[45] According to Rabbi Nehemia, there was but one act of creation and this an instantaneous one whose subsequent autonomous unfolding is recounted by Moses in terms of its readily distinguishable stages. Moses called these stages "days." Rabbi Yehuda demurred, insisting that each "day" of creation was self-contained, that there was no intrinsic dynamic moving nature from the stage of fishes to the stage of land animals, for example. On this very point, siding with Rabbi Nehemiah as it were, the Zohar[46] comments on the *Song of Songs* passage "the blossoms appear in the land," that this refers to the creation of the World, which took place by way of seeds that were planted by God which then developed and unfolded on their own.

(c) Among the Kalam, we have already noted a curiously atemporal conception of time and among the strict Aristoteleans, such as al-Farabi, ibn Sina and ibn Rushd, a conception of time that naturally lends itself to Spinoza's and Leibniz' formulations in which time is at once enobled as eternal and denied ontological significance, because in such a tight necessitarian scheme as theirs, the present is already "big with" the future and the appearance of the future is merely an artifact of limited consciousness.

(d) There is an orthodox Judaic conception of time as the primary dimension of the spiritual life of the faithful, the very word for religious festival, *moed*, having the (plausible) meaning "rendezvous in time." Time has a significance in Judaism that is thus far more central than space and that could be said to be the domain of true human being, creative being, such that it is this that underlies the great significance attributed to the Sabbath. By contrast, time in orthodox Islam is extremely static, history being divided into blocks of time defined by prophetic revelations within which nothing new happens until, from the outside, a new revelation takes place.

(e) In the metaphysics of ibn Gabirol, and of the roughly contemporary Moslem philosopher abu Zeyid, there is a well-articulated typology of time. Nature is the field of time in the Aristotelean sense. The soul possesses a temporal character, duration, but this temporality that marks the life of the soul is ontologically "above," in the sense of prior to, natural time while being "below" the mode of temporality called eternity. The intellect, in turn, has a temporal character that is "above" the temporality of duration but within the mode of temporality called eternity. God, finally, is wholly above temporality and thus God's "life" or being is "above" even eternity.

3. Are all world events in time?

For everyone, although in different ways, the answer to this question is no, that is, everyone thinks that some events at least are not in time. Even for the Aristoteleans, who hold that time and the being of the physical world are inseparable, substantial change takes place in no-time.[47] That is to say, the genesis and destruction of forms is said, by Aristotle, by ibn Sina, by ibn Rushd, by Maimonides, among many others, to take place not merely instantaneously, but atemporally. This notion is formulated by the great translator and Maimonidean commentator Samuel ibn Tibbon as follows: the action of spiritual beings is atemporal because there can be no transition from potentiality to actuality—hence no temporal process—for a being that, as incorporeal being, is already wholly actual (albeit only in a limited way, else it would be God).[48] Simplicius is quoted approvingly by Crescas[49] as having held that thinking takes place in no-time and this is yet another way of formulating the principle that substantial change is atemporal. The rational faculty in Man constitutes an incorporeal substance whose essential activity is thinking. Changes in the content of the rational faculty constitute, therefore, substantial changes and hence must take place atemporally. The upshot of this is that temporality is not the absolute horizon of the Sphere of Being for any Medieval thinker, not even for the radical Islamic Aristoteleans. This suggests an opening in their metaphysics which could naturally have lent itself to conflation of Aristotelean with Neoplatonic metaphysics quite independent of the erroneous attribution of Proclus' *Elements of Theology* to Aristotle, commonly considered to be the cause of the Medieval Islamic melange of Aristotelean, Platonic, and Neoplatonic philosophies.

Notes

1. S. H. Nasr, *An Introduction to Muslim Cosmological Doctrines* (Cambridge: Harvard University Press, 1964). Also, Emile Fackenheim, "The Possibility of the Universe in Al-Farabi, Ibn Sina, and Maimonides," in *Essays in Medieval Jewish and Islamic Philosophy*, ed. Arthur Hyman (New York: Ktav Publishing House, 1977), pp. 303–334; hereafter cited as *Essays*. Don Isaac Abravanel, in *Mifalot Elokim*, 2, 3 says that ibn Sina took the notion of the necessary *ab alio* from Plato.

2. Moses Maimonides, *Guide for the Perplexed*, 2, pp. 14–18.

3. Ibid., 2, p. 17.

4. Ibid., but see also 2, 13.

5. Ibid. 22.

6. Maimonides, *Guide for the Perplexed*, 2, 17.

7. Levi ben Gerson, *Milhamot Hashem (The Wars of the Lord)*, book 6, chaps. 10–12, 20, 21, Also, Seymour Feldman, "Gersonides' Proofs for the Creation of the University," in Hyman, *Essays*, pp. 219–243.

8. Maimonides lengthy and important critique of the philosophy of the Kalam is in the *Guide*, 1, chap. 73ff. Also, Nasr, loc, cit.

9. Access to these arguments is provided in *Averroes' Tahafut al-Tahafut*, S. van den Bergh, 2 vols., Luzac, London, 1954. See also Majid Fakhry, *Islamic Occasionalism*, Allen and Unwin, London, 1958. Malebranche is the Western occasionalist *par excellence*, of course, but Descartes' conception of time is akin to that of the Kalam as well, and thus anchors such occasionalist notions as those of Malebranche and Geulincx, for example. In *Meditations* 3, Descartes writes: "It is a matter of fact perfectly clear and evident to all those who consider with attention the nature of time, that, in order to be conserved in each moment in which it endures, a substance has need of the same power and action as would be necessary to produce and create it anew, supposing it did not yet exist, so that the light of nature shows us clearly that the distinction between creation and conservation is solely a distinction of the reason."

10. Maimonides' analysis of time, following Aristotle's *Physics* Books IV, V, and VI, is in the *Guide* 2, 3–30. Levi ben Gerson's analysis is in *Milhamot*, Book VI.

11. Maimonides, *Guide for the Perplexed*, 2, p. 15.

12. Ibid., 2, p. 16.

13. Levi ben Gerson, *Milhamot Hashem*, 6, p. 14.

14. Ibid., 6, p. 10. Aristotle wrote, in Physics 4, 13: "The 'now' is the link of time . . . and it is the limit of time." Insofar as the 'now' divides a temporal interval, it is always different. Insofar as the 'now' connects temporal intervals, it is always the same. Thus "now" is used by Aristotle equivocally to describe a certain two-valuedness of time that seems to me very close to what Gersonides was saying and on which the latter argued to the possibility of an absolute beginning to a temporal interval.

15. Rom Landau, *The Philosophy of ibn Arabi* (London: Allen and Unwin, 1959).

16. There is no lack of translations of Jewish mystical texts into English, but the opacity of the texts and the enthusiasm of some translators makes caution a byword in utilizing them. *The Book of Concealment* together with the *Books of the Greater and Lesser Assemblies (Idrah Rabbah* and *Idrah Zuttah)* employed here are available in a useful translation by Roy A. Rosenberg in *The Anatomy of God* (New York: Ktav Publishing House, 1973).

17. Midrash *Genesis Rabbah* 8:2. Also, *Exodus Rabbah* 30:9 and 34:2, and *Leviticus Rabbah* 19:1; and the *Zohar* commentary on *Exodus*, the chapter "Terumah."

18. *Genesis Rabbah* 9:2.

19. *Zohar on Exodus*, chap. "Terumah"; also *Numbers Rabbah* 12:6, 13:2, and 13.6.

20. *Zohar*, and *Zohar* commentary on *Leviticus* chapter "Aharei Mot."

21. Maimonides, *Guide for the Perplexed*, 2, 26.

22. Itamar Grunwald has published a translation and critical edition of the first part of the *Sefer Yetsirah* in *Revue des Etudes Juives*, 132 (4), 1973, pp. 475–512.

23. Judah HaLevi, *Kuzari* 3, 27; 4, 25; 5, 14.

24. Philo, *al Briath Haolam* (translated from the Greek), chapters 3ff. On Philo, H. A. Wolfson, *Philo*, (Cambridge: Harvard University Press, 1947).

25. Joseph Albo, *Sefer Halkkarim (Book of Principles)*, 2, 18, 19.

26. See, for example, *Robert Grosseteste and the Origins of Modern Experimental Science*, A. C. Crombie (London: Oxford University Press, 1962) and F. Yates, *Giordano Bruno and the Hermetic Tradition* (N.Y.: Random House, 1964).

27. Abraham ibn Daud, *Sefer HaEmunah HaRamah (The Book of the Transcendent Faith)*, I, 1.

28. This theory is well-developed in Menahem MiRecanti's commentary on the *Torah, Genesis*, reprinted in the *Sefer Levush Malkhut*, vol. 3.

29. Abraham bar Hiyya, *Hegyon HaNefesh (Meditations of the Soul)*, Book 1.

30. Sa'adia, op. cit., I, 1.

31. Sarah Heller-Wilensky, "Isaac Arama on the Creation and Structure of the World," in Hyman, *Essays in Medieval Jewish and Islamic Philosophy*, pp. 259–278.

32. Maimonides, *Guide for the Perplexed*, 2, p. 13.

33. Harry Austryn Wolfson, *Crescas' Critique of Aristotle* (Cambridge: Harvard University Press, 1925) especially the notes of Crescas' analysis of Propositions 5 (on motion) and 15 (on time) as listed by Maimonides in *Guide* 2, Introduction, 2; hereafter cited as *Crescas'*.

34. *Crescas'*, p. 289.

35. Ibid., p. 656.

36. *Midrash Exodus Rabbah*, 15:22. See also *Leviticus Rabbah* 31:7 and *Guide* 2, 30.

37. Sa'adia Gaon, *Sefer Emunoth VeDeoth (The Book of Beliefs and Opinions)*, Books 7 and 9.

38. Nachamanides, *Torath HaAdam*, Sha'ar HaGmul.

39. Maimonides, *Guide for the Perplexed*, 2, pp. 27–29.

40. Albo, *Ikkarim*, 2, p. 18.

41. On Isaac Israeli, A. Altman and S. M. Stern, *Isaac Israeli* (London: Oxford University Press, 1958).

42. See the chapter on the Ikhwan in Nasr, *Muslim Cosmology*.

43. Nasr, *Muslim Cosmology*; also *Encyclopedia of Islam*, "zaman," "Razi."

44. Judah HaLevi, *Kuzari*, 4, p. 25.

45. *Midrash Genesis Rabbah* 12:4 (carried over into 12:5), also the opening discussion in the MALBIM's commentary on the *Torah, Genesis* 1, 1.

46. *Zohar on Genesis* 1.

47. Wolfson, *Crescas'*, Wolfson's elaborate notes in Proposition 15. Also ibn Tibbon

(following note) chap. 8, and Aristotle, *Physics* 221b3.

48. Samuel ibn Tibbon, *Sefer Yikavu HaMayim, (The Book of the Gathering of the Waters),* chaps. 4, 5, and 8. In chapter 6 he gives his reading of Maimonides' problem with the quotation from *Pirke de Rabbi Eliezer* referred to above, p. 14.

49. Wolfson, *Crescas',* p. 547/548.

A Structuralist View of Biological Origins

B.C. Goodwin

Contemporary biology is structured by ideas whose proximal roots lie in the 19th century, a century dedicated to historical, utilitarian, and mechanical explanations. Time, utility, and mechanism therefore lie heavily upon this natural science. This is reflected in the dominant positions of genetics and the theory of evolution, which are regarded as the twin pillars of modern biology. Genetics concerns itself with the properties and the behavior of the heriditary material which consists of those "instructions" found by trial and error to generate useful, or successful, mechanisms (parts of organisms), while evolution describes how these instructions change in time under the action of various contingencies, described as random variation in the genetic material and selection pressures on the useful mechanisms generated by it. These selection pressures define the interaction between the organism, or more strictly its parts, and the external environment, also defined by contingencies. This interaction is interpreted in terms of the metaphors of conflict, competition, and survival of the fittest. Historical continuity, linking all organisms in an unbroken chain back to the hypothetical primordial ancestor, arises from the self-replicating property of the hereditary material which is considered to underly organismic reproduction.

This conceptual scheme is grand in scope and simple in axiomatics. It can be applied as well to pre-biological as to biological evolution. The ingenious, simple, and plausible arguments presented by Cairns-Smith (1981) in a previous paper are an eloquent witness to this versatility of the theory. So organisms, both in their origins and their evolution, become the mechanical results of contingency and utility. However, an intriguing and revealing aspect of this whole conceptualization of the biological realm is that the organism, the primary object of biological observation, has effectively disappeared. Thus Cairns-Smith focuses not upon the problem of the origin of organisms, but upon the emergence of structures capable of carrying useful information and replicating it reliably. This view of biological origins poses an interesting intellectual puzzle which is attracting the attention of some very capable scientific minds. But it is necessary to inquire whether it has any significant bearing upon basic biological problems. Is it, perhaps, a puzzle which arises by abstraction from a conceptualization of organisms which is biologically out of focus and so generates an intriguing but rather irrelevant problem? Have genetics and evolution missed the biological boat? What I shall suggest in this paper is that they have; and that this is a direct consequence of the metaphors which inform and direct these subjects. To regain appropriate focus, we need different metaphors.

The Metaphors of Modern Biology

The disappearance of organisms as significant entities from modern biology is an extremely interesting phenomenon, a cognitive event which in its context is as dramatic as the evolutionary disappearance of the dinosaurs in the late Cretaceous Age. The concept of the organism evidently did not survive the conceptual development of genetics and evolution theory in the 20th century. A recent popular manifestation of this may be found in *The Selfish Gene* by R. Dawkins (1976), in which the biological unit is quite explicitly stated not to be the organism, but the gene. Since this conceptual entity replaces the organism, it must take upon itself the metaphorical elements of the contemporary scheme, such as competitive interaction with other genes in the struggle to leave copies of itself at the expense of other genes. Hence the epithet "selfish." The metaphor of the family tree linking all genes together by lineal descent from primordial genes carries over intact in expositions such as that of Dawkins and of most contemporary geneticists and evolutionists, and obviously provides a direct connection with the origin of life as described by Cairns-Smith. The latter's genes are made out of clay rather than DNA, but are nevertheless considered to be continuous with contemporary selfish genes. The image of the potter naturally springs to mind, completing an associative sequence extending from Natural Theology, which gave Darwin his conceptualization of organisms as adapted entities (by the Great Potter's design in Natural Theology, but by natural selection in Darwin's thought), through the broken and fragmented vessel of the organism in contemporary biology to the clay particles which are regarded as the source of the whole process in a scheme such as Cairns-Smith's. The fragmentation of the organism and its replacement by a collection of genes in modern biology is an inevitable logical consequence of the 19th century view of organisms as historically continuous utilitarian machines designed by chance, contingency, and conflict. Each part of the machine is understood primarily with respect to its adaptive relations to the external environment, whereby it plays a useful role in the machine's survival and reproduction. But the self-reproducing machine is constructed by genes. So what is really important is these entities, the biological atomic units. The organism then becomes essentially an aggregate of genes which, by virtue of their capacity to direct the synthesis of the constituent elements of the organism (molecules and macromolecules) and to replicate themselves, somehow give rise to organized entities with powers of adaptation, generation, and regeneration.

What the 20th century has contributed to this story is a metaphor for this mysterious "somehow." It comes from our most useful machine, the computer. Genes in some sense are considered to define a program which generates organisms and their behavior, just as programs generate the behavior of computers. So the genetic program does what all the King's horses and all the King's men were unable to do: it puts the atomized pieces of the organism together again after its great conceptual fall. The fact that nobody has been able to show how this program works implies that it is complex and ingenious, eliciting much admiration for the genes and stimulating a great deal of research on their behavior. Thus, in addition to their metaphorical burden, the genes must carry the weight of a heavy psychological projection since they are seen as the key to a great number of man's current problems, such as cancer, aging and death, mental illness, the need to raise IQ and food production, etc.; the contemporary philosopher's stone, one is tempted to say, but in the decadent, aberrant scientific tradition symbolized by Faust, whose goal is manipulation rather than transformation.

The Metaphors of Creation and Revelation

Darwin sought to explain the origin of biological form in terms which did away with the concept of a divine or transcendental Designer. More than a century of study on this and related problems has led to the proposition that each organism has within it a designer: the genetic program. This represents a materialized essence which is ultimately responsible for all of the organism's activities. So the problem of the Designer has not gone away, it has simply come back in another, equally mysterious form. The problem of biological form and behavior has not been solved. Indeed there is good evidence that there is no solution to these problems in the terms proposed, involving the reduction of organisms to their genes (Webster and Goodwin, 1980). Genes define the potential composition of organisms, i.e., the molecular constituents out of which any particular organism is formed. But composition does not determine form, a proposition well known in chemistry and clearly true in biology. Therefore, organisms have properties, such as their form, which cannot be deduced from the properties of genes and their products. This implies that there are principles of organization of living entities which must be known if one is to understand, for example, how and why organisms of specific form come into existence. Principles or constraints of this type may be conceptualized as laws, so the implication is that there are biological laws which are manifested in the forms of organisms and are not manifested in parts such as their genes. Organisms then become the fundamental, irreducible entities of biology, and genes become necessary parts of organisms as currently known. There may well, however, have been organisms without genes in their currently-understood sense. This does not mean that such organisms were without heredity, for specific form can be inherited without genes being the vehicle of inheritance (Sonneborn, 1970). In my view, the problem of the origin of life is more relevantly couched in terms of such organisms without genes, as has been proposed by investigators such as Fox (1965), rather than in terms of genes without organisms. The reason is simply this: there is no evidence that it is possible to go, either deductively or physically, from genes to organisms; but there is clear evidence that there are non-genetic (i.e., non-DNA) methods of inheritance in simple organisms such as protozoa. The whole search for self-replicating units without organismic properties as the precursors of biological evolution is based upon the assumption that the passage from genes to organisms is a relatively trivial one, and this is without any justification.

The belief that organisms reveal or manifest laws of biological organization is a very old one, and we may consider it to be based upon the metaphor of the laws of creation: the created or the manifest universe and its parts are obedient to laws, which are revealed in its structure and behavior. In such a universe, all is not contingency; not everything can happen. An example from classical physics is given by bodies moving under the action of a central attracting force; they can move only in a particular set of distinct motions as described by Newton's laws and the inverse square law of gravitational attraction. The possible geometries of motion are defined by the conic sections: circle, ellipse, parabola, hyperbola. Which of these is actually followed by a specific body is determined by the particulars of its initial conditions of position and velocity. The planets happen to move in ellipses. Thus, observed forms are determined by both universal constraints or laws, and by particular conditions; i.e., by a timeless law and by temporal contingencies, so that eternity and history each play a role in any given process. Eternity and history here denote relative concepts, of course, the former expressing a condition of temporal invariance relative to historical periods.

Belief in the existence of laws of biological organization as revealed in the properties of organisms similarly implies that not any organism can exist, or conversely, that organisms belong to a logically defined set of possible or potential forms. An actual organism is then a realized instance of the possible, and this gives it intelligiblity or rationality. This was the belief of the rational or transcendental morphologists of the late 18th and the early 19th centuries, such as Cuvier, Geoffrey St. Hilaire, and Owen. From their detailed studies of organismic morphology, they deduced the existence of general uniformities of structure across great phylogenetic spans which, to them, revealed principles or laws of form operating in the biological realm. An example is the generalized limb form of the terrestrial vertebrates, known as the pentadactyl (5-fingered) limb. The comparative anatomists showed that the great diversity of limbs possessed by creatures such as frogs, birds, bats, horses, and humans, all have a basic uniformity of structure such that each can be seen as an adaptive transformation of an underlying generalized limb form. Thus they are homologous to one another, members of an invariant set which defines a typical form. As such, they are transformable one into the other just as the different conic sections can be transformed into one another by changes in parameters, representing contingent effects. Darwin took this abstract concept of the typical form, materialized it, and put it into history, calling it the common ancestor from which all others were descended by adaptive variation. The ancestral form, for Darwin, arose by chance, not by law, and the variations are also contingencies, so that the only necessity in biology became survival. But this is essentially a tautology, and is quite unable to explain, in a deductive sense, the existence of regularities of form above the level of the individual species such as the pendactyl limb. There are many different limb designs which would suit birds as well, if not better, than the one they have. And since the genetic designer can generate any imaginable structure, being unconstrained by laws of form according to current theory, it is not clear why organisms have stuck to certain basic patterns. What one is offered by way of explanation is a principle of conversation of developmental mechanisms, conversation being another deep 19th century belief. And so evolutionary biology, seeking to explain organismic structure, became the domain of Just So Stories of which there is an endless number since contingency is without form or bound. In myth, this irrationality characterizes the chaos which normally precedes the creation. The brilliant advances made during this century by molecular genetics and biology in describing the molecular constituents of organisms, how they are made, and interact do not save general biology from its failure to arrive at a deductive, rational theory of the organism and its evolution, for the above reasons. The error of modern biology has been to pursue the wrong abstraction, to study the historical problem of material inheritance as formulated first by Darwin and then by Weismann (1891), which is an extremely limiting view of organisms and their evolution. In the 20th century this has turned organisms into aggregates of genes and their products. This is genocentric biology, as out of focus as geocentric descriptions of planetary motion. To find a more relevant abstraction, one in which inheritance is an aspect of biological process but is not regarded as its essence, we return to the metaphors of creation and revelation.

Embryological Revelation

That part of biology where one witnesses the phenomenon of creation most directly is embryology, which studies the genesis of organisms from the egg. Here, complex, organized forms can be seen emerging from the simplest three-dimensional structure, the sphere. As argued above and in a detailed discussion in Webster and Goodwin (1980), genes and their products are not sufficient to account for this process, so it is necessary to ask what kind of spatial organizing principle may be operating in embryos. Hans Driesch (1908), the great embryologist whose experimental and theoretical work at the turn of the century established so many of the basic properties of developing organisms, argued for an essentially vitalist principle as the embryological directing agency, which he called the entelechy. More recently, Waddington (1957) elaborated a view of epigenesis in terms of the realization of particular forms from a domain of potential, the process of embryogenesis being described in terms of a construct called the epigenetic landscape. His conceptualization was very similar to that of Driesch, both involving the selection of specific realizations or manifestations of form from a potential set. Their positions could now be described as structuralist, in the following sense. Organisms are characterized as relational structures whose primary properties are wholeness and self-regulation, and their forms are all members of a single potential set defined by generative rules or laws of form. For any particular species, embryogenesis is described by a sequence of members of the potential set of forms such that one spatial pattern undergoes transformation into another until the adult form is reached.

Driesch saw very clearly that no mechanism has the properties of wholeness and self-organization possessed by embryos and primitive organisms, wherein parts have the capacity to develop into wholes. For example, each blastomere of a sea urchin after the first cleavage division of the egg can become a whole embryo; and small fragments of a unicellular organism such as *Stentor*, or a multicellular organism such as *Hydra*, or a planarian worm can regenerate the whole organism, producing a miniature but complete whole. Driesch also recognized that a major problem of embryogenesis as conceptualized above is to account for the selection procedure whereby particular members of the potential set come into manifestation. As a solution to these difficulties he proposed the entelechy, a non-material agency which conferred the relevant properties upon organisms. Although conceptually clear, this construct is no more acceptable to a scientific description than is the mysterious genetic program, which is even less satisfactory than the entelechy because of its conceptual confusion. Waddington never did propose a solution to the problem, although he leaned towards genes as the selectors of the forms emerging during epigenesis. In a recent study (Goodwin and Trainor, 1980), the structuralist orientation of Driesch, Waddington, and a few others such as D'Arcy Thompson (1916) and René Thom (1972) has been pursued to the point where it becomes clearer how a few of these problems may be consistently, and scientifically resolved. I turn now to a brief description of this.

Laws of Form and Selection Rules

The belief underlying rational morphology was that 'typical' organismic form reveals fundamental, intrinsic organizational properties of the living realm. In pursuit of this objective, we (Goodwin and Trainor, 1980) selected for analysis the earliest stage of embryogenesis, the cleavage process, which typically transforms a solid sphere (the egg, a single cell) into a hollow ball made up of many cells, the morula. Classical embryological studies have identified that which is considered to be typical in this process, and that which represents specific modification of the typical form. The basic form is defined by a holoblastic (total) cleavage sequence in which, initially, the planes of successive cell divisions show a strict geometrical pattern, as shown in Figure 1. When the cleavage planes are projected onto the original spherical egg, they give the sequence shown in Figure 2. Irregularities tend to accumulate and gradually overwhelm this order after the 5th cleavage, but the basic or typical pattern is nonetheless clear. Superimposed upon this are particular variations describing the pattern observed in different species, which fall into three main classes: radial, like

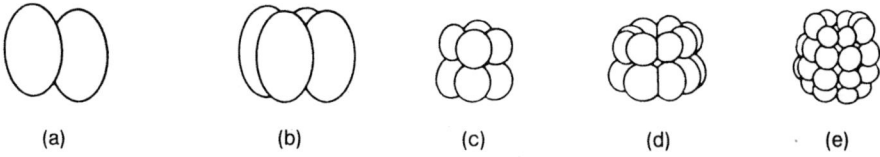

    (a)   (b)   (c)   (d)   (e)

Figure 1. An idealized description of holoblastic cleavage from the 2-cell to the 32-cell stage.

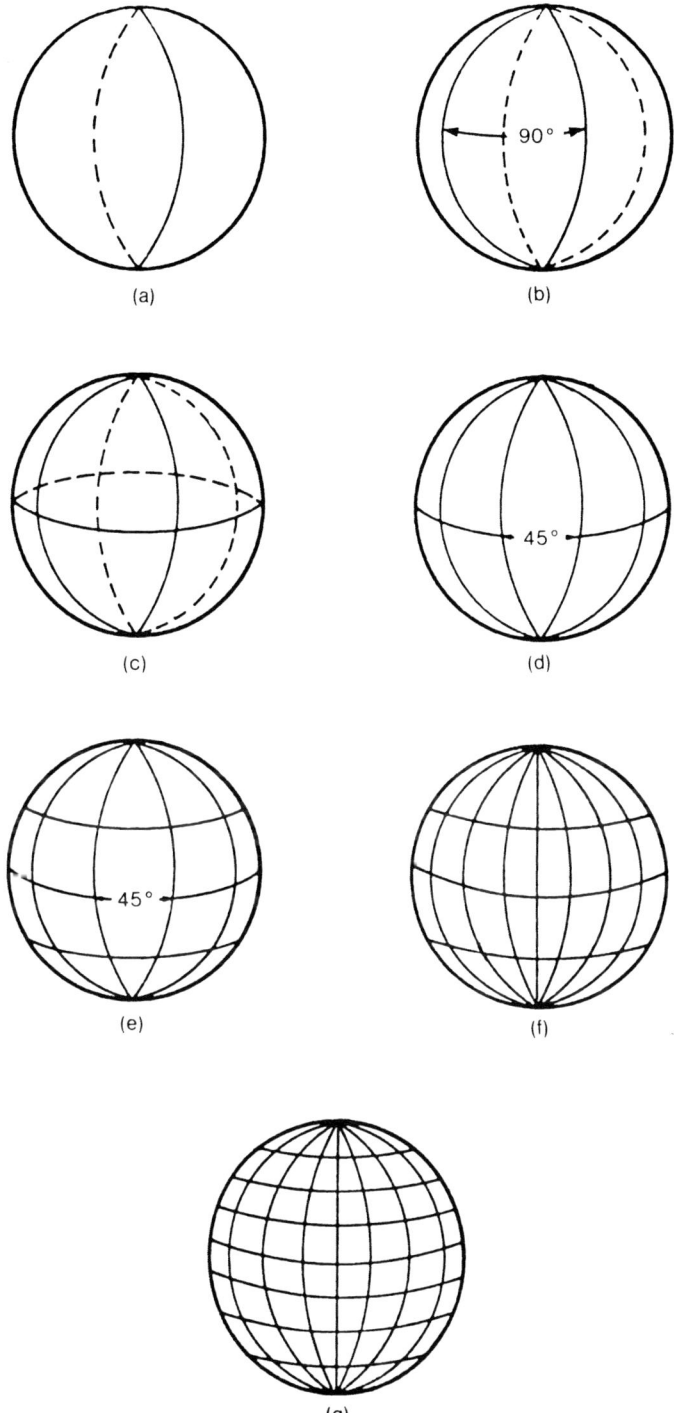

Figure 2. The typical sequence of cleavage planes drawn on the surface of the spherical egg.

Figure 1 but modified by a displacement of the cleavage planes such that cells show a systematic variation in size as one goes from animal to vegetal pole of the embryo, shown in Figure 3 for the case of *Amphioxus*; bilateral, in which there is a plane of mirror symmetry in the embryo, observed by a second eccentricity of cleavage planes along the dorso-ventral axis as it occurs in the toad *Xenopus laevis*; and spiral, in which the cleavage plans are skewed systemically along left- and right-handed spirals, as in Figure 4 which depicts the process in the snail, *Limnea*.

Since cleavage is initiated by a furrowing of the egg surface (membrane and cortex), it is natural to seek a description of this process in terms of some generalized surface energy function. Then the cleavage furrows can be interpreted as lines of least resistance on the surface, the single initial cell, the egg, being progressively partitioned into many cells by divisions involving "least action" or "work." Such a formulation of natural process is widely used in physics, and goes under the name of a variational problem. An appropriate surface energy density is defined by the function:

$$E(\theta,\phi) = A\ [(\frac{\partial U}{\partial \theta})^2 + \frac{1}{\sin^2\theta}(\frac{\partial U}{\partial \phi})^2 + \beta U^2\ ]$$

where θ and ϕ are degrees of longitude and latitude, respectively, on the sphere, $U(\theta,\phi)$ is the surface field function which will define the cleavage lines, and A and β are constants. The variational problem is then:

$$\delta \int_0^{2\pi} \int_0^{\pi} E(\theta,\phi)\ \sin\ \theta\ d\ \theta\ d\ \phi = 0$$

This leads to the solutions

$$U(\theta,\phi) = Y_{lm}(\theta,\phi) = \sqrt{2\ N_{lm}}\ P_l^m\ (\cos\ \theta)\ \cos\ m\ \phi$$

which are known as surface harmonics, the numbers l and m being integers. These functions are extensively used in the study of physical fields. What we have formulated is, in fact, a field description of the cleavage process in which the function $U(\theta,\phi)$ defines the state of the field at any point (θ,ϕ) of the surface. Let us now assume that cleavage furrows form along the nodal lines of the functions Y_{lm} (θ, ϕ), which are the lines where the functions take the value zero. Such a choice simply

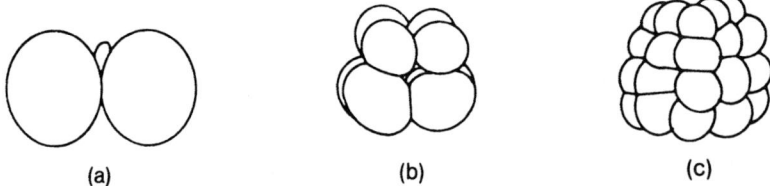

(a)   (b)   (c)

Figure 3. Stages in an actual cleavage sequence observed in *Amphioxus*, a protochordate.

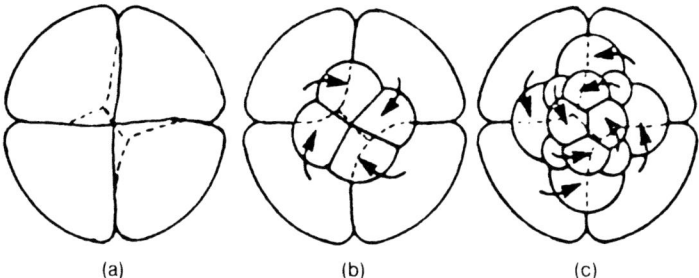

Figure 4. The spiral cleavage pattern of the snail, *Limnea*, seen from the animal pole.

uses a defined value of the field solutions as the descriptions of the cleavage lines, without committing one to a detailed description of the biological variables and forces involved in cleavage and how these relate to the functions $Y_{lm}(\theta, \phi)$. Since little is known with any precision about these processes, it is of value to proceed at a level of generality which captures the essence of the problem without attempting a detailed microscopic or molecular description. However, this is perfectly consistent with a more detailed field description which will be described later.

The nodal lines of the functions $Y_{lm}(\theta, \phi)$ are degrees of latitude and longitude on the sphere, well-defined for any pair of values (l, m). Some of these coincide with the cleavage lines shown in Figure 2, but many do not. Thus we may regard the functions $Y_{lm}(\theta, \phi)$ as providing a potential set of forms for the cleavage process, from which particular functions are selected to generate the typical pattern. What are the selection rules? Driesch, as mentioned above, suggested a non-material entelechy as the selector; Waddington intimated that genes perform this role. In terms of the field description given above, another solution presents itself. A general biological constraint is that, in the process of cell division, one cell gives rise to two. Cleavage conforms to this, so that the typical process involves a sequence of divisions of the egg such that after each cleavage division the number of cells is 2^P at the pth cleavage, where of course p is an integer. Thus in the sequence shown in Figure 2, cell numbers at successive cleavages are 2, 4, 8, 16, 32, 64, 128. In any individual of a given species, division sychrony tends to decay after some stage. But in the typical cleavage process the number of cells at the pth division is $N = 2^P$. This gives a selection rule on the functions $Y_{lm}(\theta, \phi)$. The number of cells defined by $Y_{lm}(\theta, \phi)$, taking the nodal lines to define cleavage planes, can be shown as $2m(l - m + 1)$ if $m \neq 0$, otherwise $(l + 1)$, and this must be one of the series 2, 4, 8, 16, etc. This greatly reduces the number of functions $Y_{lm}(\theta, \phi)$ which can describe the cleavage stages, but there is still some degeneracy. Thus one finds, for example, that there are four possible functions which could define an 8-cell embryo, corresponding to the four ways in which the surface of a sphere can be divided into eight parts by drawing equally spaced degrees of longitude and/or latitude on it: 8 longitudes; 4 longitudes and 1 latitude (the equator); 2 longitudes and 3 latitudes; or 7 latitudes. To get the observed pattern, we need further constraints.

It can be shown that the "energy" associated with any harmonic, Y_{lm}, varies as $l(l + 1)$, so a minimum energy criterion, which we now adopt, requires that l be as small as possible. This selects the correct description of the 8-cell embryo (4 longitudes and 1 latitude, the equator; l = 3, m = 2), but for other stages such as 16 cells there remains a degeneracy of description since (l, m) = (5, 4) and (5, 2) both give N = 2m (l − m + 1) = 16. As a final selection rule, we use the fact that all animal eggs are polarized so that the animal pole differs from the vegetal pole. In the description this means that there is a polar field, which may be used to define a final selection rule which removes the last degeneracy and generates the typical cleavage sequence. This rule is that m be as large as possible for given l, since this means the selection of cleavage lines which coincide with the direction of the polar field, namely degrees of longitude. As a result, one can draw up a table which shows the field descriptors, Y_{lm}, corresponding to each cleavage stage, as generated by the rules or constraints described above (Table 1). These come from general considerations about "energy" minimization, binary cleavage, and polarity. There is no evidence that any of these is genetically determined, although no one would dispute that gene products are involved in the cleavage process. The relevant distinction here is between necessary raw materials for construction, and design of a building. The design of the typical cleavage process, as argued by Goodwin and Trainor (1980), is a result of organizational principles intrinsic to the living organism, which in the case of embryos is, initially, a single cell. This illustrates the principles of organocentric biology, as contrasted with the genocentric view which regards cleavage to be a result of a "genetic program," despite a failure to define the nature of this program and the absence of evidence that genetic mutants can systematically alter the geometry of the typical cleavage pattern. In the absence of such evidence, one is led to the alternative view that the spatial organization revealed in the geometrical order of the embryonic cleavage planes is a manifestation of intrinsic organismic order which the above description treats as a field property. This conforms to a long embryological tradition which has used the language of fields to describe such phenomena.

The particular field description given above in terms of harmonic functions has the distinct advantage of providing an explanation of the features of wholeness and regulation which Driesch and later embryologists had identified as basic embryological properties. This is because harmonic functions, such as $Y_{lm}(\theta, \phi)$ have the property that, no matter what the size of the sphere, the same functions describe the cleavage planes; i.e., they have the property of pattern invariance with change of size, conforming to the biological observation that eggs of very different sizes show the same typical cleavage pattern. Furthermore, if the function is defined over only a part of the sphere, then it can be uniquely reconstructed over the whole. This mathematical property corresponds to the regulative capacities of embryos, that from a part the whole may be reconstituted. As mentioned above, this is a property not only of embryos but also of adult organisms of many species, which can regenerate the whole from a part. They, too, have field properties leading us to the general conjecture that organisms are fields. This contrasts with the genocentric view that they are aggregates of genes and their products.

Besides emphasizing that organisms are relational structures with properties of wholeness and regulation, the structuralist view sees any particular form revealed in an organism as a member of an invariant potential set of forms defined by generative

Table 1. Correspondence between $Y_{lm}(\theta, \phi)$ and cell number, defined by $l + 1$ if $m = 0$, otherwise by $2\,m(l-m+1)$.

l value	m value	cell number	(l, m) pair selected
1	0	2	
	1	2	(1, 1)
2	0	3	
	1	4	
	2	4	(2, 2)
3	0	4	
	1	6	
	2	8	(3, 2)
	3	6	
4	0	5	
	1	8	
	2	12	
	3	12	
	4	8	
5	0	6	
	1	10	
	2	16	
	3	18	
	4	16	(5. 4)
	5	10	
6	0	7	
	1	12	
	2	20	
	3	24	
	4	24	
	5	20	
	6	12	
7	0	8	
	1	14	
	2	24	
	3	30	
	4	32	(7, 4)
	5	30	
	6	24	
	7	14	
11	8	64	(11, 18)
15	8	128	(15, 8)

rules or laws, specific form coming into actualization or manifestation as a result of the operation of constraints or selection rules. Table I shows the potential set, $Y_{lm}(\theta, \phi)$ for the cleavage planes, generated by a variational principle or rule together with the result of the operation of particular constraints which uniquely select the functions describing the cleavage planes. Each of these field descriptions is a transformation of the other within the invariant set. Thus a structuralist description of a typical biological pattern is achieved, and the rules which generate this may be regarded as laws of form.

So far I have argued that the basic cleavage sequence is generated by biological "universals"; i.e., by intrinsic properties of the living state. However, no actual organism manifests the basic form, although *Amphioxus* (Fig. 4) comes close to it. This form is a generalization, modified by specific or contingent factors which vary from species to species or between members of a species. Some of these contingencies may be genetically determined, as in right-handed and left-handed cleavage in the snail, *Limnea*, resulting in correspondingly coiled shells. Superimposed upon the basic pattern are specific, and regular, modifications, such as the slightly eccentric cleavages of *Amphioxus* relative to the antero-posterior axis (Fig. 3). These can be described by scale modifiers (Goodwin and Trainor, 1980), which may consistently be interpreted as the effect of additional "gradients" or fields within the egg, long studied by embryologists but still not identified in terms of substantial determinants. Thus the variations on the basic pattern are described as the results of particular contingencies or factors which vary from species to species, with evidence that some of these are genetically determined. Genes have a well-defined place in a structuralist biology: they are modifiers of basic form, not its generators, so they have in fact been unfortunately named. Perhaps a more suitable term for them would be cons, deriving both from their role as carriers of contingent effects and for the confidence trick which contemporary biological thought has played on itself.

The Research Programme of Structuralist Biology

The confidence trick to which modern biology has succumbed is that from randomness and contingency one can get order. In fact what one gets is variety. The research programme of a structuralist biology is to discover the generative laws which are responsible for that order which characterizes the biological realm, and then to demonstrate how chance and contingency result in variations on this order. The organism is then revealed as an entity which transforms randomness into order by virtue of its organizational principles. Mutations in the genes and changes in the external environment provide the organism with, for example, variations in its composition, in the interaction pattern between molecules and reactions, in such variables as salt concentrations, temperature, light, and so on. Any of these may induce transformations from one member to another of the set of forms which is available to the organism, just as a change in initial conditions can induce a transformation to a new pattern of motion in Newtonian mechanics. But these potential forms are limited to a defined set, such as the functions $Y_{lm}(\theta, \phi)$ which are the potential descriptors of the cleavage planes in the analysis presented above. This particular description may and may not prove to be embryologically useful, but it does illustrate the type of analysis which characterizes a structuralist treatment of the problem, continuing the research programme of rational morphology. One goal of such a programme is to understand in rational or generative terms representative examples of typical form. This means that the organizational principles, rules, or laws which underlie various biological structures should be expressed in terms of functions selected from a potential set. For example, the pentadactyl limb as a typical form should be expressible as a set of field functions describing the basic limb structure, starting at the shoulder with a single bone (the humerus or the femur) and ending with five digits. All actual limbs must then be shown to be specific modifications of this form. It is now reasonably clear how this may be achieved for this biological structure, and it has close similarities to the treatment given of the cleavage process, despite the apparent difference in the forms involved. However, it is to be expected that the underlying field principles are essentially the same, as anticipated by embryological studies which show that similar principles of polarization and regulation are involved, as they are in all developmental processes.

One is led to the conjecture that organisms are structures defined by general principles of organization and transformation and that the set of typical forms among which transformation is allowed may be defined by the harmonic functions, solutions of the field equations of Laplace and Poisson. Since all organisms would be mathematically transformable into one another within this set, they could be compared with respect to principles of symmetry and other relevant measures of order and complexity so that a logical ordering of forms may be generated. This hierarchy of beings could provide the rational basis for a taxonomic ordering, which was another of the goals of rational morphology.

Comparative studies between organisms at different stages of development also become possible within this scheme. Thus in a recent analysis (Goodwin, 1980) it was shown that the basic form of a very simple organism, the unicellular ciliate protozoan *Tetrahymena*, can be described by the same fields as those representing the gastrula stage of an amphibian embryo, using harmonic functions which are related to but distinct from the $Y_{lm}(V)$ used to describe cleavage planes. This kind of homology appears to be the same as that used by the fervent Darwinist, Ernst Haeckel (1923), as an example of the principle which he stated in the form "ontogeny recapitulates phylogeny." He believed that any organism during its embryonic development (ontogeny) passes through the morphological forms characteristic of its ancestors (its phylogeny). Since the protozoa are the most primitive animals, an amphibian embryo should, according to Haeckel's principle, pass through the protozoan stage.

However, the embryological evidence does not suppor this so-called biogenetic law. In general, there is little correspondence between the adult form of a more primitive species, and an embryological stage of a more recently evolved one. What one sees, however, is that the embryos of different species tend to look more alike the earlier the stage at which one compares them; and this is strikingly true of species with the same basic body plan, such as the vertebrates. In terms of the perspective developed here, this is a reflection of the expectation that the further one is in embryogenesis from the adult form of a species, the closer one is to basic expressions of typical form, and hence the greater the expected similarity of form between the embryos of different species. However, the similarity, being one of homology, need not be obvious, expressible as it may be only in terms of basic symmetries and underlying field patterns whose manifest detail may differ markedly between species by virtue of different selection rules. The homology which I have suggested to exist between the form of the unicellular organism *Tetrahymena* and the amphibian gastrula is not obvious since one is a single cell and the other a multicellular embryo. What suggests the homology is that both the spatial pattern of *Tetrahymena* as revealed by the positions of its rows of cilia and its mouth, and the presumptive fate map of the amphibian gastrula together with the dorsal lip of the balstopora, are each describle by the same harmonic functions but with different curves expressed in the two cases by different selection rules. However, the amphibian gastrula field is considerably more complex than the *Tetrahymena* pattern insofar as the rules selecting the curves of the presumptive fate map are more complex in their definition. It is not yet clear exactly how these comparisons of order and complexity are to be formulated rigorously. I have discussed these questions of constraint, symmetry, complexity, and order rather more generally in another context (Goodwin, 1978), where I relate them to problems of knowledge and its use in generative processes. Evidently what is needed is a criterion of ordered complexity in relation to the constraints or selection rules which generate particular patterns from a defined potential set.

In the above example, I have been comparing unicellular organisms with multicellular forms and suggesting homologies between them, which suggests another rather radical departure from contemporary biological orthodoxy. This departure is that the cell is *not* the fundamental unit of biological organization. Once again, it is the organism with its field properties which plays the role of fundamental unit in a structuralist biology, and whether the organism is made up of one cell or of many is irrelevant from the point of view of wholeness, self-regulation, and transformational capacity. Thus cellular partitioning does not divide the multicellular organism into a collection of autonomous interacting units, as is so often assumed in theories of embryo-genesis, following the logic that wherever there is a nucleus with its genes, there is a quasi-autonomous decision-making unit, an automaton (Wolpert and Lewis, 1975). It would appear, rather, that the strategy of multi-cellularity is a means of packaging and assembling gene products, allowing the organism to generate complex local mosaics with sharp spatial discontinuities of state. But the fundamental organizing entity remains the whole organism, irreducible to any of its material parts.

Limitations and Perspectives

All perspectives have their limitations, and structuralist biology fails in at least one important aspect. Identification of some of the principles underlying biological form gives this a certain logical intelligibility, but it does not describe the nature of the dynamic which confers upon development the character of an organized process in time. A purely synchronic description, which is what has been offered in the example of typical form as revealed in cleavage is inadequate. A diachronic or time-dependent theory which shows how and why one field transforms into another is required and this is not provided by a mere description of the rules which select the relevant fields from a potential act. It is necessary also to describe an internal dialectic whereby one form changes into another as a result of forces generated by the fields themselves. We have begun to make some progress in this direction, and have again found that the type of theory which is emerging was anticipated by those we have called 20th century structuralists, particularly Waddington (1956).

Using the cleavage example, it appears that the surface fields described by the functions $Y_{lm} (\theta, \phi)$ are complemented by fields describing the internal forces which organize structures such as centrioles, mitotic spindles, and chromosomes into their characteristic spatial patterns and orientations within the dividing cells. These fields act as inducers or "evocators" (Waddington's term) of the surface fields describing the cleavage furrows along which cells divide, and the interaction between these complimentary fields then generates the dynamic which results in the state which produces a new internal field, a process for which we can use Waddington's term "individuation," since a new individualized form is thus brought into being. This then acts as the evocator of a new cleavage field, which interacts with the evoking field to generate a new individualized form, and so on. This, of course, is no more than a schematic picture employing the language of interacting fields to describe a process which is itself unitary and, it seems, essentially non-linear; but the dialectical description allows one to use quasi-linear fields in the analysis, thus simplifying the treatment considerably.

The recursive nature of the cleavage process as described above is quite obvious, and this represents an important feature of embryogenesis: the same process recurring sequentially within a systematically changing context generates ordered complexity in time. However, embryogenesis also involves the emergence of different types of order so that in terms of our field description, one type of field gives way to another. It seems that this arises by virtue of a natural decay of order in a recursive process due to the accumulation of "noise," together with a continually changing context which makes it possible for other processes to begin after certain stages. Thus the next obvious and important event after cleavage in amphibian embryogenesis is gastrulation, wherein the hollow ball of cells generated by cleavage undergoes transformation into a three-layered structure by an infolding of the cells at a certain point of the surface, the dorsal lip of the blastopore. This is possible only after cleavage has proceeded far enough for cells to be sufficiently small to invaginate (fold in) at the dorsal lip. But besides this possibility, there must be a field which defines the position of the dorsal lip and associated spatial order, the fate map in the presumptive endoderm and mesoderm. The initiation of this field seems to occur at sperm entry, the earliest event in embryogenesis, for this establishes the plane of bilateral symmetry of the future embryo, which passes through the centre of the dorsal lip of the blastopore. However, this bilateral field, distinctly different in its spatial form from that describing the cleavage planes (Goodwin, 1980), grows gradually in strength relative to the cleavage field, which becomes progressively more disordered as higher and higher harmonics come into being, and at some point the bilateral field takes over as the dominant ordering force in embryogenesis, initiating a new phase of the process. Again, this is no more than a speculative description of the type of emergent process which could underlie the structural dynamics of embryogenesis, and it is necessary to develop a time-dependent field theory to give a more adequate treatment of this problem. This is an aspect of the research programme which comes out of, but transcends, a strict structuralist description, and it is something we are currently studying.

My own belief is with the emergence of such dynamic field theories in biology, will come an understanding of the sense in which the biological process is creative. I have discussed this in relation to embryogenesis (Goodwin, 1978), where a consistent case can be made for the creative nature of this process by virtue of the very large set of possible responses to perturbation (potential set of forms), freedom of the selection process from external control, and appropriateness of the selected responses to the perturbation. A similar case can be made for evolution: new forms of organisms are generated as a result of internal (including genetic) and/or external perturbations which can cause transformations to different members of the potential set, hence new organismic forms, the transformations being dependent upon, but not determined by, the perturbation. Some of these new forms are appropriate to the perturbation: i.e., they are stable (can survive) under the new condition. Neo-Darwinism considers organismic form to be determined by genes and the external environment, there being no autonomous transformer of these into specific form (i.e., no autonomously organized organism), so there can be no creative response since this always involves appropriate selection from a set of alternatives, not determined by the stimulus but by internal criteria of order and organization. Within the new perspective, biological process may be perceived as a dialectic of creative transformation. Thus order and creativity replace contingency and opportunism as the foundations of the new biology.

Acknowledgment

I wish to acknowledge a deep indebtedness to my colleague, G.C. Webster, for the development of many of the ideas expressed in this essay.

Notes

Cairns-Smith, A.G. 1981. This volume.

Dawkins, R. 1976. *The Selfish Gene*, Oxford: Oxford University Press.

Driesch, H. 1908. *Science and Philosophy of the Organism*. London: A & C Black.

Fox, S. W. 1965. *The Origins of Prebiological Systems and of Their Molecular Matrices*. N.Y.: Academic Press.

Goodwin, B. C. 1978. A cognitive view of biological process. *J. Social. Biol. Struct.* 1: 117–125.

Goodwin, B. C. 1980. Pattern formation and its regeneration in the protozoa. *Soc. Gen. Microbiol.* 30: 377–404.

Goodwin, B. C., & Trainor, L. E. H. 1980. A field description of the cleavage process in enbryogenesis. *J. Theoret. Biol.* (in press)

Sonneborn, T. M. 1970. Gene action in development. *Proc. Roy. Soc. Lond. B.* 176: 347–366.

Thom, R. 1972. Stabilité Structurelle et Morphogénèse. Mass.: W. A. Benjamin, Inc.

Thompson, D'Arcy, W. 1917. *On Growth and Form.* Cambridge: Cambridge Univ. Press.

Waddington, C. H. 1956. *Principles of Embryology.* Allen and Unwin.

Waddington, C. H. 1957. *Strategy of the Genes.* Allen & Unwin.

Webster, G. C., and Goodwin, B. C. 1980. *The origin of species: a structuralist view.* Submitted for publication.

Weismann, A. 1891. *Essays on Heredity.* trans. A. E. Shiply. Oxford: Oxford Univ. Press.

Wolpert, L., and Lewis, J. 1975. Towards a theory of development. *Fed. Proc.* 34: 14–20.

The Origins of Time

M. Heller

> "What was so fascinating about this, Jesse thought, was its ordinary nature—the canal, the locks, the noisy water; the town itself ordinary and quiet, as if it had existed for centuries, with a profound certainty of its right to exist, no awareness of the fact that it had no reason for existing, no guarantee of its right to exist. It was here; it moved in a slow, timed orbit."
>
> Joyce Carol Oates, "Wonderland"

0. The fundamental axiom of science and of everyday experience asserts that everything has its history. Also, the universe as a whole turns out to be a "historical being"; contemporary cosmology attempts at reconstructing the cosmic history. "One of the greatest discoveries of science is that the universe also changes with time, and like living systems may well have a kind of birth and death also." (Davies, 1978, p. 74). But why has the universe its history? This apparently trivial question opens a fascinating research field for theoretical physics. Our attempt to answer this question touches an old philosophical issue—the problem of the origins of time.

Let us remark that the history is not merely the past, but a past that is—in a certain manner—contained in the structure of the present. If in the initial singularity—in the Big Bang—the Universe completely forgot what came before, one cannot meaningfully speak about the history of the initial singularity. However, one *can* meaningfully speak about the history of the Cosmos after the initial singularity. Why?

1. Contemporary physics explores reality through theoretical models that physics has itself constructed. Although an agreement between models and reality is to be controlled by experimental verifications, this seems to be a modern version of a typical Kantian problem: how to change from our constructs to "things in themselves"?

All macroscopic physical theories presuppose a certain common space-time arena. This background space-time model has a mathematical structure called a manifold structure. The concept of manifold arises as a generalization of the ordinary surface concept. The generalization goes along two lines: (i) transition from two to n dimensions (in physical applications usually n = 4), and (ii) forgetting about the space into which the original "surface" is imbedded. (For instance, O'Neill, 1966.)

However, not every four-dimensional manifold is a suitable candidate to serve as a model for the physical space-time. The necessary condition is that it should carry the so-called Lorentz structure (Lorentz metric). Physics can happen on a manifold only if it is a Lorentz manifold (Hawking and Ellis, 1973). The essence of the Lorentz structure consists in that it makes meaningful the concept of a distance in space-time.

2. A manifold carries the Lorentz structure if and only if a non-vanishing direction field is defined on this manifold, i.e., if and only if a direction is defined at each of its points (Geroch, 1971). It turns out that it is always possible to construct such a Lorentz structure so that the direction field is timelike.

In other words, a manifold is a suitable arena for physical process if and only if any of its points distinguishes between two time directions. However, two important circumstances have to be noted: first, this primitive time concept is purely local ("at any point"); second, the concept is primitive indeed—in the Lorentz structure alone there is no criterion to determine which of the two directions is past and which is future. For the "arrow of time" one must look elsewhere (Davies, 1974; Heller, 1975).

The result seems to be philosophically deep and interesting: the existence of a kind of local time—yet arrowless—is a precondition for every physics. Indeed, the possibility to make empirical predictions belongs to the very essence of physics. Without predictions physics could not exist. Predictions, in turn, presuppose at least a local time. Moreover, from the methodological point of view predictions are exactly as good as retrodictions—only psychologically are predictions of higher value for physicists. Therefore, the arrow of time is irrelevant as far as the *possibility* of physics is concerned.

3. And what about a global time? Two cases should be distinguished here. If a manifold is *non-compact** a non-vanishing direction field always exists on it, and the manifold always carries a Lorentz structure. One can always choose the Lorentz structure so that there will be no closed timelike curves on the manifold. In such a universe, time passes irreversibly

If one assumes additionally that the manifold in question is *causally stable*, i.e., if a small perturbation of the Lorentz structure does not produce closed timelike curves, then along every timelike (and null) curve, smooth monotonically increasing functions can be defined (Hawking and Ellis, 1973, p. 198−201). These functions can serve as clocks measuring the lapse of the global time of the observer whose history is represented by the given timelike curve. Surfaces of simultaneity, in general, are not yet determined uniquely. The point is, however, that on any non-compact causally stable manifold, a global *open* time always exists.

4. The case of a *compact* manifold is not so simple. There is a topological characteristic of such manifolds, called the Euler-Poincaré characteristic. To define it, take a two-dimensional "closed" surface which is compact (for instance, surface of a sphere), decompose it into an arbitrary number of "cages" (see figure), and check that the integer

$$\chi = \alpha_0 - \alpha_1 + \alpha_2$$

---

*A set S consisting of an infinite number of points is called compact if every infinite subset of S has at least one limit point in S. Compact sets are thus closed (i.e., contain all their limit points) and bounded. All finite sets are compact.

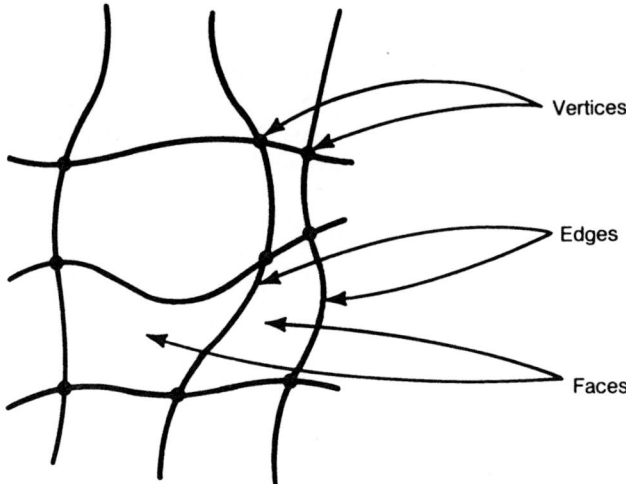

Figure 1. Decomposition of a compact manifold.

where α_0 is the number of vertices (0-dimensional figures), α_1 the number of edges (1-dimensional figures), α_2 the number of faces (2-dimensional figures), is independent of the decomposition. χ is called the Euler-Poincaré characteristic of the surface (for a sphere $\chi = 2$). This definition is naturally extended to higher dimensions. And now, every compact manifold admits an everywhere non-vanishing direction field, and consequently a Lorentz structure, if and only if the Euler-Poincaré characteristic of the manifold vanishes, $\chi = 0$.

However, any Lorentz structure on a compact manifold produces *closed* timelike curves. This means that in a compact Lorentz manifold there are always events which happen infinitely many times. Such events locally distinguish between two time directions, but globally there is no distinction between future and past: going sufficiently faraway into the future one finds oneself in the one's past. "Topologically closed time is the idea that time reenters itself upon the completion of a cosmic cycle and the end becomes a beginning—not a new beginning in linear time, but somehow the beginning which now appears to us to have been the past" (Fraser, 1978, p. 67).

Compact space-times do possess, however, certain properties which make them "non-nice" from both mathematical and physical points of view. We cannot exclude a priori that the universe we are living in has closed timelike curves, but we can say that physics does not seem to like such curves. Moreover, they are unnecessary. It turns out that a compact space-time cannot be simply connected (Bass and Witten, 1957). This means that any compact space-time M can be obtained from a non-compact space-time M' by identifying certain "parts" in M'. The non-compact Lorentz manifold M' has exactly local properties the same as a compact Lorentz manifold M, and M' always can be assumed, instead of M, to model the physical space-time.

The idea of a cyclic time is evidently connected with the determinism problem. "Those of the gods—the Stoics believed—who are not subject to destruction, having observed the course of one period, know from this everything that is going to happen in all subsequent periods. For there will never be any new thing other than that which has been before down to the minutest detail" (Whitrow, 1972, p. 17). Universal determinism of this kind seems to be foreign to contemporary physics.

5. The existence of the Lorentz structure enables macroscopic physics to happen on the space-time arena. This structure, from its very essence, permits to perform all space-time measurements; it is a *metric* structure. One might say, therefore, that the Lorentz structure is responsible for "quantitative aspects" of nature. However, the necessary and sufficient condition for the existence of Lorentz structure (the existence of the direction field, compactness, Euler-Poincaré characteristic) are topological properties. But topology does not deal with "quantity" in the ordinary meaning of the word; its objects are continuity, closeness (not necessarily metric), shape, etc. Thus, one might conclude that quantitative aspects of space and time—and perhaps of all physics—result from, or are rooted in, some deeper non-metric level which could be called a "pre-physics" of spacetime.

Perhaps this touches upon the problem called for by Whitrow: "For any theory which endeavors to account for time *completely*, ought to explain why it is that everything does not happen at once." (Whitrow, 1972, p. 132).

Notes

Bass, R. W., Witten, L. 1957. Remark on Cosmological Models. *Reviews of Modern Physics* 29: pp. 452–453.

Davies, P. C. W. 1974. *The Physics of Time Asymmetry*. Surrey University Press—Intertext Publ

Davies, P. C. W. 1978. Time Singularities in Cosmology and Black Hole Evaporations. In *The Study of Time* 3, eds. J. T. Fraser, N. Lawrence and D. Park. Springer-Verlag. pp. 74–90.

Fraser, J. T. 1978. *Time as Conflict*. Birkhäuser Verlag.

Geroch, R. P. 1971. Space-Time Structure from a Global Viewpoint. In *General Relativity and Cosmology*. Intern. School of Physics "Enrico Fermi," course 47. Academic Press. pp. 71–103.

Hawking, S. W., Ellis, G. F. R. 1973. *The Large Scale Structure of Space-Time*, Cambridge: at The University Press.

Heller, M. 1975. Global Time Problem in Relativistic Cosmology. *Annales de la Société Scientifique de Bruxelles* 4 89: 522–532.

O'Neill, B. 1966. *Elementary Differential Geometry*. New York: Academic Press.

Whitrow, G. J. 1972. *The Nature of Time*. Pelican Books. p. 17.

The Relationship Between Our New Sense of Time and Our Sense of an Ending in Tragedy

S. L. MACEY

Tragedies, which by general consensus over a long period of time are acknowledged to be great, seem to be confined to specific periods. They appear to have been composed at those times when a civilization regarded itself as a chosen people. If we use as examples the Athens of the 5th century before Christ, the France of the Roi Soleil, the England of the Elizabeth who was "Gloriana," or perhaps even the America of Eugene O'Neill, Tennessee Williams, and Henry Miller, we will quickly realize that the heroes with whom we are concerned demonstrate a progressively changing relationship with their audience. The Greek heroes are by definition related to the gods, Shakespeare's protagonists are kings or at the very least *gens de condition*—people of the highest importance in the state—whose fall involves the fate of a people. By way of contrast, the protagonists of O'Neill, Williams, and Miller are all too much like ourselves.[1] We are not here directly concerned with defining tragedy or even with relating it to death—for Oedipus and Electra tragedy does not seem to require death and in *Death of a Salesman* the death may be less a tragedy than the closure of one—but we are concerned with the influence on tragedy of an increasing awareness of clock time and of geographical place. The subject of this paper is the change in the tragic hero's function or status as well as the narrowing of the discrepancy between the time and place of the hero and that of the audience. These changes were conditioned by changing attitudes to chronology.

Aristotle maintains, "The best tragedies are constructed around a few families and are concerned with Alcmaeon, Oedipus, Orestes, Meleager, Thyestes, Telephus and any other such men who have endured or done terrible things." They are "men who are better . . . than we know them today."[2] But the histories of such men—though they relate indirectly to the audience—go far back into myth, to the times of the Trojan War and beyond. We now attempt to pinpoint the Trojan epic at c. 1200 B.C. Such dates, however, had no meaning for the Greeks.[3] Today's more precise chronologies belong to a period after what Quinones calls "the Renaissance discovery of time." They reflect the attempts of Schliemann and others to anchor in space and time the concrete relics of Greek myth. As J.T. Fraser puts it, "for the Greeks the past was the tradition of heroes. Great individuals interacted with each other and with a society of divinities in an essentially cyclic world."[4] In such a world the sense of an immediacy related to linear and chronological time was missing. This sense of immediacy is perhaps best illustrated in the English Renaissance by the clock that strikes the 11th hour in the final act of Marlowe's *Faustus*. Shakespeare's canon is full of instances reflecting the significance of clock time.

Despite the recurrence of events related to the seasons or even to the jubilee (Leviticus 25), the main Hebrew concept of time, by way of contrast with the Greeks, tends not to be cyclical. Nevertheless, the biblical chronology also had to await its modern interpreters in order to become more precisely defined. The history of the Hebrews posits a world that—according to Bishop Ussher and others—began at a definite time in 4004 B.C.; Jehovah's injunction against graven images may, however, have militated against the pictorial and dramatic representation of the Hebraic world. Apart from the *Purimspiel* that serves as an inversion of the usual codes of conduct, theatre (the traditional vehicle for tragedy) appears to have been frowned upon in areas where the Jewish, Islamic, and Puritanical influences remained strong. Despite the sense of historicity that pervades it, the Bible involves quantitative chronological measurements that strike us as mythical. Typical examples would be the 969 years of Methuselah or the 3 times 14 generations that separate Abraham from "Joseph, the husband of Mary, of whom was born Jesus." Further-more, as with Troy, the attempt to anchor the Hebraic myth in a precise chronologi-cal sequence—in this case by Bishop Ussher's *Annales Veteris et Novi Testamenti* of 1650–64—did not take place until men were more immediately concerned with chronology.

Despite the more recent developments, objective chronology is not universally accepted. Philosophers and some poets, among others, remain disturbed by the dialectical confrontation between the objective measurement of time epitomized by the Newtonian clockwork universe and the subjective measurement of time suggested by Locke's succession of impressions (*Essay Concerning Human Understanding* 2.14 and 15). Common life, however, has come to be dominated more and more by the external and objective measurements of the clock, and despite such apparent exceptions as the stream-of-consciousness technique, literature too, is increasingly controlled by chronologies related to an objectively ascertainable time and place. Though the main developments in the precise measurement of objective time occur after the middle of the 17th century, Western civilization had, of course, begun to evolve a tighter form of chronology long before the life of Newton and the concurrent invention by Huygens of the first accurate escapement for clocks. With the advent of the Julian Calendar in 46 B.C., the Romans developed a means for tabulating chronology, although not yet like our own, which was much more sophisticated than that of the Greeks.[5] Yet aesthetically the Romans seem to have lived in the shadow of the Greeks, and like the Hebrews, they never produced great stage tragedy. Certainly Senecan drama is not great tragedy.

There is a remarkable absence of great tragic drama between the 5th century B.C. in Athens and the Renaissance England of Elizabeth I. Though the Catholic church may be held responsible for long holding the theatre at bay, they eventually found it convenient to act out the tropes of the liturgy, biblical scenes, lives of the saints, and moral homilies to audiences that knew neither Latin nor the art of reading. Although in such plays devils as well as a cuckolded Joseph often provided a focus for comedy, tragedy, as we understand it was not produced.

When the audience is predominantly committed to the values of heavenly bliss, the worthy protagonist who is about to die has a prescience of eternal joy, which blunts the edge of tragedy and produces instead a martyr play or *Martyrerstück*. An example of such martyr plays is the *Katharina von Georgien* (1647) of Andreas Gryphius, in which the Christian Katharina becomes a martyr at the hands of the pagan king Abas, or Gryphius's *Carolus Stuardus* (1649), which deals with the martyrdom of the English Charles I in the very year that he died. In England, the values of Shakespeare and his audience were much more ambivalent. H.D.F. Kitto may make as much as he pleases of *Hamlet* being a religious tragedy,[6] but the main thrust of *Hamlet* is the tragic loss of a young Renaissance prince whose reputation (like his father's) must be cleansed in this world. Englishmen were in the process of becoming concerned more with time than with eternity.

Unlike both Catholic martyr plays and Greek tragedies, Shakespearian tragedy seems less overtly concerned with the gods, but Shakespeare also differs from Greek tragedy in that his protagonists are no longer heroes in the sense of being related to the gods. If we consider the range of Shakespeare's tragic heroes as extending from a king like Lear to *gens de condition* like Romeo and Juliet, they are stilll distanced in time, space, and status, but they are much closer to the average members of the audience than the Greek heroes had been.

The reduced distancing in Shakespeare begins to make us more conscious of his anachronisms. It is customary to stress Shakespeare's anachronisms—the clock in *Julius Caesar* II.i, the fact, as Ian Watt explains, that "Troy and Rome, the Plantagenets and the Tudors, none of them are far enough back to be very different from the present or from each other."[7] The word "anachronism" was not used in English until 30 years after Shakespeare's death,[8] but developments that would make such a word necessary were already under way. With regard to horology, Cecil Clutton et al., in *Britten's Old Clocks and Watches* states that "no English watch is known of a date before 1580 and only a short time before this is there any record of an English watchmaker,"[9] while in a parallel development, Ricardo J. Quinones tells us, in his *Renaissance Discovery of Time*, "In the early 1590s a spate of English works appeared with time as a vital concern."[10] This interest may be contrasted with the earlier Chaucer whose works, for historical reasons, are (apart from the special case of *Treatise on the Astrolabe*) almost without horological reference, despite the fact that he is concerned with virtually every other aspect of time, including its transitory nature, the passage of the seasons, and the impact of the planets and stars on the lives of men.

In essence, then, Shakespeare's tragic protagonists constitute a moment of transition to the modern world. They are less distanced from us in time, space, and status than the Greek heroes, but unlike the protagonists of martyr plays, their values are essentially concerned with this world. There is a finality in the knowledge that Lear will "Never, never, never, never, never," see Cordelia again, which no Christian apologist can satisfactorily evade. The sense of closure that comes with death in *Hamlet, Lear,* or *Romeo and Juliet* has a finality directly related to values that are essentially of this world. Moreover, heroic Renaissance tragedy seems also to gain from the tension between a more modern sense of clock time and an older, religious sense of heaven's eternity such as many in the audience unconsciously retain.

Though Corneille and Racine appear to choose more "heroic" subjects than Shakespeare, they are confined by "unities" that Shakespearean tragedies could still ignore. The phenomenon of the imposition of the unities of time, action, and place on the Continent and more specifically on French drama by the Académie Française after 1634 has never been satisfactorily explained. The French unities, however, were clearly an attempt to impose *vraisemblance* on an unrealistic subject, and in doing so they went far beyond the rules of Aristotle's *Poetics* that they claimed to follow.[11] In addition, the relationship of the unities with political rigidity in the case of Corneille's *Le Cid,* and the analogy with Cartesian method should not go unnoticed. Though Dryden was particularly proud of the way that he put Shakespeare's *Anthony and Cleopatra* into the straitjacket of his *All for Love* (1678), the three unities that he there employed never really captured the interest of English theatregoers.

English authors and critics, however, paid lip service to the unities until the need for such unrealistic "realism" was exploded by the common sense of Dr. Johnson in the preface to his edition of Shakespeare's *Works* (1765).[12] Yet by that time the English had developed a new approach to literature (including tragedy) that was beholden to neither the rules, the content, nor the form of previous genres. In Ian Watt's words, "Defoe and Richardson are the first great writers in our literature who did not take their plots from mythology, history, legend or previous literature." Though it might well be argued that Shakespeare's history plays have deep historical structures, Ian Watt maintains that instead of the

> ". . . historical outlook . . . which has caused the time-scheme of so many plays both by Shakespeare and by most of his predecessors from Aeschelus onwards, to baffle later editors and critics . . . . the modern sense of time began to permeate many areas of thought. The late seventeenth century witnessed the rise of a more objective study of history. . . . At the same time Newton and Locke presented a new analysis of the temporal process; it became a slower and more mechanical sense of duration which was minutely enough discriminated to measure the falling of objects or the succession of thoughts in the mind. These new emphases are reflected in the novels of Defoe."[13]

Watt might have added, of course, that the developments he describes also follow upon the invention, in the third quarter of the 17th century, of the first clocks and watches with an accuracy sufficient to satisfy the needs of modern urban man. But Watt does add, and this is clearly to our purpose, that Richardson exceeds Defoe in being "careful to locate all [the] events of his narrative in an unprecedentedly detailed time-scheme . . . we are told, for example, that Clarissa died at 6:40 P.M. on Thursday, 7th September."[14] Though some relationship with the martyr plays can be argued, *Clarissa Harlowe* (1747–48) is our first great nonheroic tragedy written in the novel form, and Richardson went out of his way to justify his epistolary technique on the grounds of relating the present activity of the characters to the sense of immediacy of the reader. He was the first author who deliberately exploited the technique of suspense through his manipulation of time.

That the novel (our one major new genre which developed in the 18th century) is more firmly anchored in time and therefore in space than previous genres—including plays that followed the unities—has been generally recognized by critics. E.M. Forster considers the portrayal of "life in time" essential to the novel;[15] Northrop Frye considers "the alliance of time and Western man" to be the novel's defining characteristic;[16] and Adam Mendilow considers that "the novel is a complex of time-values."[17] Ian Watt, whose first chapter in *The Rise of the Novel* deals with the close relationship between realism and the novel form, sees realism in terms of "the two coordinates, space and time."[18] The novel relates space and time (frequently an almost contemporaneous time and familiar space) to a degree not seen before in major literature. This can be attributed to the fact that the readers want to follow the activities of protagonists with whom they can identify. But we have already been tracing in tragedy the transition from protagonists who are heroes in the classical sense to protagonists who are *gens de condition*. The latter are considerably closer to us than Greek heroes in terms of Watt's coordinates of space and time, as well as in terms of status. With the advent of Richardson's *Clarissa Harlowe*, the difference between tragedy and comedy in terms of time, location, and status has been essentially eliminated. As Aristotle told us and as Aristophanes demonstrated, comedy has generally tended to be closer to the audience in terms of time, place, and atmosphere than has tragedy.

Aristotle noted long ago that "writers of comedy" had "constructed their plots in accordance with probability," given their characters "typical names," and dealt with "men who are inferior" (*Poetics* Chs. 9, 5, 2). The increasing emphasis on the immediacy of time and place in the early 18th century permitted a fairly natural use of this comedy convention in the six best-known journal-novels of Defoe.

The epistolary form of Richardson is even more immediate than the journal form of Defoe, and Richardson's *Clarissa Harlowe*, the only tragedy among his three novels, is, as I have noted, remarkable for showing the same immediacy in terms of time, place, and status as do his two other works. In this respect *Clarissa Harlowe* provided an important milestone in literary history. Clarissa can expect the same dowry as the Harriet Byron of Richardson's *Grandison* (1754), and she is involved with a man whose expectations place his fortune in almost the same category as Grandison's.[19] But Sir Charles Grandison is a gentleman, whereas the immediacy with the reader is weakened by Clarissa's Lovelace being an aristocratic villain whose seduction of that bourgeois young lady results in her death. In terms of both character and theme, Lovelace stands outside the familiar, immediate world of the reader. As we shall see later, this would become an important theme in German middle-class tragedy.

We have noted that when Richardson wrote *Clarissa* (1747–48), he could be remarkably specific about the time of her death, as indeed about many other coordinates of time and space. In a similar way, Fielding, in his comic novels, relates the cross-country tour of *Tom Jones* (1749) to the 1745 Rebellion, and the overseas journey of *Amelia* (1751) to the Siege of Gibraltar. Defoe attempts a comparable realism, but perhaps in part because the convention of novels "lying like the truth" had not yet been established in his time, Defoe is involved in some remarkable anachronisms. I have suggested in an article in *Notes and Queries* that the heroine of *Moll Flanders* (1722), who is said to be writing her autobiography in 1683 at the age of seventy, does in fact have a life virtually concurrent with Defoe's.[20] His *Roxana* (1724), whose protagonist is said to be a lady of easy virtue at the court of Charles II, involves anachronisms that are equally preposterous. But the bourgeois Defoe, like Richardson, writes so circumstantially and precisely about time and place that he invites attention to anachronisms.

The historically important *London Merchant* (1731) is our first tragedy involved with a member of the lower middle class. It had been written by George Lillo in the year of Defoe's death and sixteen years before *Clarissa*. Lillo, a London businessman like Defoe and Richardson, wrote his play about the tragedy of a London apprentice, who, despite the love of his master's daughter Maria, was seduced by the prostitute Millwood and murdered his uncle for his money. The scene is set in London just before the sailing of the Armada in 1588, but the story of the good apprentice Trueman and the seduced apprentice Barnwell may be compared for immediacy with a comparable story in Hogarth's 12 plates of *Industry and Idleness* (1747). In *The London Merchant* the simple prose form and the pride in City values of its merchant Thorowgood produce immediacy and realism. For over 100 years London merchants regularly sent their apprentices to see this play. George Sherburn (like Allardyce Nicoll) regrets that Lillo's *London Merchant* "stimulated practically no disciples in England."[21] When, however, the play is viewed as a successful and self-conscious announcement that the whole spectrum of middle-class protagonists, down to and including the rank of apprentice, are worthy subjects for tragedy, there seems to be no point in continuing to hammer home the message.

In an article some years ago, entitled "Nonheroic Tragedy: A Pedigree for American Tragic Drama," I showed how Lessing's *Miss Sara Sampson* (1755)—influenced by the *Kaufmann von London* ("London Merchant") and by *Clarissa Harlowe*—inaugurated a whole century of German middle-class tragedy. German middle-class tragedy, in its own turn, provided a pedigree for the nonheroic American tragic drama of O'Neill, Williams, and Miller. At the risk of over-simplifying my earlier argument, this development demonstrates that since slowly, erratically, and yet inevitably the concomitant of modern industrialization involves the franchise among men and women spreading to all classes, colors, and creeds, it is natural that more and more people are becoming sufficiently important to warrant consideration for the role of tragic hero.[22] The corollary of this argument is that tragic heroes have been coming increasingly close to ourselves in terms of place and time.

The early German middle-class tragedies, however, required a somewhat further distancing from the audience than did the English *London Merchant* and *Clarissa Harlowe*. For historical reasons, it would be a long time before German audiences could identify themselves with a middle-class tragedy in which German characters enacted their conflicts on German soil. Lessing's *Miss Sara Sampson*, Pfeil's *Lucie Woodvil*, and Brawe's *Freigeist*, which all appeared in the second half of the 1750s, have English characters of the status of *gens de condition* who tended to be removed to the neutral surroundings of an inn, while Martini's *Rhynsolt und Sapphira* has been set in the province of Zeeland at the safe distance of the time of Charles the Bold (1447–57). Whereas the classical hero's affair is between himself and the gods, and he draws much of his strength from the ability to act for himself alone, the struggle of the bourgeois protagonist is no longer with the gods, but with the values and forces of his own class. The bourgeoisie's most effective control involved withholding its hallmark of respectability from erring members of its class. As a result, the fear of disgrace (*Schande*) permeates bourgeois tragedy from Lessing to Hebbel. Indeed, there is no German play of note in the whole series in which I have not found the very word *Schande*.

The inevitability of a bourgeois girl choosing death rather than disgrace after seduction by an upper-class devil figure gave the audience a particular sense of immediacy regarding many of the middle-class tragedies. This occurs, for example, in Wagner's *Kindermörderin*, Lenz's *Soldaten*, and Goethe's "Gretchen Story," three dramas of the Storm and Stress, in which the lieutenant, Desportes, and the Mephistophelian Faust respectively play the devil figures. In Lessing's *Emilia Galotti*, the heroine, faced with an almost inevitable seduction by the prince, begs her father for release from *Schande* by means of the dagger in his hand. *Schande*, by contrast, plays a very small part in the thinking of such aristocratic females as Lessing's Orsina in *Emilia Galotti*, or Schiller's Lady Milford in *Kabale und Liebe* when they are faced with similar problems. Hebbel's *Maria Magdalene* (1843) is the last of the conventional German middle-class tragedies, but by now the social class is lower and more immediate in terms of time and place, while the devil figure, Leonhard, is from the same lower class as the heroine. Despite the fact that her seducer is no longer from the upper classes, Klara, who has tasted forbidden fruit, drowns herself rather than bring disgrace on her family.

After the 1848 revolutions, members of the European middle classes came to be treated more and more as individuals, who, in the wake of Darwin and Haeckel, were considered subjects suited to psychological tragedy on a wide variety of special problems. Since then, the *Kleinbürgerwelt* who tried to control social behaviour through the code of *Schande* has been breaking down throughout the Western world. In the new technology-based society we have almost all become enfranchised. Yet our identification with the tragic heroes of Ibsen and Gerhart Hauptmann in the nineteenth century or O'Neill, Williams, and Miller in our own has become if anything greater. They are nearer to us in terms of time, place, and status. One suspects that today most of us relate far more readily to the "Death of a Salesman" in his Chevy than to the sufferings of Oedipus or Creon in the mythical mists of Greece.

But we have also lost something. In the Athens of the 5th century B.C., as well as in Shakespeare's time, the audience for tragedy was composed of all classes. Yet concurrent with the increasing immediacy of the tragic hero and his predicament we find a blunting of tragedy through sentiment after the late 17th century (as indicated in the adaptations of *Lear*, *Romeo and Juliet*, and *Hamlet*). Furthermore, comedy becomes more popular,[23] but it also (as in the French *comédie larmoyante* and the English eighteenth-century comedy of sentiment) tends to be blunted by the bourgeois desire for sentimentalized happy endings.

There would appear to be a tension between our need for a tragedy that directly portrays the modern predicament and the pain that the immediacy of this portrayal gives to us. Shakespearian Renaissance tragedy seems to have gained from a tension between the growing sense of clock time and the older sense of heaven's eternity, but after the third quarter of the 17th century watches became not only far more accurate but also far more numerous. The horological developments relate directly to radical advances in navigation and trade,[24] which are a prerequisite to the industrial revolution and to our own clock-regulated urban life. These are precisely the developments that have made us all worthy of consideration for the role of tragic hero. Moreover, they can also be directly related to the simplified and therefore more immediate prose style first demanded by the scientists during the third quarter of the 17th century.[25]

In conclusion, after the third quarter of the 17th century, literature became more prosaic and more realistic. By the early 18th century, literature's middle-class patrons demanded that the heroes of the novel (the popular new genre directly related to strict confines of space and time) be fashioned for the most part in their own image. Such heroes are largely devoid of a preoccupation with heaven or heroism, and are circumscribed by a world in which their role is very much bounded by the implications of clock time. The nonconformist middle classes in England were descendants of the 17th-century Puritans, but the intensity of their fervor had been watered down, and the heroic intensity of Puritan ideologies had dissipated. In the nonheroic tragedies of the 18th century and thereafter (on the page or on the stage), the sense of death seems to have become not only more immediate in terms of time, space, and status, but also more final.

Notes

1. I have considered in this paper only works that are normally called tragedies. For a valuable overview on time and tragedy that also takes into account the relationship with the Theatre of the Absurd see: J. T. Fraser, *Of Time, Passion, and Knowledge* (New York: George Braziller, 1975), pp. 412–419; hereafter cited as *Of Time*. See also Tamás Ungvári, "Time and the Modern Self, A Change in Dramatic Form," in *The Study of Time, Vol. I*, eds. J. T. Fraser, F. C. Haber, and G. H. Müller (New York: Springer Verlag, 1972), pp. 470–478, for a sensitive study suggesting a new attitude towards time in major drama written since the 19th century.

2. Aristotle, "Poetics," in *On Poetry and Style*, trans. G. M. A. Grube (New York: Liberal Arts Press, 1958), chs. 2, 13.

3. Ian Watt considers that the Renaissance views time "as the shaping force of man's individual and collective history." By way of contrast, "the philosophy and literature of Greece and Rome were deeply influenced by Plato's view" that the Forms or Ideas "conceived as timeless and unchanging . . . reflected the basic premise of their civilization . . . that nothing . . . could happen whose fundamental meaning was not independent of the flux of time." Such ideas are not inconsistent with the practice of ancient Greek authors who dated historic events after 776 B.C. by naming the related winner of the foot race in the Olympics. Ian Watt, *The Rise of the Novel* (Berkeley and Los Angeles: University of California Press, 1965), pp. 21–22.

4. Fraser, *Of Time*, p. 21.

5. Samuel L. Macey, "Cronus in the Eternal City: Time in Rome and England," *Social Science*, 53 (Summer 1978): 139–146.

6. H.D.F. Kitto, *Form and Meaning in Drama* (London: Methuen, 1956), pp. 180–182 and 246–337, *passim*.

7. Watt, *The Rise of the Novel*, p. 23.

8. Herman J. Ebeling, "The Word *Anachronism*," *Modern Language Notes*, 52 (February 1937): 120–121.

9. Cecil Clutton et al. eds., *Britten's Old Clocks and Watches*, 8th ed. (New York: E.P. Dutton, 1973), p. 43.

10. Ricardo J. Quinones, *The Renaissance Discovery of Time* (Cambridge: Harvard University Press, 1972), p. 297.

11. Though Aristotle in *Poetics* calls several times for the unity of action in tragedy, he nowhere mentions the unity of place. His single reference to the unity of time is far from being a precept (ch. 5). Clearly the proponents of the three unities reinterpreted Aristotle's *Poetics* in order to justify coordinates of time and place more appropriate to the Renaissance and post-Renaissance than to the Greeks. The rigidity of the unities was more easily accommodated to the psychological tragedy of Racine than to the action-oriented plots of Corneille. In terms of duration and structure, the effect of the unities on tragedy may be usefully compared with the later effect of the stream-of-consciousness on the novel.

12. The unities were, however, questioned much earlier. See, for example, John Gay's dedication to *The Mohocks* (1712), and the last paragraph of his preface to *The What d'ye Call It* (1715).

13. Watt, *The Rise of the Novel*, pp. 14, 23–24.

14. Ibid., p. 24.

15. E.M. Forster, *Aspects of the Novel* (Harmondsworth: Penguin, 1962), p. 36.

16. Northrop Frye, "The Four Forms of Prose Fiction," *Hudson Review*, 2 (1950): 586.

17. A.A. Mendilow, *Time and the Novel* (London: Peter Nevill, 1952), p. 63; I am grateful to Professor Mendilow for help with this article.

18. Watt, *The Rise of the Novel*, p. 26.

19. *Clarissa*, letters of Wednesday, March 1, and Tuesday, May 23; *Grandison*, letters of January 6 (vol. i), and Tuesday, March 7 (vol. ii).

20. Samuel L. Macey, "The Time Schemes in *Moll Flanders*," *Notes and Queries*, 214 (September 1968): 336–337.

21. A *Literary History of England*, ed. Albert C. Baugh (New York: Appleton-Century-Crofts, 1948), p. 897.

22. *Comparative Literature Studies*, 6 (Winter 1969): 1–19.

23. Allardyce Nicoll, A *History of English Drama 1700–1750*, 3rd ed. (Cambridge: University Press, 1965), pp. 125, 137, 139.

24. Samuel L. Macey, "The Early History of Chronometers: A Background Study Related to the Voyages of Cook, Bligh, and Vancouver," *B.C. Studies*, 38 (Summer 1978): 14–23.

25. R. F. Jones wrote an important series of articles on this subject in the 1930s. See "Science and English Prose Style in the Third Quarter of the Seventeenth Century," *PMLA* (1930); "Science and Language in England of the Mid-Seventeenth Century," *JEGP*, 31 (1932); and "The Attack on Pulpit Eloquence in the Restoration: An Episode in the Development of the Neo-Classical Standard for Prose," *JEGP*, 30 (1931).

The Beginning and End of Time in Physical Cosmology

D. PARK

It appears that 15 to 18 billion years[1] ago there was an immense eruption of energy, and all the bits of matter and energy then present in the universe started to move apart at immense speed. Within minutes, atoms of the chemical elements began to form: hydrogen, deuterium, and helium first; others later. All the time these fragments of matter were rushing outward at an enormous rate which however slowly decreased, since each particle first feels the gravitational attraction of the rest and that attraction (it can be shown from Newton's theory of gravity) is directed back toward the point where the explosion took place. It is the same as the attraction that slows a ball or a rocket as it leaves the ground. A ball finally falls back again to the ground. A rocket, at greater speed, may take longer to fall back or it may move so fast that it continues on into space and never returns. The outcome depends on the initial speed. In exactly the same way, the force of gravity either will halt the expansion of the universe and bring it back again, once more to a single point, or else it will not, and the motion will continue forever. Which of the two will happen can in principle be decided by scientific measurement and we are likely to know the answer in a few years. The question is difficult only because the observed speeds place us rather close to the dividing line between continued expansion and ultimate collapse so the measurements must be more precise than is now possible.

This, in a few sentences, is the modern version of the history of the whole universe: it began, and at a time which we know reasonably well; and whether it will end, perhaps several tens of billions of years from now, or whether it will go on forever, we should soon be able to surmise.

This brief sketch will doubtless seem to many readers incredibly simple and naive. It is intended to. It raises important questions of method and philosophy. Among them are: Are space and time logically prior to the substance of the universe and its history? Where is the initial point? Why this point rather than some other? How can astronomical observations support such positive conclusions? How can our science, based on observations taken in one small region of the universe during one brief epoch, presume to make statements about the whole universe, at remote times when it was and will be very different from what it is now? I will contribute a little more to your skepticism: There are cosmologists, serious, adult men and women, who are writing papers to discuss events which took place as early as 10^{-48} second after the initial event,[2] when the sphere containing all the matter and energy of the universe that we now observe was only 10^{-5} cm in diameter. The purpose of this paper is to explain how the main conclusions are arrived at and what the makers of such theories think they mean.

Models

For 2500 years Western philosophy has struggled with the logical problem of the relation of concept to experience, of idea to object. If the object is a chair, which we can buy, sell, see, touch, and sit on, the problem is one of logical formulation more than it is a practical one. If somebody declares "the chair is just simply *there* and that's an end to it," we may not feel like continuing the conversation, but there are no facts with which we can confront him to force him to change his mind.

I think the situation is quite different in cosmology. We observe the universe through the radiations that we receive from it, and almost all of them are of one kind, electromagnetic: light and its invisible counterparts at longer and shorter wavelengths. We perform no experiments on the universe; we simply observe it. What *is* the universe, anyhow? As we see it, it is luminous dots on a dark background. As we imagine it, it is much more. What we imagine is our model of the universe. The astronomer's imagined universe contains stars, clouds, and galaxies, which have certain properties and behave in a lawful manner. The matter in it is made of atoms or isolated particles, and they also behave lawfully. Stars, clouds, galaxies, atoms, and particles are all models.

Atoms, for example, are as remote from our experience as galaxies. Observe the approximately equal ratios

distance to the remotest known astronomical object: length of my
finger $\approx$ length of my finger: size of an atom.

The reason we know so much more about atoms than about the universe is that we need not be content to observe atoms. We can perform experiments on them. Atoms are not things, at least in the usual sense of the word—for instance, under special circumstances they can be in two places at once, and I don't think that things can do that. If they are not things, what are they? Rather than become involved in an unprofitable search for the *mot juste*, let me say that an atom is known to us experimentally in terms of the results of our experiments, and conceptually in the mathemical model we have made to explain these results. Happily we have such a model and it accounts for the phenomena very well. In exactly the same way, I believe it is best to say that the universe is a mathematical model. In this way I do not have to answer the question whether I believe that the universe, at age 10^{-48} second, was literally as I say it was. I can deal abstractly rather than practically with the question of how to define time and distance in a world where there were no clocks or meter sticks. If calculation based on a model containing such concepts predicts that things will be as we in fact observe them to be, and if the natural laws embodied in the model do not have to be invented *ad hoc* but are already familiar to us from investigations in the laboratory, then I think it is reasonable to say "I believe the model is a good one." Most scientists do not need to sidestep statements of literal fact, but physicists and astronomers deal in statements so extreme that for them it is sometimes a good idea.

Please note my insistence that the natural laws used be well-known in other contexts. Such laws are generalizations immense in their scope. The laws of physics tend to take the form: If a physical system is arranged in a certain way, I will tell you what it will do next. The initial conditions are separate from the law and may be specified at will. We believe the laws because we have used them under many choices of initial conditions. But how is such a formulation to be carried over into cosmology? There is only one universe; it has only one history. If we wished to formulate a law for the universe we would have no way to disentangle it from the initial conditions, the general from the contingent. This is why we insist that the laws which go into cosmological theory are those for which we already know how the disentanglement is made. We use no laws which while pretending to be general apply only to a single case.

The Canonical Model

The model of the expanding universe with which cosmologists usually begin their discussions is founded on four pieces of observational evidence which I will now explain in very general language.

1. *Uniform distribution of matter*. Until one considers the universe on a very large scale, the distribution does not seem uniform at all. Surrounding each star or planet is a large amount of empty space; stars are clustered into galaxies separated by enormous empty expanses; there are clusters of galaxies and even some rough groupings of clusters,[3] but on the scale of our deepest observations things appear to even out. There is no general tendency of the density of matter to increase or decrease as one looks in different directions from the earth or at different distances from it.

2. *The red shift*. It has been known for 60 years that the light from distant galaxies, as well as from the newly-discovered quasars, is shifted towards the red end of the spectrum. That is, spectral lines in the ultraviolet that astronomers study in nearby objects only on photographic plates can be observed visually if they originate in objects that are far enough away. This phenomenon was interpreted from the beginning as a Doppler shift due to velocity of recession of the sources, but that does not absolutely establish the case. The general theory of relativity asserts (and we know from observation) that light from a source in an intense gravitational field will be shifted towards the red, and of course it is conceivable that other

mechanisms of which we know nothing may be at work over these enormous intervals of distance and time to produce the same effect. Indeed there is now reason to believe that the gravitational effect is operative in at least some quasars, and that estimates of their distances and velocities may need to be revised, but this does not affect the conclusion derived from studying hundreds of other objects.

3. *Background radiation.* If the universe began in a great explosion there must have been a great burst of light. Since for most of history so much of space has been empty, this light should still be around. But light exerts pressure, and as the universe expands this pressure performs work aiding the expansion. It therefore loses some of its energy, and with light this loss is observed as a shift of color towards the red end of the spectrum. The shift is very large. The observed wavelengths are a thousand times longer than they are initially; the radiation must be studied with radio receivers rather than optical telescopes, but it is there. There is lots of it. Every cubic centimeter of the room in which you sit contains about 300 photons left over from a stage of the universe so early that if one were to represent the history of the universe on a scale 1 kilometer long, this stage would be situated only about 8 centimeters from time zero.

4. *Chemical composition.* Another thing we know about the universe is its chemical composition. Spectroscopy tells us what the sun and other stars are made of, and we can also count the various kinds of quickly moving, electrically charged atoms that reach the earth's atmosphere from outer space, the cosmic radiation. Except for slight variations which are understood, the relative abundances of chemical species in cosmic radiation are found to be the same as those in the typical star. Essentially, the universe is made of about 75 percent hydrogen, 24 percent helium, and 1 percent all the rest of the chemical elements. Any theory of the universe must account for the relative abundances of all the nuclear species, especially the smaller atoms which are the most numerous, and this is what the theory of the Big Bang does.

In this theory it all happens very quickly. In the earliest stage (more of this later) space is filled with a mixture of all the many kinds of elementary particles, but even by the end of the first second after time zero the more exotic species have disappeared and most of what is left are electrons, neutrons, and protons, the ordinary constituents of atoms everywhere. By the end of the fourth minute, the ratio of helium to hydrogen is almost as we observe it today, and a few heavier nuclei are beginning to form.[4] The details of the process depend very delicately on the properties of the various nuclei that are formed, as well as on the rate of expansion, and when one has looked at the calculations and seen how well the results agree with observation it is hard to believe that the basic model of the Big Bang is a big mistake.

The Big Bang

If the universe started out with a big bang at time zero, then it is not unreasonable to assume—on the principle of sufficient reason or simple economy of thought—that equal amounts of matter were projected in every spatial direction. If all the matter had started off at the same speed it would now be in the form of a spherical shell with the starting point at the center. But this is not what is observed. Space is, on the average, uniformly filled, and this requires a miracle. It requires that not only the distribution in direction but also the distribution in velocity be exactly uniform: that there must have been as much quickly-moving as slowly-moving matter at the outset. This is inexplicable from the point of view adopted here; we shall see below how it finds its natural explanation.

If we assume the miracle, we can explain at once a highly significant observational fact which I have not yet mentioned. If the matter started out moving away from the starting point at different velocities, then now, after many years, the matter that moved fastest at first will have travelled furthest: the faster it is moving, the further away it will now be. And this is exactly what is observed. We can measure velocities by the Doppler shift and estimate distances by a complicated chain of observation and reasoning, and the proportionality of velocity to distance, on the average, all over the observed universe, stands even firmer today than when it was first announced 65 years ago.

Distance travelled equals velocity multiplied by time of travel. We estimate distances and measure velocities of the remote galaxies. Dividing the first by the second gives the time of travel. All remote galaxies give about the same figure, roughly 18 billion years. The true age will be a little less. This is because gravitation is slowing the expansion; things used to move faster and the time was correspondingly shorter. A good estimate might be about 15 billion years.

We can estimate the age of the universe but not, if it is finite, its size. All we can do is to sample the universe, and whether the sample observable in our telescopes is relatively small or relatively large we do not know. Since the sample we see is so uniform from one end to the other, we are encouraged to believe it gives a fair picture of the universe as a whole, and all our theories are based on this assumption.

Curved Space

Einstein's general theory of relativity (1915) has revolutionized our ideas of space and time, but it was not Einstein who first really understood its revelance for cosmology. The one who did it was a remarkably versatile Russian mathematician, hydrodynamicist, and meteorologist named Alexandr Friedmann who in 1922 published a paper "On the Curvature of Space."[5] Most of the observable consequences of general relativity are very small—tiny modifications of Newtonian physics, but on the cosmological scale they may be vast. Einstein's theory treats space and time as if they were elements of an actual physical system, spacetime, which is affected by the presence of matter and energy and satisfies its own dynamical laws. Friedmann's contribution was to propose that the expansion of the material universe is simply an observable sign of the expansion of the underlying space, and he showed that the phenomena as known then could be explained as a consequence of Einstein's spacetime dynamics. Friedmann's equations are central to the modern theory.

Friedmann's curved space can be either finite or infinite in its total content. A consequence of the theory is that if the spatial and material content are finite, the universe has a finite history: having started with a Big Bang it ends in a Big Crunch as all the matter in it, pulled inward by gravitational attraction, rushes back together again. If the content is infinite, gravity slows the expansion but never reverses it and the galaxies will go on forever getting further apart. Note that general relativity does not predict which will be the case; it allows either behavior, and which our universe follows depends on the initial conditions. About them the theory has nothing to say.

In my descriptions I shall talk as though the universe were finite and started out as a single point. To me a point is aesthetically more satisfying than infinity (if one has to choose one or the other), and besides, from the phenomenological point of view it does not make much difference. Of course, if we were watching the universe from outside* the difference would be conspicuous, but we are not; we are inside it. Even if we imagine ourselves present at a very early epoch we could not survey all the universe; then as now we could only sample it, and who is to say, outdoors and looking up at the sky with or without a telescope, whether what we see belongs to a finite or an infinite world?

If the universe has a choice of being finite or infinite in its spatial extent and in the length of its history, which course is it actually following?

The theory provides a criterion that is in principle easy to apply: if a certain number Ω determined by the mean density of matter in the universe and the speed of its expansion is greater than 1, the universe is finite in time and space; if Ω is less than 1, the universe will expand forever.[6] The only trouble is that some of the observational quantities in Ω are hard to measure. We see stars and galaxies that glow and gas clouds that emit or absorb light, but to evaluate Ω we must know how much matter there is that we do not see. That is difficult but not impossible. In 1979, a good value for Ω was about 0.01. Today it is moving towards 0.5. I think that many astronomers and physicists have a taste for finiteness and both hope and believe that it will finally turn out to be greater than 1. In another ten years we should be reasonably sure.

The Einstein theory is now very well confirmed by observation, and it seems likely that the cosmological model based on it is a good one. It answers a few questions that may have been bothering the reader:

1. Where did the Big Bang take place? (One ought to erect a monument—.) It took place everywhere. If matter was infinitely dense at the beginning, we must not imagine it all clustered together at some point in space but rather only that the entire region available to it—all of space—was smaller.[7] We can now explain at once the miracle of the uniform distribution of matter in the observed universe: the expanding space is simply, on the average, uniformly filled. How else would you have it be?

2. What is beyond the edge of the universe? Whether it is open and infinite, like an infinite plane or other curved surface, or closed like the surface of a sphere, it has no boundary and nothing is outside it.

---

*These words make no sense if we consider the universe as "everything there is," but at this point I am thinking of it as a mathematical model. I mentally embed this model in a space of higher dimension and contemplate it from outside just as Claudius Ptolemaeus mentally situated himself at a point thousands of miles up in the air to draw a map of Europe.

3. Why did not all the radiation from the big bang simply leave the scene of the explosion? Why is there any of it left here now? The universe is full of radiation and always has been. There is an interesting way to explain this radiation. I have mentioned that as we look outward in space we look backward in time. The light we receive from distant objects shows us a picture, much reddened by the Doppler shift, of them as they were long ago when they first emitted the light. Look out still further, to a time before there were any stars. Look out in any direction. What is the remotest thing in time and space that you can see? The Big Bang. It is the blinding light from that which, greatly dimmed and reddened, bathes us and is detected as the background radiation.

In these answers we encounter a huge conceptual break with the past. Newton, standing on the shoulders of his predecessors, imagined space and time as logically prior to matter, and I have used this convention in my descriptions of the particles released in the big bang rushing out into previously empty space. Leibniz, in his correspondence with Samuel Clarke, argued that the material content of space and time is logically prior to space and time—that geometry is defined by the totality of spatial relations between solid objects and time by the totality of temporal relations between events. It seems at first sight that one cannot really settle the Newton-Leibniz question, but the Einstein theory has settled it for our time in favor of Leibniz: matter is considered as prior and determines, largely or completely, the spacetime it inhabits. The equations of the general theory of relativity describe how this occurs. The "largely or completely" goes back to my discussion on the importance of disentangling general laws from the contingent conditions under which they are applied. In applying Einstein's laws to the universe as a whole there are still some obscurities as to how the disentanglement should be done.

Matter determines the form of spacetime, and spacetime then determines how matter shall move under the force of gravity. (I do not discuss here generalizations of Einstein's theory to take account of other forces. Gravity is the force which shapes the cosmos.) Matter defines fixed scales of distance and time—we shall see below what facts authorize me to say this—and so if I speak of the expansion of the whole universe I have a scale in mind: the universe expands but stars and galaxies do not get any larger. Of course, I could say that the universe remains the same size while stars and galaxies shrink. There is no observational difference, but to do so would load the equations of dynamics with unnecessary complications, and they are intractable enough as they are.

After the Bang

So far I have spoken mostly about the history of the universe as a whole, but physicists and astrophysicists have also considered the evolution of the various forms of matter which it contains. We know what forms it *now* contains, and furthermore we have studied in elementary-particle laboratories some of the particles which we believe it once contained but which are now very rare.

I sketched the history of the universe (our model of it) at the beginning of this paper. I now sketch it again, filling in some of the details provided by studies in nuclear physics, astronomy and geology.[8]

Time 0 to 1 second. A chaos of particles of different kinds, doubtless including many unknown to us. At the end of this period the temperature is about 10^{10} degrees C, and the density about one ton per cubic centimeter.

1 second to 3 minutes. Collisions between particles become less violent and a few light nuclei such as deuterons and alpha particles begin to form. At the end of this period the temperature is about 10^9 degrees C and the density is about 40 kilograms per cubic centimeter.

After this point nuclei must form very rapidly. They are formed by the capture of neutrons that are flying around, and neutrons only last for a few minutes before they decay. At the end of 10 minutes there are almost no more free nuetrons and the chemistry of the universe is fixed except for relatively small changes which later occur inside stars.

For about a million years the universe is full of hydrogen and helium nuclei and tiny amounts of a few other elements, as well as electrons, neutrinos, and photons. At the end of this period the universe has cooled to about 10,000 degrees and atoms can begin to form without being immediately broken up by the violence of collisions with other flying particles. Soon most of the free electrons are captured into atoms and the world is full of hydrogen and helium gas. It becomes transparent to light as it is now, and it is the light from the epoch immediately before, when the universe was still opaque, which we see as the background radiation. The Doppler shift, or the expansion of the universe, has now cooled it from 10,000 degrees to about 3 degrees above absolute zero.

Meanwhile galaxies form into great flat discs, slowly rotating, containing clouds of hydrogen and helium gas. Out of them stars form by gravitational contraction. Inside a star, nuclear reactions coalesce hydrogen and other light nuclei together into heavier ones, a process which produces the energy to keep the star shining. When the hydrogen is gone the star collapses and then, if it is a particularly massive one, it explodes. Then, if there are hydrogen clouds or other debris in the neighborhood, a new star may form containing some material from the old one. The sun is such a second-generation star, and our planets seem to have been formed from the remaining debris of the ones which preceded it. All the atoms of the earth and us who inhabit it, except for some of the hydrogen, were once cooked together in a star which vanished about five billion years ago. Some half billion years later the sun and our planets had formed; and a little after that, about three billion years ago, life began on earth. Figure 1 shows a geologist's view of the history of mankind. (Remember though that the indications of time are inexact, since everything changes very gradually.)

Endings

Leaving the human episode for others to consider, I pass on to the future. The sun is about in the middle of its life. For about another four billion years it will continue to radiate heat and light; then its nuclear fuel will start to give out and it will start to

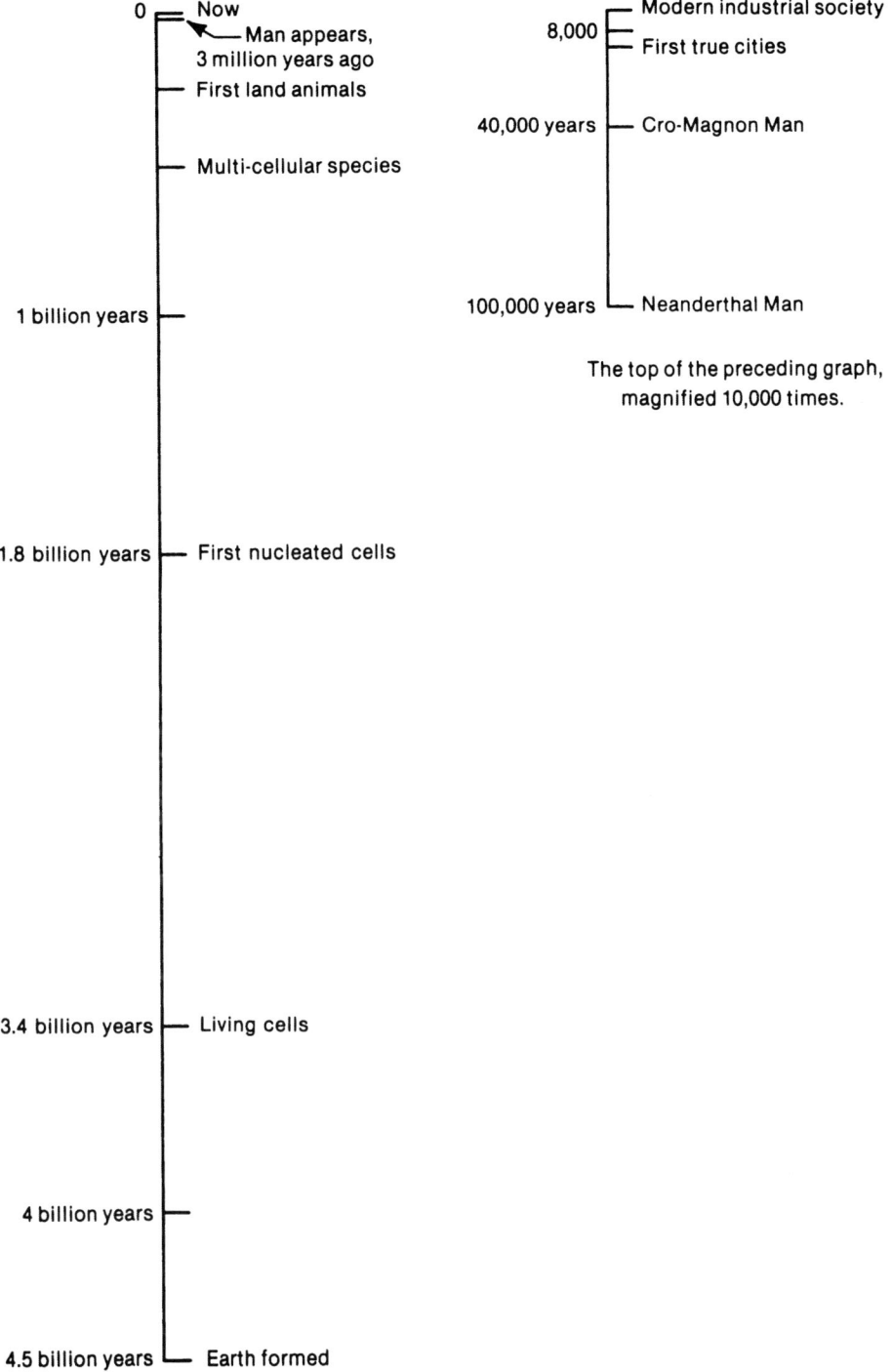

Figure 1. A geological view of the history of life on Earth (as of 1979).

cool. As it cools the core will contract; this will go on for perhaps a billion years. Then the sun will be a small, dense core surrounded by an immense tenuous envelope of gas, glowing a dim red. Astronomers call such stars red giants. After some more changes the sun will contract until it has become a white dwarf, its huge mass compressed into a volume about the size of the earth's. Then very slowly the white dwarf will cool to black. Nobody knows how many black dwarfs are drifting around in space at present, since they can not be seen, but in another 10 billion years there will be many more of them.

If the destiny of the universe is to collapse, we are not yet halfway there, since the universe is still expanding and collapse will take nearly as long as expansion. It is a reasonable guess, though no more than that, that 50 to 100 billion years from now it will all be over.

If the universe expands forever the stars will go out, the smaller ones ending as black dwarfs or neutron stars (pulsars) and the larger ones as black holes. Gradually the galaxies will lose their structure. A few stars will evaporate into outer space; the rest will gradually spiral towards the center and be swallowed up in immensely massive spinning black holes.

It now seems likely that protons and neutrons, the main constituents of solid matter, are unstable with a life of something like 10^{32} years. After that there will be only black holes, large and small, and a tenuous gas of electrons, positrons, photons, and neutrinos. Gradually the electrons and the positrons will annihilate each other, thereby producing more photons.

One of the triumphs of modern theoretical physics is a proof, or almost a proof, given by Stephen Hawking in 1975,[9] that black holes are inherently unstable, that they will finally decay by the emission of particles.

Post-Terminum

If there comes a big crunch, what happens next? We cannot say, and it may be impossible to say. The reason is that the canonical model starts from a point at time zero and ends in a point. Physicists answer questions like "what happens next?" by referring to causal laws; another name for them is equations of motion. These are the laws that contain all we know about how things change and move. All the equations of motion we know are expressed in terms of the underlying spacetime structure, but when the universe is reduced to a point there is no such structure, for a point, said Euclid, is that which has no parts, no structure. Therefore there is no physical law either, no causal principle leading to the next moment, whatever that phrase may mean in such a situation. If there were some physical background structure even more fundamental than spacetime, against which the changes of spacetime could be measured, then one might imagine a theory that treated spacetime like a bouncing ball. But our spacetime geometry is intrinsic, defined in terms of operations carried on inside it, since that is how we know the world. A curved spacetime in four dimensions can always be considered to be embedded in a flat spacetime of higher dimensions, but the embedding of spacetime is a mathematical trick, and we have no physical laws applying to the space of higher dimensionality. It is deceptively easy to imagine events before the big bang or after the big crunch, but in physics there is no way to make sense of these imaginings.

Extrapolation

The laws of stellar and planetary dynamics were formulated by Newton in the 17th century and modified by Einstein in the 20th. The laws of atomic dynamics have been known for only about 50 years. Newton's laws were directly verified in experiments on pendulums and on balls rolling down inclined planes. The laws of atomic and nuclear physics came out of experiments performed in laboratories constructed in our own time. In constructing a model, said Aristotle, we may assume what we please, but should avoid impossibilities. In the canonical model it is assumed that the laws as we have formulated them apply without change under extreme conditions of pressure and temperature, a million times those in an exploding atomic bomb, at very remote epochs in space and time. Aristotle perhaps allows us to make such assumptions—there is nothing inherently absurd about them—but in doing so are we doing anything more than playing a game? This question has been very carefully discussed in recent years, and I can devote only a few remarks to it here.

Physics at High Pressures and Temperatures

From the atomic point of view, pressure and temperature are simply the effect of the impact of atoms. Under extreme conditions the impacts are both hard and frequent, but each such event is only an impact, and we have had many years to study them, one by one or a few at a time, in the laboratory. Modern accelerators enable us to study collisions at energies up to 20 billion electron volts, corresponding to a temperature of 2×10^{14} degrees C. In the canonical model this temperature occurred at 10^{-11} seconds after time zero. For times earlier than that the theory must tell us what to expect.

Physics of Remote Events

First, events remote in space. There is no observation suggesting that the laws of nature vary from place to place at a given time. The signals we receive from stars, galaxies and gases in outer space are essentially the same whichever direction we look. Based on this fact, it is a fundamental postulate of the canonical model that in any epoch the universe is the same everywhere. (As I suggested earlier, the model is based firmly on Sufficient Reason.)

There are three ways in which the laws of physics could fail to apply over long intervals of time: a) They could contain slight errors which are magnified as large numbers are put in; b) the laws themselves could change over time; c) the laws could have the right form but the so-called constants of nature: the numerical values of the gravitational constant, Planck's constant, the speed of light, the elementary electric charge and so on, could be not quite constant. These three ways are not logically distinct, but it is helpful to consider them separately.

a) The laws of nature as we understand them are expressed in terms of well-defined mathematical structures that we consider both simple and beautiful. Complexities, and there are indeed many, arise from the application of these laws in complex situations. Corrections must be applied, but they arise from theories which are simple. There is little room for adjustment, and where there is we consider the theory incomplete. There is no way in which the general theory of relativity can be slightly adjusted. It is either right or wrong. Cosmology is a testing-ground for many theories, but we believe it will help us to accept or reject them *in toto* and not to make little changes.

b) This is a very attractive idea. Why should natural law be logically prior to the universe? Why should it not arise out of the universe in some way, modifying continually as the universe changes? If this were true we might finally be able to derive natural law from observation rather than needing to postulate it *a priori*. What we would then have to postulate is a super-law connecting natural law as we know it with the state of the whole universe, and it is this super-law that ought to be simple and beautiful. Difficult, perhaps possible, but it seems there is no reason to do it. Studies of the radiations from distant objects show that atoms accurately obey the same laws, over immense distances of space and back to remote epochs of time, as they do in our laboratories. A recent observation[10] shows that a relationship involving both the electronic and the nuclear structure of atoms has undergone no change larger than a few parts in 10,000 during the last four-fifths of the history of the universe.

c) As I have said, the distinction between this possibility and the preceding one is only one of language, but the language is convenient. One can ask, Has the mass or the electric charge of electrons changed over time, and is the speed of light constant, as these quantities enter the laws of physics? Here geological data are of good use, though since the earth is only 30 percent as old as the universe, we cannot draw conclusions about the beginnings. Still, this is more than offset in most people's minds by the great precision possible in the measurements made on materials we can hold in our hands.

There is a methodological point: how should we define quantities like mass and length in the early universe? Mass, for example, is typically defined in terms of operations with rigid apparatus. But how to define it for an epoch in which there were neither operators nor any solid matter? The question deserves more thought than it seems to have received in the literature, but there is a way around it. Even though we may be uncertain how to define the scale of masses a fraction of a second after time zero, we can consider quantities like the ratio of the electron's mass to that of the proton which are independent of the scale used. Similarly, the quantity $2\pi e^2/hc$, known as the fine-structure constant (e is the electron's charge, h is Planck's constant, c is the speed of light) has a value close to 1/137 independent of all the scales of measurement used as long as they are defined consistently. From a very clever argument[11] based on geological data we know that the fine structure constant has changed by less than one part in 10^8 during the last two billion years, and there are other similarly definitive results.[11]

That the laws of nature do not vary over time and space is not surprising if one believes that these laws (whether or not we know what they are) are to be expressed as simple and definite mathematical structures, with no room for little changes. One could have been wrong in that belief. One could still be wrong, of course, but now it seems less likely.

With these insights we can return to my earlier remark that in our theories matter and its metrical properties are considered to determine the metrical properties of spacetime. The diameter of an atom is given approximately by the expression $h^2/4\pi^2me^2$ where h is Planck's constant and m and e are the mass and charge of an electron. There is no *a priori* reason to suppose that h, m and e may be considered as independent of the place and time at which they are measured, but it seems that this is the case. We have therefore some justification for assuming that the earth and we and our measuring instruments are not expanding as the universe expands and that our instruments may reasonably be used to measure that expansion. Similar considerations apply to the measurement of time and to the structure of stars, most of whose matter is not in the form of atoms.

All this makes it seem that we are assuming that the physical laws governing the universe either came into existence at the moment of the big bang or, if one does not like that, existed before it, which one may not like any better. Here it is important to remember that the universe, which we have been discussing as if it contained stars and galaxies not to mention ourselves, is in reality a mathematical model, and the laws belong to the model. It is convenient, and justified by observation, to set up scales of measurement and laws of dynamics in this model as I have described. Questions such as whether the laws were in force before time zero or why they are as they are and not some other laws are good scientific questions which may well have answers within some later and more extended model, but they have no answers, and indeed they cannot even be asked in an exact way, in terms of the model we have.

The Large and the Small

Among the dreams of physicists there has long been the dream of a theory that unites the large and the small—a theory to show us that the properties of elementary particles are necessary consequences of those of the cosmos. The dream is shattered: the cosmos changes; particles, it seems, don't. But there is another connection, a historical one. In the very early stages of the universe, the smallest fraction of a second after time zero, the course of events was determined in great detail by the particles present and their interactions. In that sense the properties of the universe now depend on the laws of microphysics, and conversely, by studying the universe now we may be able to learn facts about particles which we would otherwise not know. This is true for the following reason: what we can learn experimentally about particles is limited by the energies at which experiments can be carried out and the densities of the particle beams we can produce. Energy shows up as mass in particle reactions, and very massive particles require very large energies to produce them, while beam density determines the number of events observed. Both the energy and the beam density available at accelerators are limited, now and in the future, by money, available space, and possible technology, but in the early universe, if one goes back far enough, there is no limit either to energy or to density. "The early universe," said Yakov Zel'dovich, "is the poor man's accelerator"—and the man need not be so poor.

The great passion of contemporary particle physics is unification—to find a theory which connects all the different kinds of particles and their interactions in a necessary way by relating them to a single, unalterable mathematical structure. (By unalterable I mean that if a theory turns out to be wrong in part it must be abandoned entirely and not touched up.) There are now some candidates.[12] None is anything like a complete theory, but they begin to tell what kinds of particles exist and how they interact. Among the particles that have been predicted are some with masses 10^{15} times that of neutrons and protons. To test this prediction by laboratory experiments is impossible, now and in the future. There is no reason to think there are now any such particles in the universe, or will be before the big crunch. But it turns out that in the canonical model, at about 10^{-43} seconds after time zero, the conditions existed for producing them and that their existence and properties at that moment had large and lasting effects on all of subsequent history.[13] For example, they determine the ratio of the number of photons of background radiation to the number of protons in an average volume of the cosmos today. The ratio, about a billion, is explained by this theory and by no other that I know of.[14]

There are several other such results, all recent and all tentative. I mention only the suggestion of William Press[15] that the formation of galaxies begins as a sort of curdling process in the enormously dense mass of particles that existed at a time about 10^{-39} seconds after time zero. Figure 2 shows the history of the universe represented on a kind of scale appropriate to these considerations. Note that the precise moment of the big bang cannot be shown on such a scale.

Absurd? Possibly these new ideas are wrong, but the test of a model, beyond Aristotle's criterion mentioned earlier, is that it should be based on known facts and explain known facts. It is hard not to believe that the beginning of the universe is within our reach to understand, even as we may never be able to talk about events before that beginning, and that this understanding must arise out of, and contribute to, our understanding of the physical world at its most fundamental level.

Notes

1. A billion, in United States English, is a thousand million.

2. 10^{-48} means the number represented by a decimal point followed by 47 zeros and a one. 10^{48} means a one followed by 48 zeros.

3. See for example P.J.E. Peebles, E.J. Groth, M. Seldner and R.M. Soneira, "The Clustering of Galaxies," *Scientific American* 237 (Nov. 1977); p. 76.

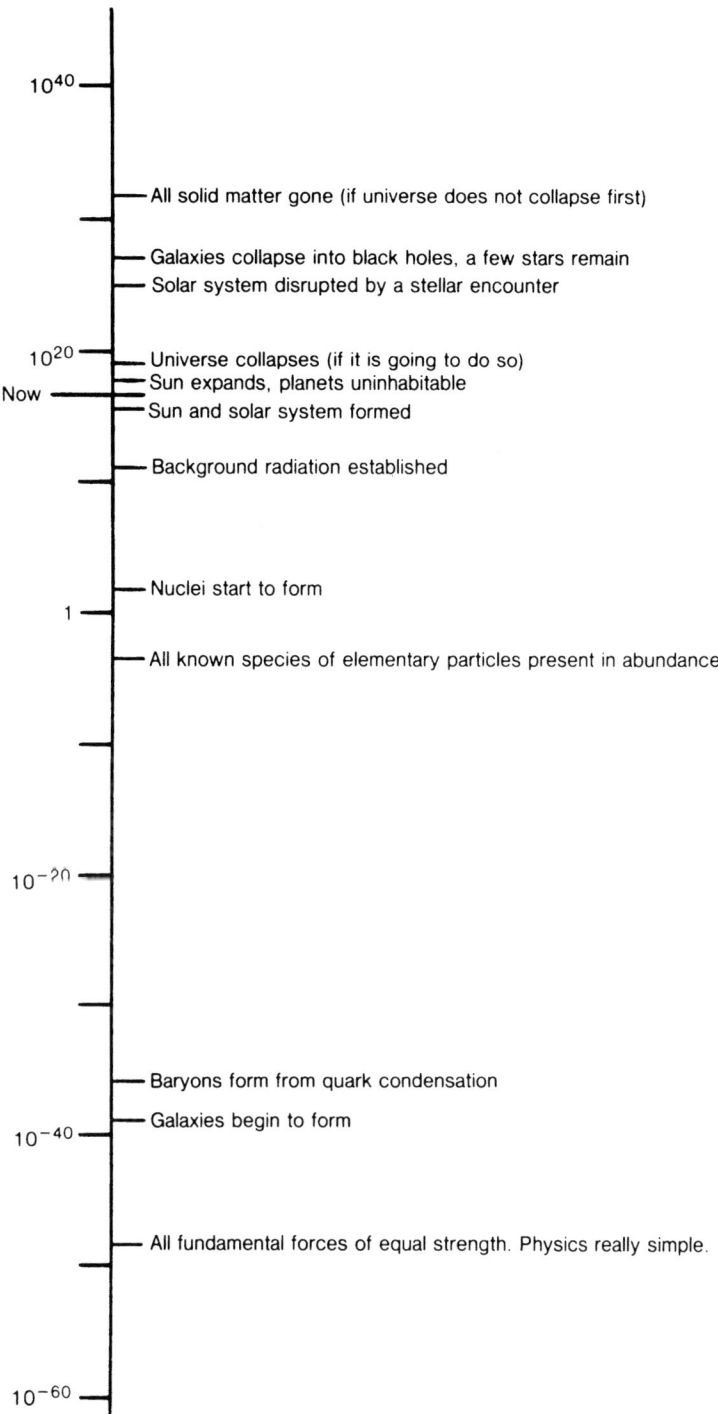

Figure 2. A physicist's view of world history.[16]

4. A very readable account is Steven Weinberg, *The First Three Minutes* (New York: Basic Books, 1977).

5. *Zeitschrift für Physik,* 10 (1922): 377. ibid, 21 (1924): p. 326.

6. Let G be the Newtonian gravitational constant, ρ be the average density of matter in the universe, and H (Hubble's constant) be the constant in the proportionality of recessional velocity v to distance d in distant objects: $v = Hd$. Then

$$\Omega = \frac{8\pi}{3} \frac{G\rho}{H^2}$$

7. If space is now and was initially infinite, it is difficult to say that space was smaller then, even though mathematicians are used to saying that one infinity is smaller than another. Let us just say that matter was initially more densely packed in the infinite space than it is now.

8. Prospects for the remote future are discussed by F. J. Dyson, "Time Without End: Physics and Biology in an Open Universe," *Reviews of Modern Physics* 51 1979): p. 447.

9. S. W. Hawking, "The Quantum Mechanics of Black Holes," *Scientific American* 236 (Jan. 1977): p. 34. J. B. Hartle, "Particle Production by Black Holes," in *Particles and Fields—1977* (New York: American Institute of Physics, 1978).

10. A. D. Tubbs and A. M. Wolfe, "Evidence for Large-Scale Uniformity of Physical Laws," *Astrophysical Journal* 236 (1980): p. L105 and references cited there; see also reference 11.

11. F.J. Dyson in *Current Trends in the Theory of Fields* (New York: American Institute of Physics, 1978).

12. S. Weinberg, "Unified Theories of Elementary-Particle Interaction," *Scientific American* 231 (July 1974): p. 191. S. L. Glashow, "Quarks With Color and Flavor," *Scientific American* 233 (October 1975): 38.

G. 't Hooft, "Gauge Theories of the Forces Between Elementary Particles," *Scientific American* 242 (June 1980): p. 104.

13. D. N. Schramm, "Cosmology and New Particles," in *Particles and Fields—1977* (New York: American Institute of Physics, 1978).

G. Steigman, "Cosmology Confronts Particle Physics," *Annual Review of Nuclear and particle Science* 29 (1979): p. 313.

14. A.D. Sakharov, "The Baryonic Asymmetry of the Universe," *Soviet Physics JETP* 29 (1979): *p. 594.*

E. W. *Kolb and S. Wolfram,* "*The Development of Baryon Assymmetry in the Early Universe,*" *Physics Letters* 91B, (1980): p. 217, and other work cited there.

15. M. S. Turner and D. N. Schramm, "Cosmology and Elementary Particle Physics," *Physics Today* 32 (Sept. 1979): p. 42.

W. H. Press, "Spontaneous Production of the Zel'dovich Spectrum of Cosmological Fluctuations," *Physica Scripta* 21 (1980): 702.

16. I first saw a diagram of this kind at a talk by A. Zee at the University of Maryland in November 1979.

Beginnings and Endings: Hesse and Kawabata

M. PILARCIK

The idea of beginnings and endings suggests a number of intriguing and often paradoxical questions about the nature of time and the human perception of events, change, and movement within the limitations of a temporal framework. From one standpoint, beginnings and endings provide a means of ordering events within a coherent unit, thus imbuing them with a certain definition and significance. For example, Huizinga, in his seminal work on the play element of culture, recognizes time restrictions as a chief characteristic of play: "Play begins, and then at a certain moment it is 'over.' It plays itself to an end."[1] These boundaries create a temporary sphere of activity with its own order and separate it from the confusion of ordinary life. Aristotle also uses a linear and progressive time sense to define tragedy as a whole and complete action, " 'Whole' means having a beginning, a middle and an end. . . ."[2]

Nature itself, however, does not consist of such clearly defined beginnings and endings, nor does the human being always perceive reality as well-ordered, linear movement within distinct temporal limitations. For example, Goethe wrote the following lines in his poem, "Permanence in Change":

> Let the beginning and the end
> Draw themselves together into one.[3]

Eliot in his *Four Quartets* also expressed a similar view. In "East Coker":

> In my beginning is my end.

And in "Little Gidding":

> What we call the beginning is often the end
> And to make an end is to make a beginning.
> The end is where we start from.[4]

The idea of an all-embracing unity behind the multiplicity of things and the ultimate relativity of beginnings and endings appears in the following quotes from the writings of the Taoist sage, Chuang Tzu: "Who knows that life and death, existence and annihilation, are all a single body? I will be his friend!" "The Way is without beginning or end, but things have their life and death—you cannot rely upon their fulfillment."[5]

On one hand there is a tendency to recognize order, logic, causality, and laws in the external world by referring to the beginning and ending of events, days, years, life, the universe itself, or to create an order within the temporal bounds of a piece of literature, or music, or a game; but there is also the subjective experience of time as a fluid, illogical process where modalities of beginning and ending, past, present, and future can overlap, intersect, become indistinguishable, or vanish in a vision of timeless unity. The discrepancy between the objective and subjective time-sense is a prevalent theme in modern literature, often referred to as an "obsession."[6] Thus, structural beginnings and endings based on progression and linearity in external time often conflict with an inner private time-sense.

The concept of beginnings and endings and its implications are a vital force both thematically and structurally in two masterpieces of modern world literature, Hesse's *The Glass Bead Game* and Kawabata's *The Master of Go*. Analysis and comparison of the two novels reveal that each author in his own unique fashion uses a literary means to embrace and integrate both the sense of external, temporal beginnings and endings in time and the dissolution of such distinctions in cyclical continuity and the inner experience of subjective time. Although the works are quite different, they have common elements that focus on the time issue: the nature of a provocative game that assumes cosmic and mystical significance, the life and death of a renowned master of the game, and the structure of the novel itself.

In Hesse's *The Glass Bead Game* these elements generate an intricate pattern of paradoxical time relationships. The novel takes the form of a compilation of material documenting the history and development of the Glass Bead Game, the life and legendary death of Magister Ludi Knecht and Knecht's early writings. The introduction and most of Knecht's biography are written in chronicle form by an anonymous Castalian historian who has access to the archives and faithfully reports the facts, based on the materials and documents available to him. Since Knecht dies outside of the Castalian borders, the last chapter, depicting his death, is told in legendary form as passed down in the oral tradition by his students. The thirteen poems and three autobiographies are the collected personal writings of Knecht from his student years.

The chronicler writes from the perspective of approximately 2400, placing Knecht's life at around 2200. Therefore, he relates past events while simultaneously evoking a sense of the present as well as the future. From the reader's point of view, the events lie well in the future, and the 20th century, referred to as the Feuilletonistic Age, becomes a vital part of the historical process which results in the development of the Game and the province of Castalia. Ironically, the chronicle form and the historical approach taken by the narrator are used to trace the development of a game that by its nature seeks to transcend time and space, and a province that chooses to ignore history and remove itself from the world of time and historical process. Knecht's life and his death provide the link between the aesthetic, timeless ideal of the province and the outside world of change and flux.

Although the Game came into existence at a specific point in time, the narrator makes it quite clear that the idea of the Game can be recognized much farther back in many past ages and in cultures where men sought to find the *unio mystica* of all knowledge. There is no identifiable origin of the idea itself, since it transcends the bounds of time: "For like every great idea it has no real beginning; rather, it has always been . . . ."[7]

The Glass Bead Game is, therefore, one particular fictitious manifestation of this eternally present idea. It embodies the visionary ideal of bringing all intellectual and spiritual values together into a harmonious synthesis. Through the manipulation of abstract laws and a system of universal symbols or ideographs, the player can integrate the most disparate themes and the highest spiritual values into the simultaneous present. The structure of the Game, or *lingua sacra*, thus provides the means to formulate new relationships of universal truths and thereby to penetrate the imperishable universal oneness behind all multiplicities and oppositions. Despite the great amount of discussion given to the Game in the novel, it remains a highly elusive and abstract construct which ultimately defies containment in finite, temporal reality and is inimitable in the world outside of the novel.

The followers of the Game turned away from the world and its sensual inclinations in order to preserve and cultivate the eternal and permanent values of pure mind or *Geist*. They established the educational province of Castalia as a realm removed from the flux and change of nature and human affairs; and within this self-contained sphere the Castalians are able to devote their attentions to intellectual pursuits and spiritual contemplation. The Castalian dedication to the unity and harmony of conflicting principles is purely intellectual, concerned only with spiritual, immutable values. As such, it creates a crystalline, timeless state in opposition to the natural world of time and sense, or *Seele*. The Game, which is the chief activity of the province, represents the most extreme point of this antithesis. Although the Game attempts to mirror the cosmic order by bringing all opposition into harmonious interaction, it cannot accommodate transient phenomena, subjective sensations of individuals, or anything, for that matter, which cannot be abstracted into a universal, communicable essence.

The timelessness and sense of "flat present" evoked by the Castalian Province are intersected by the movement of Knecht's life as he passes through the hierarchy of the Order and beyond it to the outside world of nature and the flux of time.[8] Through his experiences and encounters with other figures, he becomes the focal point of a universal dynamic that embraces all oppositions, similar to the configurations of a Glass Bead Game but on a level that includes the Game itself. The juxtaposition of three figures in particular is instrumental in shaping the time-sense that emerges in the novel: the Music Master who represents the secular Order of Castalia and its ideal spiritual state of cheerful serenity or *Heiterkeit* unaffected by conflicts in time, Elder Brother who secludes himself from Castalia and the world to live in harmony with the cyclical processes of nature and time, and Father Jacobus who represents the religious order and the role of commitment to the world and history.

Although the Music Master epitomizes the best of Castalia, his art is temporal by nature and, therefore, must have a solid foundation in concrete reality. In comparison with the Glass Bead Game, which deals only with the abstracted essence of music, this art relies on a sensuous basis and the physical nature of sounds through time to give it content. As Knecht explains in one of his lectures, one cannot know music from the abstracted distillations used in the Game: "We make music with our hands and fingers, with our mouths and lungs, not with our brains alone. . . ."[9] The Music Master's art exists in time. To deny time would be to deny its very existence; therefore, its spiritual essence is realized only through physical sensations in time. Music provides the first experience of the spiritual for Knecht, opening his mind to the sublime clarity and beauty behind the sensuous content. When the Music Master dies, it appears to him as though the Master's physical decline is a steady process of dematerialization and the spiritual transfiguration of the person into the essence of the art he loved. There is no sense of abruptness or finality, but rather a shift to another level of reality beyond time. The image that remains is a radiant look in the eyes, indicative of harmony, serenity, and timelessness.

The recluse, Elder Brother, rejects Castalia and dedicates himself to the study of Chinese thought in a small, Oriental world that he fashions for himself in a bamboo grove. His attention is directed to the mystical Taoist writings of Lao Tzu and Chuang Tzu, rather than to the more rational Confucian classics. It is his involvement with the oracular "game" of the *I Ching*, however, which is particularly significant. This ancient text is based on combinations of solid and broken lines representing the dual nature of the universe. The hard or solid lines are known as *yang*, the soft, broken lines are *yin*. In combination they create a circular pattern of sixty-four hexagrams which expresses all of space and time in a never-ending process of change contained within one primal unity. Change is the universal law which nothing can escape and which accounts for the endless variety of phenomena in the world. Through change the unity becomes multiplicity without ever losing its fundamental oneness.[10] Knecht's acquaintance with the *I Ching* necessitates a confrontation with nature as finite, multitudinous forms caught in the unending flux of time and change. Castalia attempts to transcend time by divorcing the Game from the phenomenal world, but the *I Ching* reveals that the mystery of eternity exists not *beyond* change but *in* it. The eternal flux of time with its continuous patterns of creation and destruction, beginnings and endings, is an expression of the movement of the *Tao* or the interplay of *yin* and *yang*. When Knecht suggests his plan to incorporate the *I Ching* into the Glass Bead Game, Elder Brother compares it to trying to bring the world to a bamboo grove.[11]

Father Jacobus, a Benedictine historian at Mariafels, introduces Knecht to an understanding of history, morality, and individual responsibility to the world in time, and teaches him to perceive his life as an historical reality and himself as an actual part of that historical, living, organic process. Father Jacobus attacks the Castalian sense of history as "bloodless and lacking in reality," nothing but abstract laws and formulas like mathematics, but "no reality, no good and evil, no time, no yesterday, no tomorrow, nothing but an eternal, shallow mathematical present."[12]

He accuses the Castalians of naiveté and ignorance due to their isolation in an artificial, amoral realm, worshipping not a transcendent Godhead, but a Game that is itself a mere form created in time and space, and which has no room for genuine knowledge of man in all his aspects, spiritual as well as bestial. The Game is fascinating play, a fine cerebral occupation, but its players have a repugnance for anything which clouds the rarefied atmosphere of their scholarly existences.[13] They ignore current events, politics or the uniqueness of events and individuals in reality and time. To study history, on the other hand, is no irresponsible game. One must willingly submit "to chaos and nevertheless retain faith in order and meaning."[14] History is matter, flesh, blood, life, death, and conflict in time. To ignore this is to ignore one's individual responsibility to actively participate in, and to shape, the course of events.

It becomes obvious to Knecht that both the Game and Castalia, as creations in time, are subject to the very laws of time and change which they reject. Although they serve the timeless ideals of *Geist*, they are themselves not timeless, but historical phenomena with a beginning and ultimately with an end. His recognition of the relativity of these institutions and his eventual decision to leave Castalia do not deny the value of eternal or spiritual ideals, but rather they affirm a vision of total reality, which includes both time and timelessness, *Seele* and *Geist*. When Knecht becomes Magister Ludi he experiences the possibilities of the Game and his position to the fullest, but also their limitations. He is ready to embark on a higher game.

The idea of beginnings and endings is extremely important for an understanding of the final episode of Knecht's life. Before leaving, he has a discussion with Tegularius about a poem he once wrote, originally titled "Transcend," but later altered to "Stages." There are certain lines which spontaneously spring into his mind and lead him to recall this particular poem:

> In all beginnings is a magic force
> For guarding us and helping us to live. . . .
> So be it, heart: bid farewell without end![15]

These lines refer to moments of his life which he calls "awakenings." They are powerful and direct experiences of reality that greatly affect his perception of things. He describes them as "shaking him down in his unconscious mind like a spring storm"[16] or as a "spiritual shock."[17]

His explanation of the awakening experience is reminiscent of the Zen "*satori*" (Chinese "*wu*").[18] In a fictitious letter written some years after the publication of the novel, Hesse uses the voice of Knecht to describe "awakening" as similar to the Zen experience, referring to it as a union with the All that is experienced as a reality throughout the soul and body.[19] The noted D. T. Suzuki describes *satori* as a "mental cataclysm," a "fiery baptism . . . one has to go through the storm, the earthquake, the overthrowing of the mountains, and the breaking in pieces of the rocks."[20] In an instant, one gains awareness of all oppositions and contradictions as an organic whole. This sudden experience of looking at things anew is an intuitive perception that cuts through analytical and rational thought and leads to a rebirth.

Knecht is struck by the intense reality of these moments, which he also describes in powerful analogies: "They are tremendously real, somewhat the way a violent physical pain or a surprising natural event, a storm or earthquake, seems to us charged with an entirely different sort of reality . . . ."[21] Unlike the sublime sense of oneness and harmony he feels in meditation or when involved with the Game, these experiences strike him personally and necessitate action and decision. They provide the impetus to progress from one stage in life to the next in a continual process of transcending and leave-taking. As each stage is nearing its end, there is an eagerness for death and new beginnings.[22] The experience takes place in an instant of time and opens his eyes to the immediacy of concrete reality rather than universal laws. Castalia, by denying time, denies the richness, the change, the eternally new and revitalizing beginnings of reality. Without time there is no change; there is no growth.

At the point of his departure from Castalia, however, Knecht recognizes that the awakenings are not leading him in a linear, upward progressive path towards a goal, but rather to continually new situations and to new transformations, which appear more like a circular or spiral movement: "Straight lines evidently belonged only to geometry, not to nature and life."[23] The position of Magister Ludi is not the end goal towards which his life is aimed, but rather an experience that in its turn has to be transcended. Knecht's life returns to the world of time and sense which gave him birth and which will accept him in death.

In the act of leaving Castalia, and in the subsequent ending of his life, Knecht becomes the focal point in which the timeless ideal of Castalia and the time-bound world of nature are brought together. This synthesis occurs within the flow of time, and yet it is related in a chapter entitled "Legend," as a mythical, almost mystical, episode beyond a specific time and space. The historical tone of the narrator's account is thus balanced by this shift into a different realm and time frame.

Knecht meets his death high in the mountains, removed from ordinary life and situated away from the routine affairs of the everyday. The rugged harshness of the primitive world strikes him as strange, and uninviting, but with a certain "silent and cold grandeur."[24] Despite his feelings of physical discomfort from the unfamiliar environment, he is ready to meet and interact with his new pupil, Tito. The radiant light of the morning sunrise and the dazzling impact of nature move the young boy to a rhythmic dance in celebration of the surging life-giving forces. The paganistic dance has a sense of self-abandonment and reverence similar to the sacred rituals of the Glass Bead Game. It elevates him to a "priestly solemnity."[25] For Tito, this is a spontaneous, totally unconscious response to the moment, to nature, to the destiny awaiting him. After he wakes from the trance of this sensuous worship, however, he self-consciously attempts to shatter the memory of the ecstatic moments. His face, which had a look of timelessness, an ageless expression, suddenly becomes childish, and he is obsessed with time.[26] As though in haste to do something important, he quickly suggests a race with time, in a swim to beat the sun as the mountain shadows move across the lake.

Knecht responds by deliberately leaping into the icy waters after Tito. The extreme cold proves to be too much for him, and he succumbs to the hostile elements. The chance and suddenness leading to his death make it such a disturbing issue in the novel. One expects the natural ease and harmony that seemed so characteristic of the accomplished Master's life in Castalia to continue. In entering the world of nature, however, Knecht opens himself to chaos, to risk, to the accidental and the transitory. He gives himself just as fully to this new situation, with its pain and its death, as he did to the serenity of meditation and the sublimity of the Glass Bead Game. Unlike the highly spiritualized death of the Music Master, Knecht's death stresses the physical. He experiences weakness and fatigue and the penetrating sensation of the icy cold waters that simultaneously feel like leaping flames.[27]

This ending gives completeness to the life of Knecht and defines it within a temporal framework, but it is not meant to be the end of the novel. The last image in Knecht's biography is of Tito who, terribly shaken by Knecht's disappearance in the lake, becomes conscious of his responsibility in this event, and he experiences a spiritual awakening or new beginning. The antithesis of Castalia and the natural world are brought together by Knecht's act of plunging into the water at the same spot where Tito entered. Only Tito emerges from the water to continue the cycle of beginnings and endings, but on a higher level of synthesis.

The water symbolizes a return to the primal source of life. As Roger Norton in his study of the novel points out, "It is this central 'mother ground,' rather than life in the ordinary sense, which reveals itself in this death scene to have been the goal of Knecht's journey out of the Order."[28] His physical death results in Tito's spiritual birth.

The poems following this final episode of the biography were written during Knecht's student years. They remove the sense of ending in time by circling back into Knecht's life to a period when he defied Castalian rules and gave personal and lyrical expression to his inner thoughts and doubts. The poem, "Stages," appears in its entirety and speaks directly to the sense of rejuvenation contained in Knecht's death:

> Even the hour of our death may send
> Us speeding on to fresh and newer spaces,
> And life may summon us to newer races.
> So be it, heart: bid farewell without end.[29]

The three fictitious autobiographies that Knecht was assigned to write also look back on his inner imaginative life. In these shorter pieces Knecht projects himself into other lives at other times in the indefinite past. The consecutiveness of time thus becomes a pattern of intersecting cycles and eternal renewal in an unending process of beginnings and endings that unifies all three parts of the novel.

When the work is examined as a whole, the events and encounters do not lead towards a specific ending at all, but are all contained within one point of view that perceives the beginning in the end and the end in the beginning, as well as the dynamic movement within this basic unity. The movement of the novel, therefore, emanates from the self-transcendence of this one consciousness through the experience of continual beginnings and endings in time. The consecutive form of the novel is thus converted into a lyrical process [30] in which external and internal time are brought together in a unified vision of reality and the self.

Kawabata's *The Master of Go* also creates an intricate pattern of time relationships as perceived from a single, lyrical point of view. The expectations of narrative to move in linear fashion from beginning to end give way to an inner rhythm that permeates the novel from the outset and shapes its course. The novel begins and ends with the death of the Master Shūsai. Within this framework where beginning and ending are one, Kawabata moves easily forward and backward in time, allowing the succession of images and memories to replace causal development and to interrupt the consecutive progression of events. The novel takes shape organically and reflects the fluidity of time within a patterned timeless whole. Like *The Glass Bead Game*, Kawabata's novel is not concerned with plot or the surging of action towards a climax, but rather with the gradual unveiling of what already is.

The Master of Go is based on an actual championship Go match played in 1938 between the aging, aristrocratic Master, Shūsai, and his youthful contender, Kitani Minoru, renamed Otaké in the novel. At the time, Kawabata was asked to cover the classic match for the sponsoring newspaper, the *Tokyo Nichinichi Shimbun*, now known as the *Mainichi Shimbun*. By 1954 he had transformed his perceptions of the game and players into what he referred to as a "*kiroku shōsetsu*" or chronicle novel. "*Kiroku*" refers to the recording of facts and events, and the "*shōsetsu*" form resembles the novel but is somewhat more flexible, personal, and essentially nonfiction or autobiographical.[31] *The Master of Go* is thus a blending of documented events and a subjective interpretation and experiencing of those events. In one respect this creates an underlying tension between the linear sequence of events in time and an internal time sense that is not bound by the causal and linear. The intermingling of objective and subjective time gives the novel its internal dynamism and movement. In addition, the historical events and detailed record of the match are seen in a much broader context that clearly transcends the bounds of faithful reportage and personal reveries. Through this Go match, Kawabata captures the elegaic mood of ephemeral beauty as a splendid, ancient tradition comes to an end and an elegant spirit fades in the wake of a new, coarser age influenced largely by Western values and a progressive, scientific rationalism. The game and its players are transformed into symbolic images of time in transition, where ending verges on beginning in a fluid, continually changing universe.

The Go game serves as the focal point of this transition from the past to the modern age, and also functions on a variety of levels as a highly evocative symbol of time and timelessness. It assumes an air of permanency and continuity through time because of its ancient origins, dating back approximately to 4,000 years ago.[32] In contrast, the lives and affairs of man seem fleeting and insubstantial. Although its exact origins and purposes are somewhat obscure, we know that the Chinese conceived of the game, and that it was associated with politics and military strategy, but may also have been developed as an intellectual exercise. There are also strong indications that the game had close connections with magic, the art of divination, and mystical enlightenment.

Ancient Chinese and Japanese legends tell about the power of the game to suspend time for those who become absorbed in it. For example, a woodcutter believes he spends a few hours watching two old men play Go by a mountain roadside, but after the game the men disappear, and the woodcutter finds that his hair has turned white and his axe handle has rotted with age. The game is sometimes called by the name *ran-ka* or rotten axe based on this tale. There is a similar legend about two boys who happen across the entrance of a cave in which two fairies are playing Go. While they watch, a hare jumps up and down at the feet of the fairies. Each jump coincides with a complete revolution of seasons in the outside world. By the time the game ends, years have elapsed. In the Japanese puppet play by Chikamatsu, *The Battle of Coxinga*, this almost magical power of the game plays a significant role. As the character, Go Sankei, watches two strange old men playing Go, his perception of time is considerably altered. Concentration on the patterns created by the stones enables him to see through time and space and to witness the events of five years as though in one simultaneous instant. As one old man explains,

> If it looks like a go board to you, it is a go board, and for the eye that sees go stones, they are merely go stones. But there is a text which likens the world to a go board. For those who see with their minds, the center of the universe is here.[33]

For those who understand its mysteries, the game can penetrate the hidden depths of the universe where primal forces are at play and fate works itself out:

Go Sankei: And the result of your contest?
Old Man: Does not the good or bad fortune of mankind depend on the chance of the moment.[34]

The board with its 19 vertical and 19 horizontal lines produces 361 intersections which could represent the days of the year; the 4 corners could suggest the 4 seasons; and the interaction of the black and white stones could mirror the eternal pushing and yielding of the primal and cosmic forces, the *yin* and *yang*. Kawabata refers to the " 'road of the three hundred and sixty one,' which the Chinese had seen to encompass the principles of nature and the universe and of human life."[35] The game thus functions as a symbol of the universe in which the interplay of two opposing forces gives rise to all of time and space. The first stone placed on the empty board is the first definition in space and a limitation in time. The next stone of the opposite color, when placed on the board, generates a dialectical tension reminiscent of the *yin-yang* duality. The mystical aura surrounding the game derives from the universal symbolism it evokes. As Kawabata says, it is "a game of abundant spiritual powers."[36]

When the Master and his opponent, Otaké, meet over the Go board, reality assumes different proportions, and the feeling of other-worldliness is pervasive. Stones striking the board have an "unearthly quality about them, as of echoing in a chasm"[37] or "through another world."[38] The game also takes on the particular characteristics of the players who give it life. Otaké's game is described as having a "dark" quality with something cheerless and oppressive about it, like a "strangled cry." The Master's game seems to unfold of its own accord "like the flow of water or the drifting of clouds."[39] The two styles of playing reflect different experiences of time which confront each other through the game. One is an extreme consciousness of chronological clock time, and the other is a complete detachment from objective time. Within the course of the game the two extremes struggle against each other in the process of inevitable change and flux. It is the lyrical consciousness of the narrator, however, that sees beyond the struggle to the ultimate relativity of both positions.

Kawabata, in his role as the narrator Uragami, stresses the players and their relationship to time, the game, and each other rather than the technicalities of strategy and maneuvering. Both players have followed a course through their lives which has led them to this particular encounter. They are products of two different eras, and despite their individual personalities and eccentricities are representative of vaster forces "being carried on by the currents of history."[40]

The Master's tie to the past is established from the start of the novel, where he is identified as twenty-first in the Honnimbō succession. The first Honnimbō, Sansa, had been a Zen Buddhist monk, but later became instrumental in the formation and development of the Go-dokoro or Go Academy. The Master took holy orders on the three hundredth anniversary of the death of Sansa, in the Nichiren sect of the Jakkoji Temple. He is also the last to bear the title, Honnimbō, as part of the traditional succession of Go Masters. His life is thus firmly rooted in the traditions of the past and also spans a period of time from the Meiji Era (1868–1912) until shortly before World War II. Uragami explains that he "brought the game to its modern flowering,"[41] not only because of his excellence in playing, but because his life and his art are one. In his person he embodies a panorama of history, an era, and a way of life that is also the way of Go.

For the Master, Go transcends the bounds of mere game and is even more than art and ritual. In the spirit of Zen, he immerses himself completely in the game like the painter who becomes bamboo after painting bamboo; or the poet who learns about pine by being pine; or the archer who knows no distinction between self, arrow, and target. His moves in the game result spontaneously from a meditative state in which his sense of self dissolves, external time ceases to matter, and his mind is emptied of purpose, ambition, and calculation. He enters "a universe of the spirit in which everything communicates freely with everything, transcending bounds, limitless."[42] Player, opponent, and game are experienced as one, making winning or losing meaningless. In this state, characterized as *mushin no shin* or "mind of no mind," the self is unconscious of itself, unattached to any thought, unconcerned with the past, and does not anticipate the future.[43] Focus is placed on the *now*, or timeless instant that disregards the ticking of the external clock. This explains why the Master is often confused about time measurements and the meaning of fast or slow in regard to his playing. He plays intuitively and spontaneously when the time is right. In Go, for example, he plays with remarkable speed, but in other games, such as mahjong or billiards, he is "the despair of his adversaries for the time he spent in thought."[44]

At one point the Master attributes his ability in the game to having "no nerves" or to a kind of "vagueness" as in a painting.[45] This vague quality that implies lack of definition, unobtrusiveness, or emptiness is a vital element in Japanese aesthetics evoking a sense of limitlessness and timelessness. The Master's intensely spiritual nature can be characterized as "vague" in this sense. While meditating on the Game he is unconscious of time and self: "He suggested some rarefied spirit floating over a void."[46] On the other hand, he seems to have a rather tenuous hold on everyday reality. Remaining impervious to the world around him and even to the severity of his own physical condition, he constantly seeks other games and diversions in which he can lose himself. This desire for release from the world of time and differentiation assumes the form of an "addiction" or a "disquieting obsession" that in itself makes the meditative state increasingly fragile and vulnerable to pressures from the outside. Within the course of this match, the forces of time ultimately lead to the collapse of his detached serene spirituality. His mind cannot continue to deal with the vast discrepancy between his internal intuitive time sense and the rationality of objective time, nor can his body continue to combat the processes of age and physical deterioration.

The challenger, Otaké, differs from the Master in many respects and represents the active forces of the new, modern age with its rationality and heightened time consciousness. Instead of the cold, severe dignity of a sublime spirituality, he manifests the warm amiability of a world-oriented personality. This is even apparent in the fact that he is physically sturdier, has a lively household with many students, and enjoys joking and carefree activity. In his playing, the physical and rational dominate, rather than the spiritual and intuitive. Unlike the Master, who "lost in vast distances"[47] is oblivious of his discomforts, Otaké is consumed by a nervous tension that translates into various disorders, including enuresis, diarrhea, restlessness, fidgeting, and absentmindedness. The calm and quiet composure of the Master unnerves him, and he is unable to abandon himself to the harmony of play. His game reflects conscious effort, careful planning, and a focus on the future end rather than on the *now*. Although he is clearly dedicated to the game, it assumes a different dimension for him. Go does not dominate his life; nor has he been "bled" by it because it remains within the realm of competitive play, bordered by distinct temporal beginnings and endings which do not merge with his everyday reality.

Seen from Otaké's perspective, the game acquires a more rigid, orderly structure. He insists on adherence to a set of new, clearly defined rules that insures equality by imposing such restrictions as time allotments and the sealed play. These are designed specifically to insure fair competition and to prevent the kind of arbitrary, irrational playing for which the Master is notorious. It is irrelevant that the Master's art is forged by spontaneity, eccentric whims, and special consideration due to his rank; he "could not stand outside the rules of equality."[48] In former times the title of *Meijin*, meaning Master, famous person, or genius, had broad connotations beyond just excellence in the game. It was given to a player who had attained the highest rank of proficiency (*ku-dan* or ninth rank), and it reflected on his person and his way of life. The courtesy and respect that accompanied the title easily worked to his advantage in the game.

The importance given to a distinct and binding time schedule, objectivity, and equality seeks to eliminate this advantage and reflects the nature of the new age with its emphasis on vigorous competition, personal victory, and advancement in rank. It also indicates an essential shift from inner profundity to external formalism, from the finesse and mysterious elegance of Go as art to the regulation of Go as technique.

> It may be said that the Master was plagued in his last match by modern rationalism, to which fussy rules were everything, from which all the grace and elegance of Go as art had disappeared, which quite dispensed with respect for elders and attached no importance to mutual respect as human beings. From the way of Go the beauty of Japan and the Orient had fled. . . . One conducted the battle only to win, and there was no margin for remembering the dignity and fragrance of Go as an art.[49]

Unlike the Master who loses himself in the game and allows it to flow spontaneously from him, Otaké relies on the power and manipulative strength of his rational mind and will. For the Master, external time is not a conscious factor, since he is not given to logical, sequential deliberation. As he says, "I'm not much of a thinker."[50] Otaké's playing, however, results from a concentrated mental effort that is constantly aware of the passage of time. It drives him, torments him, and permeates the nature of his playing. For example, his total time expenditure is nearly twice that of the Master and almost exhausts the allowed limit set for the match, despite the fact that forty hours per player is unprecedented in the history of the game. At times he gives the impression of deliberately prolonging his decision in order to gain a psychological advantage, but at other times it appears as though he cannot make up his mind until the very last instant, and must have the seconds read off to him until the time allotment is exhausted. He functions within established segments of time which are defined by concrete, linear beginnings and endings, and does not trust an intuitive, internal time sense that transcends temporal boundaries. Once the limitations are set, he cannot free himself from the knowledge that he is working against an external time measurement. Therefore, he easily succumbs to the pressures and anxieties of a time-obsessed consciousness and a powerful ambition inhibiting the ability to lose himself to the natural flow and movement of the game. It makes no difference how much time were available to him; he would still feel the race against time and be plagued by indecision and mental frenzy similar to that in the following passage:

> "Very odd," Otaké muttered, as if in a trance. "I'm running out of time. The great man is running out of time, forty whole hours of it. Very odd. Nothing like it in the whole history of the game. Still wasting time there, are you? Should have played in one minute, no more."[51]

The introduction of the time restrictions and sealed play has a significant effect on the course and outcome of the match. Although they are meant to insure the sanctity of the match, they impose a new element into the very fabric of the game itself. Time, which traditionally was not a tangible factor, now becomes objectified and externalized and thus assumes a real and concrete form. The Master, who does not perceive nor comprehend time in this way, has difficulty foreseeing its implications. Otaké, however, is quick to recognize the potential for new tactics and strategy. The decisive turn in the game is precipitated by a clever manipulation of time.

The crucial interchange begins when Otaké is unable to arrive at a decision about a battle climaxing in the center of the board. In order to escape the pressures of the time allotment and to gain extra time to think about a strategy, he seals a play in the upper reaches of the board that is unrelated to the tense conflict. He, therefore, finds a way to circumvent the limitations of the temporal boundaries by making them tools of his own cunning: "When a law is made, the cunning that finds loopholes goes to work."[52] There is nothing officially dishonest about such a maneuver since it conforms perfectly to the rules, yet it is quite unexpected and abruptly shifts the flow of the game. The Master sees this as a trick which vulgarizes and ruins the work of harmony that they are creating together. He expects the game to move naturally from the egoless interaction of the two players, as they create a masterpiece of black and white stones together, but the manipulation of time betrays the intrusion of self-interest into the game. This provokes a crucial response from the Master, who yields to his anger and frustration and makes the fatal move resulting in his defeat.

Nevertheless, there is a sense of inevitability about the way in which the game evolves. Both players are acting in accord with the historical forces they represent and become part of a larger scheme of things that transcends them personally. They are being carried along by the currents of time and the inevitability of change. The Master is a fragile remnant from the fading past who no longer fits in with the world around him. He evokes a profound feeling of loneliness and sadness, much like the mood captured in the following haiku by Bashō:

> Loneliness—
> Standing amid the blossoms,
> A cypress tree.

As the narrator says in a particularly moving passage, "In that figure walking absently from the game there was the still sadness of another world."[53] The Master cannot withstand the flow of time nor sustain the extreme delicacy and spirituality of his art in the wake of a coarser modernity.

The novel is permeated with the sad, nostalgic awareness of ephemeral beauty that the Japanese characterize as *mono no aware*. The autumnal mood emanates from the sense of ending that is embodied in the Master. For the Japanese, beauty exists in the transient phenomena of a world in time. Therefore, Kawabata may lament the fading tradition and sensibility that was once part of Japan, but he also is acutely sensitive to the poignant beauty of its last moments. In the voice of his narrator Uragami, Kawabata projects a resigned acceptance of man's ultimate helplessness in the eternal flux of nature and life, but also a heightened awareness of the evanescent beauty of the moment.

The novel as a whole is shaped by the blending of external and internal time perspectives within the lyrical consciousness of the narrator. The careful recording of moves, dates, and time expenditures provides the sense of linear chronology within the framework of the game; however, the narrator's memories of the Master and the match flow backward and forward in time, giving the novel a time-shape created by the juxtaposition of images from various points in time. In the first chapters, for example, the movement threads chronologically backwards from the Master's death on January 18, 1940, to his final days, then to the last play of the match at precisely 2:42 on the afternoon of December 4, 1938. It then reverses direction to the day after the match, skips ahead a few days, and ends a month after the funeral. References to the beginning of the match appear in Chapter Four. The fragmentation of chronology gives the novel the impression of a free flowing movement, but simultaneously an underlying pattern of unity is created by the fact that the ending with the defeat and death of the Master merges with the beginning of the work. The movement can be compared to the unfolding of a picture scroll in which the sequence of images is not causal and yet part of a greater whole. Events do not surge towards an ending, but attain significance in themselves and in the internal dynamism they generate.

In addition, there is a subtle though pervasive sense of cyclical movement through

time as the seasons change and the weather tenses and eases as though in harmony
with the players and their game:

> The sky was dark with the squall Otaké had called a tempest, and the lights were on.
> The white stones, reflected on the mirrorlike face of the board, became one with the
> figure of the Master, and the violence of the wind and rain in the garden seemed to
> intensify the stillness of the room.
>   The squall soon passed. A mist trailed over the mountain, and the sky brightened.
> . . . Otaké played Black 73. Once more the sky was lightly clouded over.[54]

Kawabata does not deny time in the realization of his art, nor does he exclude the
temporal world of life and nature. For him, as is typical of the Japanese *Welt-
anschauung*, the beginnings and endings of all finite phenomena reveal the source of
ultimate reality itself. This is what he seeks to convey in the wistful beauty of his
novel. The sense of timelessness emerges from a sensitivity to the eternal process of
change and the ability to perceive the infinte in the finite.

Kawabata's *The Master of Go* is quite distinct from Hesse's *The Glass Bead Game*
in both style and content, and yet this element of time and the concept of beginnings
and endings play a vital role in many similar contexts. Throughout the novels is an
interweaving of various temporal perspectives. The historical, linear passage of time
in the external world is juxtaposed with the inner, subjective experience of fluidity
and stillness and a cosmic sense of ultimate unity within all temporal distinctions.

Both the Go game and the Glass Bead Game serve as a focal point for the
apparent dichotomy of time and timelessness, change and permanence, the trans-
ciency of life and the eternity of spirit, chronological time and the subjective,
internal time sense. On one hand, they function as highly evocative symbols of
cosmic unity and harmony, providing the means to experience the timeless oneness
in all opposition and difference. In addition, the antiquity of their traditions imparts a
sense of permanence and continuity through time against which the ephemeral lives
of the characters are projected. Nevertheless, as finite forms, they are not immune to
the process of time, change, and historical transition. The novels depict the games at
a point of transition when they have been cultivated to a heightened spiritual level,
but face the dangers of over-refinement and potential decay. The lives and deaths of
the renowned Masters, Knecht and Shūsai, bring about the inevitable process of
change and also reveal that beginnings and endings in time are the substance of
reality and the source of the timeless.

Hesse creates the Glass Bead Game and the province of Castalia as timeless,
spiritual ideals, removed from the transient phenomena of the natural world and the
affairs of human beings. In order to preserve the purity of this spiritual idealism, the
Castalians seek to nullify all things of a temporal nature: births and deaths,
beginnings and endings, chance, uniqueness, history, creativity, private sensations,
etc. What emerges is a sublime means of experiencing the timeless unity of all
immutable values and transcendent absolutes, but not of the totality of life. The
Game faces the danger of sterility and degeneration into an effete intellectual exercise
severed from the world which gave it existence. Magister Ludi Knecht brings the
Game to new heights; however, he is acutely sensitive to its limitations and

one-sidedness. His defection and death provide the ultimate synthesis of nature and spirit, world and Game, and reveal that beginnings and endings in time are part of an eternal process which is necessary for continual self-transcendence and becoming.

Kawabata perceives the Go game also at a point of extremity where it has become a highly spiritualized art, fragile and in danger of being vulgarized by the worldly forces of the new modern age. Master Shūsai brings its long tradition to a final flowering through his life, skill, and spiritual depth; however, he has sacrificed so much of life to the game that he no longer has a hold on everyday reality. His intense spiritual involvement is tainted by the indications that he has a psychological need to escape the world and dwell in the rarefied realms of the spirit. His defeat and death do not point to a higher synthesis and transcendence, as is the case with Knecht, but rather to a total collapse of his serene spirituality and the ending of a splendid tradition. The game incorporates the spirit of the new age and continues through time. In the eyes of Kawabata, this is a natural inevitability in the eternal flux of time.

The process of change permeates both novels; however, the distinct difference in tone and mood generated by the authors arises from a different perception of this process. Hesse acknowledges the pain and sadness of leave-taking and ending, but he expresses an eagerness for the resurgence of new beginning that is self-transcendence, progress, and spiritual rebirth. Kawabata captures a small segment of the flow of time at a point of new beginning, but focuses on the poignant, wistful beauty of ending.

Notes

1. Johan Huizinga, *Homo Ludens: A Study of the Play Element in Culture* (Boston: The Beacon Press, 1955), p. 9.

2. Aristotle, *On Poetry and Style*, trans. G. M. A. Grube (Indianapolis: Bobbs-Merril, 1958), p. 11.

3. Lass den Anfang mit dem Ende, Sich in *eins* zusammenziehn! ("Dauer im Wechsel") Johann Wolfgang von Goethe, *Werke*, ed, Erich Trunz, 8th ed. (Hamburg: Wegner, 1966), 1, pp. 247–248.

4. T.S. Eliot, *The Complete Poems and Plays: 1909–1950* (New York: Harcourt, Brace and World, 1962), pp. 123 and 144.

5. Chuang Tzu, *Basic Writings*, trans. Burton Watson (New York and London: Columbia University Press, 1964), pp. 80 and 103.

6. See A. A. Mendilow, *Time and the Novel* (New York: Humanities Press, 1965); Hans Meyerhoff, *Time in Literature* (Berkeley and Los Angeles: University of California Press, 1960).

7. Hermann Hesse, *Magister Ludi (The Glass Bead Game)*, trans. Richard and Clara Winston (New York: Holt, Rinehart and Winston, 1969), p. 7; hereafter cited as *GBG*.

8. See Hilde D. Cohn, "The Symbolic End of Hermann Hesse's *Glasperlenspiel*," *Modern Language Quarterly*, 11 (1950): p. 348; hereafter cited as *Glasperlenspiel*.

9. Hermann Hesse, *GBG*, p. 77.

10. Max Kaltenmark, *Lao Tzu and Taoism*, trans. Roger Graves (Standford: Stanford University Press, 1969), p. 45.

11. Herman Hesse, *GBG*, p. 117.

12. Ibid, pp. 150–151.

13. Ibid., p. 174

14. Ibid., p. 151.

15. Ibid, p. 343.

16. Ibid., pp. 342–343.

17. Ibid., p. 345.

18. For a discussion of Hesse's interest in Zen see Adrian Hsia, *Hermann Hesse und China* (Frankfurt am Main: Suhrkamp Verlag, 1974), pp. 122–126.

19. Hermann Hesse, "Josef Knecht an Carlo Ferromonte," in *Materialien zu Hermann Hesses 'Das Glasperlenspiel'*, ed. Volker Michels (Frankfurt am Main: Suhrkamp Verlag, 1973), I., p. 336.

20. D. T. Suzuki, *Zen Buddhism* (Garden City: Doubleday, 1956), I, p. 83.

21. Hermann Hesse, *GBG*, p. 365.

22. Ibid., p. 368.

23. Ibid., p. 350.

24. Ibid., p. 389.

25. Ibid., p. 391.

26. Cohn, *Glasperlenspiel*, p. 351.

27. Hermann Hesse, *GBG*, p. 293.

28. Roger C. Norton, *Hermann Hesse's Futuristic Idealism* (Frankfurt am Main: Peter Lang, 1973), p. 118.

29. Hermann Hesse, *GBG*, p. 411.

30. For a detailed discussion of *The Glass Bead Game* as a lyrical novel see Ralph Freedman, *The Lyrical Novel* (Princeton: Princeton University Press, 1963), pp. 96–114.

31. Edward G. Seidensticker, trans., *The Master of Go*, by Yasunari Kawabata, (New York: Alfred A. Knopf, Inc., 1972), p. 5; Hereafter cited as *MG*.

32. For a history and theory of the Go game see O. Korschel, *The Theory and Practice of Go*, trans. Samuel P. King and George C. Leckie (Rutland: Charles E. Tuttle, 1965).

33. Chikamatsu, *Four Major Plays*, trans. Donald Keene (New York: Columbia University Press, 1961), p. 115.

34. Ibid.

35. Yasunari Kawabata, *MG*, p. 118.

36. Ibid.

37. Ibid., p. 28.

38. Ibid., p. 88.

39. Ibid., p. 136, p. 128.

40. Ibid., p. 144.

41. Ibid., p. 54.

42. Yasunari Kawabata, *Japan the Beautiful and Myself*, trans. E. G. Seidensticker (Tokyo: Kodansha International, 1969), p. 56.

43. D. T. Suzuki, *Zen and Japanese Culture* (Princeton: Princeton University Press, 1973), p. 111.

44. Yasunari Kawabata, *MG*, p. 78.

45. Ibid., p. 76.

46. Ibid., p. 64.

47. Ibid., p. 109.

48. Ibid., p. 54.

49. Ibid., p. 52.

50. Ibid., p. 76.
51. Ibid., p. 150.
52. Ibid., p. 54.
53. Ibid., p. 63.
54. Ibid., p. 87.

Perspectivity and the Principle of Continuity

C. M. Sherover

Everyday speech, as much learned discourse, often refers to particular things as first coming into existence and then later expiring. The reality of the particular entity or event is then discerned, identified, and even explained, in terms of the quantified temporal distance between these identifiable termini of its duration. This common manner of speaking and thinking presumes to understand particular entities and events as though the identity of each is somehow contained within its determinate temporal boundaries; we identify particular people, things, and events by means of their bi-terminal dates. And we speak of a person, a thing, an event in terms of its beginning and its end, as though these two chronological notations were preeminently intrinsic to the nature of its being.

Philosophic thought, like the sciences it has spawned, has generally declined any obligation to take a common mode of speech or thought at face value. Indeed, responsible philosophic thinking has served a critical function: the distillation and reformulation of those validities which lurk in the unreflective everyday use of language, with frequent misleading metaphors that too often mask perspectives and insights seeking expression. It does this in at least two ways: (1) by an essentially literary critique which seeks precision of meaning; and (2) by a metaphysical critique which exposes those presuppositions upon which such insights depend for their authenticity of meaning.

The intent of this paper is to point out some of the conceptual problems that arise when the rubric of 'beginnings and endings' is used in any literal sense. Three groups of separable but closely related confusions are engendered. Collectively they demonstrate the very limited metaphoric utility which the rubric of 'beginnings and endings' may serve in any discussion involving time or temporality.

I.

To approach an object or event primarily in terms of its beginning or its end is to imitate the unimaginative surveyor diligently, but narrowly, pursuing his assignment; he cannot describe the land he measures but only the boundaries which mark its external limits. His employer, as a serious investigator, is concerned with terminal boundaries only because of his interest in what lies between them.

Interest in a particular event, examination of a particular phenomenon, attention to a particular person, are each excited by what it does, the functions it serves, the role it plays, the activity in which it engages—not supposed points of genesis and termination, but what lies between them is what engages attention or focuses investigation. This temporal 'between' of an entity or event is its durational being, its lastingness, its span of activity; it is this 'between' that gives meaning to its discerned terminal 'points.' To examine the entity or event primarily in terms of its beginning and end is not to discern *it*, but its limits. Its time is *its* duration; only within its durational extent is it itself to be found.

We generally mark the bi-terminals of a particular durational being by dating them. Yet we seemingly need a continuing reminder of the literal emptiness of this standard artifice; it is authentically descriptive of no existent entity. Rather, this descriptive vacuousness distracts our attention from what it allegedly 'locates' and points us away from the centered 'between' to its supposed temporal limits.

Dating indicators generally treat time as though it were some sort of container within which things happen— 'at this point *in* time.' The dating system then serves as a sort of spatial locution to 'locate' the 'place' of an event. Time is thus treated as a kind of spatiality and 'locating' an event transforms temporal description into a kind of temporal geography. Such spatializing thinking, epitomized in a geography of dating, presumes a 'container' model by which to think about time; as such it leads directly to the many incongruities and paradoxes which mark the literature at least since Zeno (ca. 450 B.C.). The many resulting absurdities and puzzlements which ensue from treating time as something it is not, are signs of a refusal or inability to treat time and temporality *in their own terms*.

When we regard time as uniquely *sui generis*, as Lotze and Peirce among others have urged,[1] we quickly see that a date is literally a term without any reference. It is intentionally vacuous just because it literally points to no point at all. It claims to 'locate' a temporal 'point' which it names as a particular non-extensive 'now'; but such a 'point,' as Aristotle had already observed,[2] merely marks the junction of a specific 'before' and a specific 'after' and is literally nothing in itself, is not 'in' time, and is certainly nothing temporal; it is but the mark of the boundary of a particular time—or, as we may now say a bit more precisely, the boundary-edge of a particular duration or lastingness of being. The dating's only literal claim is to represent a fictional point on a fictional line. Useful as this may conceivably be, its literal descriptive force—as claiming to 'locate' a specific 'when?'—is fiction compounded and literally meaningless.

A date is nothing temporal. Descriptively false because it literally does not do what it claims, it nevertheless is extremely useful when not so misconceived. A date does *not* 'locate' a mythic point 'in' time. It *does* relate simultaneities and sequences of existents and events; it 'locates' an event *qua* related to other events. It tells us what else was happening when this was happening. It serves to relate this particular event with others—'at the same time,' 'overlapping in duration,' 'that was dying as this was being born,' 'this occurred before that began.' The value of a dating system, then, is intrinsically relational. Its prescription for testing the truth of a date ascription is to relate the event it 'locates' to other proximate events—as when one 'transmutes' the 'numbering' of an anniversary from, say, the Julian to the Gregorian or the Chinese calendar.

As relational, a dating system illumines a particular entity or event as involved in a context of being. Its prime force is then to throw focus, not on the boundaries around, but on the temporal context within which the entity or event endures, manifests its activity of existing, and reflects its existential colleagues within itself. As relational, a date demonstrates the contextual character of the existent to which it refers, the ways in which that entity or event reflects and builds its context into itself. As relational, a date forecloses the legitimacy of abstracting any entity or event out from the temporality of its existing and then treating that entity or event as self-contained. As relational, it points up the durational nature of what it 'locates' and the ways in which that durational activity is in a continuity of relational mutuality with those other entities and events which share its temporal neighborhood.

When a dating system is taken as a metaphoric mode of describing, not terminal points, but a complex of dynamic relationships, it points to the matrix of relations inherent in any particular durational existent. When the function of a dating is seen as relational description, it precludes the legitimacy of severing an entity or event from its context; it urges that any entity or event can only be fathomed when seen as inextricably involved with, and expressive of, a continuity of temporal being.

II.

When a dating is regarded, not as a mythic point on a mythic line, but as a sort of shorthand for a durational complex, we no longer regard a beginning or an end as more than a proximate limit: in a real sense, a beginning prepares itself, and an ending transpires over 'moments' and carries on after it is over. As with any specific birth or death, we are immediately pointed to a process of alteration and change, to antecedents and consequents. We see ourselves looking back into a past that is no longer but is somehow still manifest in the entity or event before us. Our understanding of its activity reaches into the future that is already somehow manifest in the way we see what is before us as inviting possibilities yet to be actualized. When an entity or event is seen as a durational complex, we understand that its being is inseparable from its time, and its time is a continuity of being with all that came before and with what follows.

For any durational entity is itself an event without a precise moment of genesis or termination. As Santayana noted, "Events are changes, and change implies continuity and derivation of event from event: otherwise there might be variety in existence, but there could be no variation, since the phases of the alleged changes would not follow from one another."[3] We are not, then, bound to Hume's dictim that we can only "pronounce one thing not to be another."[4] Attentive observation of any particular entity or happening immediately points out its intrinsic temporal continuities with other things and events, and its own temporal spread no matter how precipitous it may be. Rather than speak of a 'beginning' or an 'end' we might better think in terms of progressive emergence and progressive decline. In this sense, the Greek terms of 'generation' and 'corruption' were more faithful to the facts of our experience.

Any presumption of efficient causality makes this clear. No particular thing is its own cause of being. Its present state manifests ties to an earlier condition, a process out of which it gradually developed and, perhaps some separate precipitating event which brought about that special transformation of its antecedents that makes it appear as new. Even the 'big bang' cataclysmically initiating *this* material universe 'took time' and was presumably the radical transformation of an earlier matter/energy complex, conceivably initiated by some separate precipitating event. Even the French Revolution, often described as the 'beginning' of modern Europe, did not begin at a precise moment, carried with it the heritage of prior problems and crises which precipitated its cataclysmic appearance, shaped the ways in which it developed and the ways in which it fed into what came after it died down. Unless one posits a pervasive doctrine of spontaneous and uncaused generation, one is bound to a view of nature or society, involving any motion of efficacious causality as a process of temporal continuities and as carrying past conditions in a transformed way into what comes after.

Any discerned and discriminated event, then, is a sign of both antecedents and consequents. Newness and variety—the new contours of Mt. St. Helens, the birth of a baby, the inauguration of a new governmental administration—whether in physical nature, biological emergence, or social history, arise only as transformations of what has already been, as developmental continuities from previous states and antecedent conditions, as the actualization of possibilities already available. What we regard as specific, efficacious, or precipitating causes of particular transformations are themselves dependent for their being on antecedents. Any particular entity or event, then, is initially complex in its developmental history and its pervasively historical continuities in the world.

If the other entities and events with which it interacts are likewise enmeshed in their own temporal continuities, then the temporal continuities of any particular entity or event do indeed stretch far beyond it, involving it in a dynamic matrix that is essentially organic in character. Any particular entity or event, as durational and developing, is then inherently involved with, and truly inseparable from, the continuing temporal continuities that characterize the existent world as such.

If time is the mode whereby one entity or event is related to the others, to the world in which it is acting out its own state of developmental being, then such time-relations cannot be accidental or nonessential to it; they cannot be merely external or arbitrary appendages attached to, but removable from it. The time-relations of an entity or event are *intrinsic* to it, to what it is and how it is—not only as existent, but as exis*ting*, develop*ing*, chang*ing*, matur*ing*, declin*ing*. Any entity that *is*, any event that occurs, is then constituted by its time-relations—which have made it what it is be*ing*, activate what it is do*ing* and the functions it is serv*ing*. Its time-relations are then constitutive and, in the traditional jargon, are relations *internal* to the essence of its individual identity. Because internal time-relations are inherent in definition and in structure, no entity or event can be understood or explained while its essential temporal constitution is ignored.

If time is inherently relational—whether externally or internally considered—then time cannot be exclusively linear in character. Time-relations cannot be simply characterized in terms of before-and-after sequentiality, although that sequentiality is certainly ingredient to them. As a network of functioning relations, time involves any particular entity or event with others. As a metaphoric shorthand—but at risk of gross over-simplification—we may discriminate time-relations as either sequential or simultaneous; but to take these terms as clear and distinct differentiations with full exclusve force is to treat them as accidental attributes and not as constitutive ingredients; to do so would also be a reversion to the naive simplicities inherent in a temporal geography that 'locates' the 'place' of a thing in terms of precise points within a spatially conceived box. To reduce time-relations, then, to mere sequential-ity and simultaneity in an exclusive way is to preclude the possibility of comprehend-ing the being of an existent entity, the meaning of a real event.

If time is inherently relational of real entities and events, and as such intrinsic to them, then time-relations must be conceived as real. However we may evaluate the descriptive competence of human reason's cognitive claims—whether we believe that human cognition may somehow penetrate to the heart of reality itself, or we find that human cognition is necessarily restricted within the apriori confines of its perspectival outlook—clearly any cognitive claim is predicated on the presumption of some kind of temporal realism. Any knowledge claim, restricted as it may be in its cognitive reach or certitude must engage the presupposition that time-relations are somehow real, as real as the things and events we seek to comprehend, real at least as they truly appear to human awareness. One may or might not believe that objective reality fully appears within the outlook of human cognitive competence; but even the skeptic here must presume in his cognitive investigations that the temporal continuities which appear to human cognitive reason seem to have an objective ground independent of the modes of human thinking, as objectively independent as the things and events that appear to be constituted by them. And both must recognize as well that the activity of human thinking, reasoning, cognizing, is itself temporally constituted—for any awareness of external change is itself a temporal event in the activity of the apprehending consciousness.

If particular entities and events are constituted and bound together by temporal ties—manifesting themselves as continuities of dependence, causality, purpose, function, interaction—then there can be no gaps in time or nature. Whatever discontinuous changes seem to appear, we necessarily presume that such discon-tinuities reflect a limitation of our understanding, set a task for investigation; that such discontinuities are not of the real order of events to which understanding is directed but in the understanding itself. The notion of a timeless gap between two distinct sequences is a contradiction in terms and devoid of intelligibility or ontologi-cal meaning.

As Leibniz urged, when formulating the law of continuity, we necessarily involve the premise that time is the form of the "continuous temporal modifications in the universe."[5] In carrying this forward, Kant had argued for a primordiality of temporal continuity and unity: the concepts of succession, simultaneity, and duration, he pointed out, all presuppose one time order insofar as "no time is apprehended except as part of one and the same boundless time. . . . [For time] is the principle of the laws of continuity in the changes of the universe."[6] More primordial even than the laws of reasoning, continuity is at least a 'first postulate' of the notion and experience of time.[7]

Whether we speak of the ways in which we can understand nature, or of the nature that is to be understood by us, if in any sense, we are *in* nature, then the requisite continuity is not only the primordiality of temporal contintuity of nature but also of us. As Kant was quick to note, our possible knowledge of 'all changes and successions' depends, first, on temporal continuity in the observed world; it equally "presupposes the perdurability of the subject."[8] Indeed, William James effectively carried this forward by urging that time is *the one* form of being common to both knower and known, to man and nature, and thereby the ground of the possibility of man's knowledge of nature.[9]

The principle of temporal continuity, then, has at least three subordinate principles: (1) the continuity of the phenomena constituting the natural order; (2) the continuity of the human subject who seeks to understand it; (3) the continuity of their relationship in a mutuality of temporal being permitting knower and known to meet together in a common temporal field. The principle of temporal continuity is, then, a first principle for the understanding of external objects *and* of our own selves.

What we term a 'beginning' or an 'ending' cannot then be, within this one world at least, *ex nihilo* or *in nihilo*—out of nothing or into nothing. Beginnings and endings of particular entities or events can only be taken as noticeable transformations which activate potentialities already present, actualize possibilities truly genuine, and are continuous with and inseparable from them.[10]

III.

The judgment that a particular transformation is sufficiently dramatic to occasion designation as a beginning or an end is itself a change, an event, in the consciousness of the beholder regardless of its descriptive truth. No judgment is a merely passive report of what is transpiring 'out there.' The most passive spectator kind of judgment, that does not disturb the continuing flux to which it refers (as the unobtrusive spectator does not affect the drama he witnesses), is already a selective report on what is claimed to be transpiring before one's eyes. One cannot claim to describe the infinite detail of what is before him in its totality; he thus necessarily focuses attention and selects aspects of it that excite his interest and respond to his ensuing questions.

To speak, then, of a beginning or end of a particular entity or event is to interpret specifically selected sets of continuities—the changing color but not the fragrance or the sound—regarded as important or significant to the reporting interest; but 'important' means 'important for something,' and 'significant' likewise refers to and invokes a standard of evaluation; to regard a set of continuities as important, then, immediately implicates the valuational categories of the speaker. His judgment of a particular beginning or ending is the voicing of a particular perspective and its operating judgmental standards; the judgment has its cognitive meaning, then, only within the terms of that perspective. When considering the truth of any assertion, then, one needs to implicate the value structure—the criteria for significance, importance, focus, selectivity—of the perspective from within which the assertion is made. This perspective is, itself, a dynamically particular continuity of interpretive activity; as such it is durational, cannot be reduced to momentary states, and is inextricably involved with the context of its being.

If any judgment or knowledge-claim is inherently selective, if it depends in any way on its own (perhaps unspoken) criteria or values, it is no merely passive report; for it probes the presentational flux with its own interests, standards, and its questions; its report, which is the response to these probings of interest or concern is an inherently interpretational activity. The selectivity of perceptional focus arises out of the needs or interest of the observer's attentive thinking; his judgment is structured by the modes of thought invoked to respond to the questioning interest that initiated it and brought it forth. Thinkers as disparate and yet similar as Leibniz, Kant, Peirce, Royce, and Heidegger—and Plato, too—have presented cogent reasons for this thesis that perceptual attention and consequential judgments constitute interpretive activity arising out of the particular thinker's peculiar finite perspective. [11]

Interpretive thinking is itself a process of temporal continuity that is, perhaps logically but not really, separable from the continuities to which it claims to refer. An act of thinking is a response by the thinker to what he sees and not merely a reaction to a perceptual stimulus; an act of thinking is animated as much out of the needs, concerns, interest, desires of the particular thinker as out of the object about which he speaks; an act of thinking cannot, then, be reductively explained as the mere causal effect of the observational field.

A specific act of thinking, the interpretive understanding of what is transpiring before one's eyes, is oriented forward in terms of problems to be solved, confirmations of expectations to be secured, answers to questions being asked, purposes to be pursued, discerned lacks to be filled—in terms of conceived anticipations which beckon one onward. Within the activity of what is being observed, functions are discerned, purposes are seen as being frustrated or fulfilled, experiments are conducted, activities are planned, expectations are tested. These are discrete modes of human thinking—all of which look ahead to what is conceived as possibly yet to come; they feed on a vision of possible futurity and feed it into the present activity, taking up only those specific causal chains from the past which seem pertinent to the conceived task at hand. These are human activities which presuppose a new kind of continuity—a reciprocity between present and future.

Within the standpoint of the present, I can look onto the past from which the present arose, and discriminate from that complex the specific sequential chains of developmental events which I understand to be *the* causal strands leading into the current focused matrix—as the weaver can discriminate those strands being taken into the particular pattern he is now developing from those that have already been used or will only come into other areas of his rug.

Thus I can explain the present in efficacious causal terms as being dependent for its being on specifically selected sequential chains from the past. But I cannot comprehend my own activities of planning, experimenting, or anticipating in this simple reductionist manner; for these current directive activities presuppose a vision of yet unrealized possibilities that are taken as genuine and that as such feed my present decisions and acts—in a kind of running back-and-forth of continuities between future and present that are nourishing each other.

Present directive activity presumes a view of the future as somewhat yet open and thereby somewhat dependent for its specific development upon my present judgment, decision, activity. Mere chronology is thus confused in any description of such mental activity just because it inherently involves the mixing of the not-yet futurity of anticipation with the now of the immediate present. But, clearly, any deciding and planning presupposes firm conviction in the possibility of counting on continuing continuities into the future that is not-yet. And, clearly, actional judgment necessarily steps beyond cognitive certitude as it advances (by an invocation of rational faith) into its future which is, in any strict sense, unknown (and unknowable) just because it is not, by definition, presently determinate. What is involved in this reaching-into of futurity that is necessarily present in thinking activity certainly requires extensive development.[12] But what is crucial here is just this: the principle of continuity is not to be seen solely in the causally sequential chains from the past that are presently manifest; this principle of continuity works forward in both the human act of interpretive understanding *and* in the developmental activity of the things we seek to understand, use, transform. Projection of continuity between present and the not-yet-but-still-can-be is thus intrinsic to any act of thinking as it seeks to deal with selected aspects of the world.

If perceptual and interpretive activity are selective, problem-oriented and judgmental, they depend upon the evaluative standards which constitute the continuity of the outlook and modes of thought of the particular human subject; for it is by means of these evaluative criteria or standards that individual attention is focused and directed, judgments made, decisions resolved, activities committed. These evaluative standards, which constitute any individual perspective, are then primordial to the selectivity of perceptions, the idiosyncracies of individual tastes and biases and commitments.

Any assertion, then, of a beginning or an ending is a selective interpretation of the continuing flux from the particular viewpoint of the particular speaker. Any such assertion can only claim to be a report of a discerned transformation of a particular aspect of the continuing flux *as* selected for focus in terms of the predilections and questions which the individual brings to his observation or investigation of it.

The primitive witch-doctor and the typical western physician will describe the same patient in terms of different kinds of transformations, report different 'discontinuities' as a result of diverse examinations, anticipate different pathological development, and prescribe dissimilar remedial courses of action. Any attribution of efficient causality, as any directive activity, presupposes the principle of continuity; but the specific continuities selected out for discernment, evaluation, investigation, the specific continuities projected as the 'henceforth' presupposition of deliberate activity, are dependent on the evaluative standards constituting the individual outlook of the person concerned.

Thinking is then itself a creative, interpretive structuring of presumably open situations with a problematic orientation and a point of view; it ties past into present by means of its structured focus and future into present by means of decisions and ensuing activities. Both taken together, in a synthetic whole, constitute the particular perspective from within which it reports, evaluates and decides. The particular strands of continuity one abstracts from the presentational flux with, and in which he is already involved, are judgmental interpretations from the outset; they finally depend at least as much on the values, interests, categorial structures and physical dispositions of his outlook, as they do on what is actually transpiring in the reported field. The claim of a new beginning is then inherently fallacious if it either suggests denial of literal continuity with progressively changing antecedents, or pretends that the particular idiosyncratic perspective of the speaker is irrelevant to his claim.

Different qualified observers often interpret the same presentation in dissimilar ways. Interpretive thinking cannot be explained as merely the caused effect of the observed presentation. If thinking is selective and interpretive it is not reducible to prior causes, mechanistically conceived, just because its activity consists in questing for reasons to justify belief. It is thereby itself selective and thereby insofar free. Interpretive thinking presupposes the continuity of process and connection in what it thinks about, but also its own selectivity of focus, the weighing of reasons, the invocation of evaluative criteria and their examination and refinement. Interpretive thinking thereby presupposes the ability freely to engage in such discriminating and self-directive acts.

If I am to escape error in judgment—as Descartes had already clearly argued[13]—if I am to test my hunches, explore alternatives, evaluate evidence, I must be able to direct my attention, control my thinking, govern its investigations, and refrain from premature suspension of questioning—until my own evaluative criteria for the fulfillment of my quest have been satisfied. Responsible thinking is, then, a free project which is itself constituted in terms of its own temporal continuities.

Most considerations of human freedom have been in terms of moral endeavor and practical reason. What is suggested, however, is that any conception of perception *qua* selective, any appraisal of thinking and judgmental decision *qua* evaluative and interpretive, bespeaks not only the primordiality of perspective; insofar as an individual's perspective is selectively interpretive, it is forward-looking in terms of evaluatively discerned possibilities not yet actualized. That resultant evaluation cannot be explained as neatly reducible to what already has been or to what is literally present, just because it represents a step into the problematics of a future not yet resolved. Evaluative selective thinking thus presupposes the continuity of freedom in the durational existence of conscious evaluating beings. Acknowledgement of the functioning of freedom in the activity of thought only underlines the primordiality of the individual perspective in the assertions the individual makes concerning aspects of the changing world in which he finds himself.

IV.

We cannot, then, legitimately speak of any beginning or ending in any literal or absolute sense. Any discerned beginning is but a new transformation that itself is a process—whose precise genesis can hardly be dated with exactitude, and whose nascent temporal context can only be proximately pointed out. Any discerned ending is a transformation that carries into what follows or arises out of it; its further implications, like the ripples in a stream, can hardly be definitively delineated. The assertion of a particular beginning or ending is a selective, interpretive judgment from within the perspectival framework of the speaker, and brings to bear both his focus of interest as structured by the evaluative categories he uses and the freedom that is manifest in the reasoning he does in applying them. And that dynamic reasoning is itself a temporal process that cannot be severed from the temporal continuity of his biographical becoming.

Such judgments and outlooks, then, presuppose the essential historicity of dura-tional activity: the temporal continuity of the speaker and his concerns with what he speaks about, the separable but not separate continuities of each, and the historical continuity of the cultural matrix from within which he has developed that individual distillation of his cultural ethos into the outlook he has made his own. On each of these variegated levels is presupposed the continuity of what is no longer with what is not yet but may yet be.

To assert a beginning is to point out a process which itself reaches out to what was and what may yet develop; it is a process which reflects an environment of currently transpiring interaction. It is to bring to bear on that complex happening an evaluative framework which justifies the speaker from his own judgmental outlook, to speak of a birth rather than an emergence, a new turn rather than the maturation of an earlier activity.

V.

What, then, of the utility of such a term as 'beginning' or 'ending'? By them, we ought to name the prominent part of a process and not a mythic point: a procession that ties the separable continuities of transformational complexes, represented in what is seen and in the evaluating observer, together into the event that includes them both. Our assertions concerning what we claim to describe but give voice to the values incarnate in the outlook which chooses to speak. We thus describe, not the process being discussed, as the grammar might indicate, but the response to questions of selective evaluative import, the testing of expectations whose own continuities are only somewhat dependent on, and logically prior to, what is being allegedly described. To pronounce a judgment of beginning or ending is a proclamation of a new transformation that is, in a sense, taken as an interpretive 'discontinuity' from within a continuous field.

This essentially metaphoric locution cannot legitimately serve as a framing concept that directs attention to the borders of the entity or event. When locutions such as "beginning" or "ending" are taken as relational and referential, they point to the durational reality within those borders and equally to what lies beyond; they regard such temporal boundaries as artificial selective impositions called for by the speaker's purpose to direct focus, and not as truly descriptive of the entity he is focusing on; such locutions, authentically comprehended, point in at least two directions at once: to that to which they claim reference *and* to the perspective (and its reasons) which voices them. Their authentic function is to point to a general area of interactional existence and not to a particular determinate and allegedly self-contained event. Their function is not to frame, not to locate precisely, but to point out, to direct attention toward, the generally temporally constituted contextual area within which an interpretive set of evaluative canons is claimed to have some specified degree of pertinence.

Notes

1. Charles S. Peirce, "Issues of Pragmaticism," in *Selected Writings of Charles S. Peirce*, ed. Wiener (New York: Doubleday Anchor, 1958), p. 223; R.H. Lotze, *Metaphysic*, trans. B. Bosanquet, 2nd ed. (Oxford University Press, 1887), sec. 149, in Charles M. Sherover, *The Human Experience of Time* (New York: New York University Press, 1975), p. 203; hereafter cited as *HET*.

2. Aristotle, *Physica*, 219a–220a.

3. George Santayana, *Scepticism and Animal Faith*, (New York: Dover Publications, 1955), *p. 230.*

4. *David Hume, An Enquiry Concerning Human Understanding*, 12, 3.

5. G.W. Leibniz, in Sherover, *HET*, p. 134.

6. Immanuel Kant, "Inaugural Dissertation," sec. 14.4, in Sherover, *HET*, p. 146.

7. *Ibid.*, sec. 14.6, p. 149.

8. *Ibid.*, "Scholium," p. 152; N.B. The original Latin reads: *"supponit perdurabilitatem subjecti."*

9. Sherover, *HET*, p. 351. N.B. Although James does not cite Kant, this thesis follows directly from the conjunction of two key sections of Kant's First Critique, viz., "The Schematism" and "The Highest Principle of all Synthetic Judgments"; see Kant, *The Critique of Pure Reason*, trans. N.K. Smith, (New York: St. Martin's Press), 1929, 1965, A137–B176–A147-B187, pp. 180–87 & A154-B193–A158-B197, pp. 191–94.

10. Since Aristotle, the concepts of 'potentiality' and 'possibility' have been generally confused. A current project of mine is an attempt to disengage their separate meanings. Briefly for the present purpose, one might take 'potentiality' in the sense of defining capability (whether or not realized), e.g., 'An acorn has the potentiality of becoming an oak tree'; in contrast, a 'possibility' would refer to a future set of conditions which must be met if that potentiality is to be realized, e.g., 'Because this acorn is being eaten by a squirrel, it does not have the possibility of growing into an oak tree'. In these terms any present actuality would be seen as the merging of potentiality and possibility, in Leibnizian terms of the 'necessary' and the 'sufficient,' in Heideggerean terms, of the 'ontological' and the 'ontic.' But note that both within the standpoint of the present point to ways in which futurity is to be understood.

11. See, e.g., Plato, *Theatetus*, pp. 183–86; J. Royce, *The World and the Individual*, 2, (New York: The Macmillan Co., 1901), esp. II–VI; M. Heidegger, *Being and Time*, trans. Macquarrie & Robinson (London: SCM Press, 1962), esp. sec. 41.

12. The 'existential analytic,' i.e., Division One of Heidegger's *Being and Time* , is a rigorous working-out of what this primordial future-orientedness involves in terms of an understanding of human activity; Division Two of this work seeks to explicate the temporal dimensions that are involved.

13. René Descartes, *Meditations on First Philosophy*, esp. Meditation IV.

Closure and the Shape of Fictions: The Example of "Women in Love"

Readers have known that endings are important ever since men began to read or even to listen to stories. We pay tribute to the importance of endings in numerous ways. For example, even though knowing how a work ends makes it easier to analyze on a first reading, few of us *really* want to lose the suspense which accompanies a first reading. Therefore we commonly ask others not to "give away" the ending. When we're baffled by a book, or consumed by curiosity and skip ahead to read the ending, we obey a contrary impulse, but one which also testifies to the importance of endings, to their ability to clarify just what a given fiction is "about." When reviewers of films or novels almost ritualistically note and evaluate the ending of the work and its success or failure, they too acknowledge the first fact with which I begin: the importance of an ending to the form and meaning of a narrative work.

Critics have known this basic fact as long as there have been critics. Hence, our earliest articulation of the importance of endings comes in our earliest known critical text, the *Poetics*. Aristotle's idea that the relationship of beginning, middle, and end crucially determines the form and meaning of a work of art, and indeed, makes art art, has been frequently reiterated for the novel, which is the form of fictional art I shall discuss here. Henry James maintains, for example, that what he calls "stopping places" in fiction are never entirely natural and never easily found; instead, the artist must carefully choose the appropriate "stopping-place" by what James (in a marvelous bit of Jamesian rhetoric) calls "a difficult, dire process of surrender and sacrifice."[1] In *Aspects of the Novel*, E. M. Forster makes the same point: a completed novel contains "pattern and rhythm," internal connections which give it meaning and make it art.[2] More recently, Frank Kermode explores the same idea in *The Sense of an Ending*. In novels, what look like random "pieces of information" assume significance via the ending; Kermode puts it:

> "The beginning implies the end . . . all that seems fortuitous and contingent in what follows is in fact reserved for a later benefaction of significance in some concordant structure."[3]

At this point, a few definitions will be helpful. Two terms in particular need to be defined and distinguished: endings, and (the word in my title), closure. The word "ending" straightforwardly designates the last discernible unit of a work—section, scene, chapter, page, paragraph, sentence—whichever seems most appropriate for a given text. Closure, on the other hand, designates the *process* by which a novel reaches an adequate and appropriate conclusion. Closure is thus a much more inclusive term than the term "ending"; since closure often begins with the first sentences of a novel, the study of closure requires inquiry into the total shape of fictions (into, that is, the relationship of beginning, middle, and end), rather than just into a work's last significant section. My use of the term "closure" corresponds to what Barbara Herrnstein Smith in *Poetic Closure* calls the "integrity" of a lyric and what David Richter in *Fable's End* calls the "completeness" of an apologue—a sense that nothing necessary has been omitted from a work.[4]

In the practical consideration of endings and closures, many critics have, of necessity, moved from consideration of the text as an organic whole to consideration of the effect of the text on the audience or reader. Thus, for Aristotle, endings not only complete an artistic whole, but also "purge" the audience via the cathartic experience of "pity and fear." Also, for E. M. Forster, while perception of "pattern and rhythm" may be essential to appreciating a text, the process of reading a text depends on the reader's fundamental impulses of curiosity. As Forster notes, we are all "like Scheherazade's husband [in that] we all want to know what happens next." In several comments in his prefaces, James agrees with Forster that the sense that fictions will end and our desire to know *how* they will end nurtures our desire to read novels.[5]

The weight of these venerable comments about endings suggests that endings and closures are not just important in art, but in the experience of the reader, and in life itself. Most critical discussion of endings has, however, emphasized almost exclusively the ways in which the presence of endings makes the experience of fiction *different from* the experience of life. Now, emphasizing the differences between life and fiction caused by endings is too narrow and too partial a view of the matter, for reasons I will get to shortly. Yet the view that the presence of absolute, printed, fully knowable endings in literature makes the experience of novels different from the experience of life clearly has some validity and should be briefly explored.

One reason both ordinary readers and literary critics attribute such importance to endings is, for example, the fact that we think about endings in fiction in much the same way that we think about and organize our own past experiences: in both instances we look back over a certain collection of facts from a significant vantage point which determines, in part, the meaning of those facts. In literature, this process is usually called retrospective patterning. But if retrospective patterning is one method by which we give meaning to our lives and to our fictions, only in our fictions are mortals granted the privilege of knowing an *absolute* ending and of contemplating a *completed* whole. Because of chance and mortality, the full ability to establish "pattern and rhythm," to make a "selection and comparison" of discrete facts, does not fully belong to the living being contemplating his own existence, but to the historian, the biographer, the novelist—and their readers. In novels, endings are privileged moments precisely because they eliminate the possibility of radical reversals and terminate the tentative perception of pattern which normally accompanies both living and reading.

The perception that endings confirm the pattern of a life or text but are unknown for lives in progress is far from modern. Herodotus shows awareness of it in the history of King Croesus when, after the fall of his empire and his capture by the Persians, Croesus understands the truth of Solon's words that "no man could be considered happy until he was dead."[6] Greek tragedy similarly exploits this distinction between living life and telling or writing about it. A biography of King Oedipus would be a very different document from Sophocles' *Oedipus Rex*. Events and words misunderstood in the play by Oedipus would become portentous and receive emphasis in a retrospective narration. Similarly, the dramatic ironies easily perceived by an audience familiar with Oedipus' story are in part a function of knowing its ending.

In our own day, Jean-Paul Sartre explores some of these issues via the meditations of Antoine Roquentin in *La Nausée*. Using some of Sartre's ideas as a point of departure, Frank Kermode's *The Sense of an Ending* raises an important question for the discussion of closure in fiction: if fictions have endings and structures while life does not, are fictions then a distortion of experience? One many answer this question in several ways. First, one can justifiably maintain that the question is bogus: even the most mimetic of fictions includes, after all, *harmonia* (arrangement), and is to some extent stylized. We accept and even expect such stylization as a convention of the reading experience. Second, one can point out, as Kermode does in *The Sense of an Ending*, that fictions are "mental structures" by which we "make sense of experience." We should not, therefore, underestimate the healthy functions that fictions serve, not only in literature, but also in history, philosophy, the sciences, and the other disciplines. Third, and most radically, one can challenge the very notion that fictions distort experience because they have structures and endings. One can explore analogies between the shape of fictions and the shape of lives caused by the presence of endings. It is this last strategy that I would like to explore in the remainder of this essay, first in general terms and then using the example of a complex novel, D. H. Lawrence's *Women in Love*.

One may explore analogies between living and reading caused by endings and closures in several directions. One possible direction is the drawing of pertinent analogies between the two experiences overlooked by many critics. The process of reading without knowing endings is, for example, very similar to the process of living day to day. On both first readings and in ordinary living, we make tentative guesses at direction and meaning by applying our experience of what the data we encounter *usually* lead to and mean. Thus, in literature, we expect a given love story or mystery to develop and to end in ways comparable to those of other love or mystery stories we have read. Thus too, in life, we expect to get a taxi at the airport if we usually do, or guess that we will get well from a minor illness because people rarely don't. What I am suggesting is a distinction between first readings and subsequent readings in a consideration of the relationship between living and reading. Since first readings involve the continuous making and revision of guesses, first readings are rather like the process of living in the present. Second or subsequent readings, however, resemble the ways in which we retrospectively organize experience or experience the past. Upon rereading, "pattern and rhythm,"—connections between beginning, middle, and end—may be more easily discerned and more fully understood by the reader. Appreciating such connections via retrospective patterning provides the primary pleasure of rereadings, just as reliving the facts or perceiving the patterns in our lives forms the basis on which we regard our pasts.

A related direction one can follow in challenging the notion that endings distort experience is the direction charted by Roy Pascal in a recent article called "Narrative Fictions and Reality."[7] Pascal correctly points out that life is neither entirely ordered nor entirely contingent (as Kermode seems at times to maintain). Indeed, reflection reveals many ways in which lives, like texts, include crucial senses of endings. When we have completed one stage of life, and look back on the past in light of "how things turned out," we use our sense of endings in life to organize experience. Such stages are frequently marked by an event or ritual, like marriage, illness, or death. When we comment on historical events or on present trends in light of past events, we do the same. Again, when we observe the lives of others and project endings for them, or when we speculate about our own futures and postulate certain goals or endings, we once again use our sense of endings and closures in real life experiences. Stating the case as broadly as possible, one can maintain that all conscious experience (whether of life or literature) includes some sense of an ending, some attempt to discern patterns of closure. Even as we recognize that what we may call life's "provisional endings" are less consciously controllable and are frequently more abrupt than what James calls art's "visibly appointed stopping places," we need to reaffirm that endings in fiction correspond to and serve human experiences of the most ordinary kind.

My point may be clarified by a hypothetical experiment in which individuals are asked to summarize their lives.[8] Consider a narration of the following kind, from a hypothetical, rather uncomplex college freshman:

> "I grew up in a large family on a farm in the midwest. During vacations from school, I worked on the farm. In grade school and high school I did well and was considered a bright student. My social life was good too—I've always had lots of friends and boyfriends. Now that I'm in college I have to work harder to keep up my grades. But I also want to be a well rounded person and have a good social life. After I graduate I'd like to work for a while and then get married and have a large family, like the one I grew up in."

Superficially, this speaker organizes time around education: being in school versus being on vacation or having a "social life"; a regular tick-tock rhythm of grade school, high school, college, and what is to follow college. We can safely say that this young woman largely experiences life through the "intrinsic genres" associated with being a student.[9] But rivalling education's central place in her life and ultimately displacing it as the focal point of her utterance is the idea of marriage and a family. The ending she most anticipates is not graduation from college or putting her education to work, but becoming a wife and mother. This submerged but crucial sense of the "intrinsic genre" of her life makes central rather than peripheral certain aspects of the speaker's narration, like the opening reference to her family and her insistence on having a "good social life" and "lots" of boyfriends. Such facts seem incidental on a first reading, but central upon rereading.

Consider an equally uncomplex but radically different narration:

> "'My parents wanted me to be a doctor and I always knew I would be one. I worked my way through college, majoring in biology. Then I went to medical school and completed my internship while my wife worked as a teacher. Now I'm in private practice and teach at a well-known university affiliated hospital. My work takes up all my time, but it's satisfying and I couldn't imagine doing anything else."

In this hypothetical narration, every sentence is work oriented. Professional life provides the sole context within which this individual experiences his life, and "being a doctor" constitutes the intrinsic genre of all his activities. The speaker begins and ends with the idea of being a doctor. He introduces his schooling and marriage only as they pertain to that idea. Congruently, and in marked contrast to our hypothetical freshman, the speaker's current occupation is his only conceived "ending," and his satisfaction with the present precludes all speculation about future endings and new beginnings.

If our experiment were real rather than contrived, the number and complexity of ways that individual narrations might be organized would be unlimited. One can easily imagine and construct narrations around each of the following provisional points of ending and new beginning in human experience: marriage, divorce, the death of a spouse; recovery from mental or physical illness, descent into mental or physical illness; movement from one place to another, movement from one job to another; discovery of a sense of purpose in life, disillusionment; and so on. The precise units in which an individual conceptualizes or tells about his life would sometimes reflect simply the context in which experience or narration occurs. Someone who has just moved from one part of the country to another, for example, might well tend towards formulations of the "while I lived in New York City" kind. If the context did *not* control the unit chosen, however, the markers used by an individual to describe life would make a statement about how that individual experiences life. The more consistently or exclusively an individual chose markers of a specific and distinct kind, the more certain we could be of that individual's primary orientation, of the things most important in his life.

Analogously, a novelist tends to end his works at points which seem significant or natural to him as ones of conclusion or new beginning. As was the case for our two hypothetical speakers, he is likely to provide a more than casual connection between the beginning and end of his narrative. Similarly, each portion of that narrative will bear some relationship to the ending, whether or not it is easily perceivable on a first reading. And the points at which a novelist chooses to end his fictions, particularly if similar in many of his works, may reveal much about an author's sensibility, values, and views of experience. In actual novels, factors like point of view and irony might modify the extent to which an ending directly reflects an author's views on life. But as long as analyses of individual texts consider such complicating factors, the experiment just outlined may serve as a model for the way in which anticipated endings can shape a narrative and express the teller's preoccupations.

Two important factors which I will not be able to explore very thoroughly here should still be mentioned: audience reactions and cultural assumptions. In each of the preceding hypothetical narrations, a number of possible audience reactions might be anticipated: some feminists might wish to warn the college freshman against her only half-conscious emphasis on traditional female roles; some listeners might similarly see the doctor's focus on his career as obsessive and, ultimately, as dehumanizing. Readers of novels similarly react in a number of possible and often predictable ways to the endings of novels; sometimes, indeed, an author writes with an anticipated reader response very much in mind (such is the case, for example, when an author wants an audience to accept an unhappy ending when it deeply desires a happy one). Moreover, in each of the preceding narrations, a number of cultural assumptions come into play. In most societies, the family bias of the college student's narration would suggest that the speaker was female, even if the narration was censored to remove all indication of gender. Similarly, the doctor's narration would, until very recently, almost definitely have indicated a male speaker. Numerous and various cultural norms similarly interact with the ending of fictions and determine authorial attitudes and reader reactions: endings may express the norms of the culture within which they are written, or they may attack them, or they may modify them, and so on. Each fictional ending needs to be considered in a network of reader reactions and cultural contexts as well as in terms appropriate to the text itself and to its author.

When we turn from controlled hypothetical narrations to actual novels, the discussion of closure and the shape of fictions obviously becomes far more complex. In the remainder of the paper, I would like to outline an approach to closure for D. H. Lawrence's *Women in Love*. First, let me remind you of the way in which the novel actually ends. In the penultimate chapter, Gerald comes upon Gudrun and Loerke romping in the snow. He strikes Loerke and begins to strangle Gudrun. He relaxes his grasp when Gudrun loses consciousness and Loerke taunts him. In a daze, Gerald walks off into the snow and, exhausted, passively commits suicide by going to sleep. In the last chapter, Birkin and Ursula return from Italy after Gerald's death to the village in the Tyrol. They find Gudrun as cold as the setting. Birkin views Gerald's body, both deeply moved at his death and deeply grieved that that death has come about *instead* of the spiritual union Birkin had offered Gerald. Gudrun leaves for Germany and disappears from the book. But the novel's ending presents a final scene between Birkin and Ursula in their English home, which I will reprint in full:

Ursula stayed at the Mill with Birkin for a week or two. They were both very quiet.
"Did you need Gerald?" she asked one evening.
"Yes," he said.
"Aren't I enough for you?" she asked.
"No," he said. "You are enough for me as far as a woman is concerned. You are all women to me. But I wanted a man friend, as eternal as you and I are eternal."
"Why aren't I enough?" she said. "You are enough for me. I don't want anybody else but you. Why isn't it the same with you?"
"Having you, I can live all my life without anybody else, any other sheer intimacy. But to make it complete, really happy, I wanted eternal union with a man too: another kind of love," he said.
"I don't believe it," she said. "It's an obstinacy, a theory, a perversity."
"Well——" he said.
"You can't have two kinds of love. Why should you!"

"It seems as if I can't," he said. "Yet I wanted it."
"You can't have it because it's false, impossible," she said.
"I don't believe that," he answered.[10]

Without doubt, the ending is startling in its abruptness, brilliantly so. On a first reading, the ending is a shocker, as it is in the Ken Russell film of the novel. On second or subsequent readings, the ending (as I will explain shortly), makes the reader more attuned to conflict in the novel as something which will not be resolved, but will persist to the last sentence and beyond. Surely, the ending underscores for the reader two key themes, and the connections between those themes: first the theme of male-female relationships and how to achieve what Birkin calls a perfect "star equilibrium," and, second, a more subversive theme, the theme of male intimacy and how to achieve it. The first theme is insistent enough throughout the novel. The second theme is also insistent, but less so; certainly its importance is dramatically emphasized by the ending, which refuses to allow Ursula or the reader to gloss over Birkin's unusual, and (to some readers) unpleasant and perverse desire for intimacy with Gerald.

One effect of the ending on the reader may be clarified by introducing the notion of *defamiliarization* (a term introduced in 1917 by the Russian Formalist Victor Shklovsky). In a famous essay, Shklovsky speaks of literature as revealing the "stoniness of a stone," by which he means the essence of a particular thing or a particular experience.[11] In novels, endings in effect "freeze" certain moments representing human experiences, thus defamiliarizing them for the reader and making their essence truly felt. Now the experience "frozen" by the ending of *Women in Love* is in its general contours, common enough. We may not have quarreled with our mates or with someone else very close to us about the need for male intimacy (as Birkin does here), but we have all participated in deeply felt quarrels about matters which have been quarreled about before and will be quarreled about again, matters which, as in this case, frequently can have no satisfactory resolution. Moreover, we have all, like Birkin, had recourse to obstinate and desperate refutation when faced with distasteful and frustrating truths: "I don't believe that," says Birkin—the perfectly irrefutable, but also perfectly impotent cry. The ending of *Women in Love* is, then, important in the reader's experience of the novel and to his perception of themes. It is also relevant to the reader's experience of life since those themes illuminate genuine life experiences.

The ending is also, of course, highly relevant to Lawrence's own experiences, as readers familiar with Lawrence will immediately recognize. It treats one of Lawrence's most obsessive concerns—the question of perfecting male-female relationships. Here, as elsewhere in Lawrence's writing after World War I, perfect harmony is *almost* but not quite achieved: here, because of an irresolvable and recurring conflict; in later works like "The Man Who Died" and *Lady Chatterley's Lover*, because of an enforced separation which ends the novel before the lovers can dwell in what (in Lawrentian terms) is paradisal bliss. The ending also reflects Lawrence's discovery of and attempt to cope with his own yearnings towards male friends (specifically John Middleton Murry), yearnings which might be called homosexual, but which are by no means entirely or even directly sexual. It reflects too, Lawrence's failures in this to him crucial area of life, and his deep frustration at those failures. Finally, the ending reveals Lawrence's most characteristic definition of marriage, a definition which had come to include continuing elements of conflict. Indeed, Lawrence had even come to elevate conflict into a *condition* of a desirable union, perhaps as a way of idealizing his troubled life with Frieda, his wife. In *Women in Love*, Lawrence believes that intimacy and harmony will inevitably at times break down, to be reformed and reexperienced, and then once again to break down. Even if we knew very little of D. H. Lawrence's other works or of his biography, the ending would provide reliable clues to many of Lawrence's experiences and ideas. D. H. Lawrence's sense of an ending informs his narration and reveals his preoccupations, just as the narrations of our hypothetical freshman and doctor revealed their senses of endings and preoccupations.

The effectiveness of the ending, its rightness as the conclusion to this novel is evident. Many critics have noticed the novel's startlingly symmetrical and patterned treatment of four primary relationships: those of the two sisters, Ursula and Gudrun; of the two men, Birkin and Gerald; and of the two pairs of lovers, Birkin and Ursula, and Gerald and Gudrun. The importance of the ending largely depends upon the relationship between Ursula and Birkin, and on that between Birkin and Gerald. The remainder of my discussion of closure in the novel will treat first the relationship of the lovers, and then that between the friends, as it influences the outcome of the love affair between Birkin and Ursula.

From the time that Birkin and Ursula begin to recognize their attraction to each other, the history of their relationship is one of progress towards greater intimacy, towards the star equilibrium of which Lawrence so often speaks. Each stage in that progress is marked by a quarrel, followed by a reconciliation. Their first significant encounter, in Ursula's classroom, has Birkin insist that a hesitant Ursula teach her botanical lesson by marking in vivid reds and yellows the female and male parts of a catkin. It ends with Ursula deciding, with some hostility, that Birkin "sounded as if he were addressing a meeting" (p. 37). Repeatedly in the course of the novel, Birkin and Ursula engage in spirited and sometimes angry debate; or, it might be more accurate to say, Birkin expounds his views, with Ursula a kind of petulant straight man. Instances include the discussion of the male cat Mino's domination of the female stray, the nocturnal dialogue in "Moony," and the quarrel over Hermione in "Excurse." In fact, the narrative voice speaks with admiration of Birkin's and Ursula's ability in "rousing each other to a fine passion of opposition" (p. 118). As I have already noted, Lawrence incorporates quarrels and opposition into his ideal of male-female relationships: he believes conflict can be desirable in that it allows the lovers to be like two *separate* stars constellated in equilibrium.

As the novel develops, however, the reader develops a somewhat deceptive sense about the direction Birkin's union with Ursula will take. The novel persistently contrasts the love of Birkin and Ursula with the love affair of Gerald and Gudrun; indeed, this is the most basic structural principle in the novel, one noted by almost all its commentators. The two couples' first meaningful sexual encounters are, for example, described in important parallel, yet contrasting terms. Birkin and Ursula experience what Lawrence describes as "vital, sensual reality" in clearly mutual enjoyment. Particular emphasis must be placed on the word *vital*, as the progress of their love is a progress towards life amidst a civilization in love with death. They are to each other following sexual union "the immemorial magnificence of palpable, real otherness" (p. 312)—Lawrentian terms of approbation if I've ever seen them. In the very next chapter, however, significantly titled "Death and Love," Gudrun experiences "the terrible frictional violence of death" while making love to Gerald. And Gerald feels as if he is once again made whole by lovemaking, "suffused . . . as if he were bathed in the womb again." The language used to describe Gerald's experience *might* be positive but in the novel simply isn't: lovers should come together as two separate wholes, not *to make each other whole*. And the image of Gerald as in the womb is a regressive one when compared to numerous and more vital birth images used for Birkin and Ursula. To a considerable extent, then, the reader is led by the novel to view Birkin's and Ursula's love as an ideal relationship to be emulated; the first-time reader and even sophisticated readers and critics come, moreover, to expect the lovers to fully reach the state of harmonious "paradisal bliss" they approach at moments in the novel. [12]

In the love scenes between Birkin and Ursula during the second half of the novel, they seem very, very close to this harmonious, paradisal state. I am thinking here of scenes like that in the Inn when they decide to marry; like that aboard ship as they cross to the continent; and like that in their room after the dance in the Tyrol. Let me give two good examples of moments almost paradisal between the lovers. First, aboard ship in "Continental":

> They seemed to fall away into the profound darkness. There was no sky, no earth, only one unbroken darkness, into which, with a soft, sleeping motion, they seemed to fall like one closed seed of life falling through dark, fathomless space. . . .
>
> In Ursula the sense of the unrealized world ahead triumphed over everything. In the midst of this profound darkness, there seemed to glow in her heart the effulgence of a paradise unknown and unrealised.

Second, a passage from Ursula's point of view later in the chapter, after a sexual encounter:

> They might do as they liked—this she realised as she went to sleep. How could anything that gave one satisfaction be excluded? . . . How good it was to be really shameful! There would be no shameful thing she had not experienced. Yet she was unabashed, she was herself. Why not? She was free, when she knew everything, and no dark shameful things were denied her.

In these quotations, it is evident that the relationship between Birkin and Ursula has progressed to an intimacy sometimes highly sexual and sometimes beyond mere sex, an intimacy untouchable and serene.

Like Scheherazade's husband, we all want to know "what happens next." And I expect that at least on a first reading, many of us are persuaded by the moments of bliss experienced by Birkin and Ursula and hope that what will happen next is the state extolled by fairy tales: "And they lived happily ever after." But Lawrence does not really want us to believe that all will be fine and harmonious between Birkin and

Ursula "ever after." By ending his novel with a scene of intense conflict, he prevents the "progress" of Birkin and Ursula from going too far, from being overly optimistic and even rather saccharine. Indeed, we can perhaps best sum up the true direction of the relationship by a quotation from very near the end of the novel. The narrative voice says the following:

> between two particular people, any two people on earth, the range of pure sensational experience is limited. The climax of sensual reaction, once reached in any direction, is reached finally, there is no going on. There is only repetition possible, or the going apart of the two protagonists, or the subjugating of the one to the other, or death. (p. 443)

Certainly, this is a dismal prognosis. Gerald and Gudrun choose the last option—death. More life-affirming characters, Birkin and Ursula choose the first, but still rather depressing option—repetition.

There is, then, a saving humility about the ending, a sense of discontent and failure despite the intimacy enjoyed by the lovers, which is just right at the end of *Women in Love*. In a world poised on the brink of apocalyptic destruction—as is the world of the characters—the desire to escape to life, to the "paradisal bliss" envisioned by Lawrence must be acknowledged, but must also not seem too good to be true.

Because of the structural emphasis on the parallel relationships of Birkin and Ursula/Gerald and Gudrun, the attraction between Birkin and Gerald implicitly receives less attention in the novel than do the heterosexual attractions. When the novel does confront the developing love of Birkin and Ursula with the fact of Birkin's need for an intimate male friendship, however, the quarrels between Birkin and Ursula end not in reconciliations of the kind exemplified in "Moony" and "Excurse," but in knotty, persistent conflict. I direct your attention to the end of the chapter called "A Chair," shortly before Ursula's and Birkin's marriage:

> His face was full of real perplexity.
> "Don't I?" he said. "It's the problem I can't solve. I know I want a perfect and complete relationship with you: and we've nearly got it—we really have. But beyond that. Do I want a real, ultimate relationship with Gerald? Do I want a final, almost extra-human relationship with him—a relationship in the ultimate of me and him—or don't I?"
> She looked at him for a long time with strange bright eyes, but she did not answer.

In effect, this unharmonious exchange is a dress rehearsal for the scene which ends the novel, and is, in fact, very close in spirit and dialogue to that scene. By the end of the novel, Birkin's love has become totally hopeless, however, and Ursula's anger and Birkin's insistence and frustration have proportionately grown in intensity.

In the book as it stands, Birkin's desire for Gerald is a kind of ant at the picnic, a fly in the ointment, an annoying yet inevitable block to perfect harmony between Birkin and his wife. The ending reasserts the importance of the issue of male/male relationships to Birkin and to Lawrence; the issue cannot be dropped, even when its fulfillment becomes impossible. The note of discord it produces persists after the ending of the novel.

Even within the novel as it stands, then, the problem of establishing male intimacies remains real enough and problematic enough. However, around 1916, Lawrence wrote a prologue to the novel, which he later cancelled. The prologue was first published in the 1960's, with an introduction by George H. Ford.[13] It is a very interesting document, particularly in terms of closure. In it, Lawrence establishes as the novel's first issue not male/female marriage (the subject currently discussed by Ursula and Gudrun in what became the finished novel's opening chapter), but love between men. Let us look at a few examples from the prologue:

> All the time, he [Birkin] recognized that although he was always drawn to women, feeling more at home with a woman than with a man, yet it was for men that he felt the hot, flushing, roused attraction which a man is supposed to feel for the other sex. Although nearly all his living interchange went on with one woman or another . . . yet the male physique had a fascination for him, and for the female physique he felt only a fondness, a sort of sacred love, as for a sister.
>
> In his mind was a gallery of such men [men of the Nordic and dark-skinned types]: men he had never spoken to, but who had flashed themselves upon his senses unforgettably, men who he apprehended intoxicatingly in his blood.
>
> He asked himself, often . . . would he ever be appeased, would he ever cease to desire these two sorts of men. And a wan kind of hopelessness would come over him, as if he would never escape from this attraction, which was a bondage.
>
> For he would never acquiesce to it. He could never acquiesce to his own feelings, to his own passion.

As these examples will indicate, the prologue is extremely frank in raising the issue of homosexuality. Given cultural strictures in the 1920s against homosexuality or anything remotely like it (strictures which Lawrence interestingly shared),[14] Lawrence would have been inviting disaster in using this prologue. Moreover, if the novel began by posing Birkin's attraction to other men as his central problem and ended with his quarrel with Ursula over the attraction, the focus of *Women in Love* would be rather skewed. The circular exploration of the theme of male love would give certain moments in the book greater emphasis than they currently have: moments, for example, like the end of the chapter called "In the Train," and, of course, moments like the nude wrestling match. In fact, the prologue would, I think, make the issue of male/male relationships dominate the issue of male/female relationships which is really the heart of the novel. Among other reasons for removing the prologue, then, Lawrence might have rejected it in order to achieve the book's final balance and focus.

At the same time, however, the ending restores to the novel some of what was clearly Lawrence's original intention. It prevents us from thinking at any level that the "problem" of Birkin's love for Gerald has been "solved" by the fruition of his love with Ursula. It ends the book with an appropriate and significant conflict and underscores the fact that *Women in Love* explores relationships between men and women, and men and men, which are thoroughly and unrelentingly problematic.

Lawrence's most characteristic belief about the novel was that it must be moral art, art relevant to the ways men and women think and to the lives men and women lead. He makes good that claim everywhere in *Women in Love,* but he emphasizes it dramatically during closure, and in the ending. In turn, closure in and the ending of *Women in Love* provide a clear, but not atypical, example to support my own claim: the presence of endings and closures underscores analogies between the shape of fictions and the shape of lives, analogies basic to our impulse to read fictions and to our sense of their importance.

Notes

1. "Preface to Roderick Hudson," rpt. in *The Art of the Novel,* ed. R. P. Blackmur (New York: Scribner's, 1907, 1962), p. 6.

2. E. M. Forster, *Aspects of the Novel* (New York: Harcourt, Brace and World, 1927, 1954), pp. 149–69.

3. Frank Kermode, *The Sense of an Ending* (New York: Oxford, 1966), p. 148.

4. Barbara Herrnstein Smith, *Poetic Closure* (Chicago: Univ. of Chicago Press, 1968); David Richter, *Fable's End* (Chicago: Univ. of Chicago Press, 1978).

5. Forster, *Aspects of the Novel,* pp. 26–27; in James, see p. 6.

6. *The Histories,* trans. Aubrey de Selincourt (Baltimore: Penguin, 1965), p. 48.

7. "Narrative Fictions and Reality," *Novel,* 11 (Fall, 1977), pp. 40–50.

8. Dr. Gordon Bevans called my attention to experiments in clinical psychology analogous to the experiment outlined below, which I composed independently. See L. A. Zurcher, Jr., *The Mutable Self* (Library of Social Research: 1977).

9. The intrinsic genre is that sense of the whole which allows us to appreciate the role of any part in the unity of the whole. I borrow the term from E. D. Hirsch Jr., *Validity in Interpretation* (New Haven: Yale Univ. Press, 1967), p. 86.

10. D. H. Lawrence, *Women in Love* (New York: Penguin, 1920, 1977), pp. 472–73.

11. "Art as Technique," rpt. in *Russian Formalist Criticism: Four Essays,* trans. Lee Lemon and Marion J. Reis, (Lincoln, Nebraska: Univ. of Nebraska Press, 1917, 1965), p. 32.

12. The idea of the ending to *Women in Love* as a kind of wish-fulfillment for Lawrence, one in which he projects true happiness for himself and Frieda via Birkin and Ursula, has proven surprisingly persistent. Its latest appearance is in a favorably reviewed major study of Lawrence's life during World War I. See Paul Delany, *D. H. Lawrence's Nightmare: the Writer and his Circle in the Years of the Great War* (New York: Basic Books, 1978), p. 226.

13. George H. Ford, "An Introductory Note to D. H. Lawrence's Prologue to *Women in Love,*" *The Texas Quarterly* (Spring, 1963), pp. 92–97.

14. Lawrence's letters express dislike for the homosexuality of the Bloomsbury group, several of whose members served as models for characters in *Women in Love.* See Delany, pp. 87–89.

Issues of Music and Time

Hindu-Buddhist Time in Javanese Gamelan Music

J. BECKER

> "Music is not only an art that employs or occupies time: I suggest that music is also a model of time and that the rhythm of music gradually comes to reflect cultural ideas on the nature of time."[1]
>
> —Lewis Rowell

We have all heard it said that "music is a universal language." What is meant by this statement is not that *any* music is universal, but that western classical music is universal. The distress and awkwardness of a music-loving, educated Indian at a symphony concert is matched only by the discomfort of an intelligent American at a Chinese opera performance. Musical events are profoundly culture-bound. Given enough time, one can come to appreciate the arts of an alien culture (partly by superimposing one's own set of values and aesthetics upon the listening act), but one cannot be taught to hear music as someone from another cultures hears it. Too much cultural background, too many unstated, often unstateable presuppositions are embedded within the situation of music making and music hearing. Listening to a musical events from another culture is the same kind of act as reading a poem from another culture. One may comprehend all the words (notes) and yet somehow miss the meaning. Because cross-cultural understanding is ultimately impossible does not mean that one should not try. This paper is such an attempt, an effort to briefly sketch out a few of the many underlying assumptions which provide the cognitive context, the source of richness of meaning for a performance of Javanese gamelian music.

The gamelan ensemble consists of anywhere from five to thirty bronze gongs and xylophones, with a drum or drums to regulate tempos (Figure 1). Since the 14th century when Islam became the predominant faith in Java, up until the present-day, there has been an uneasiness, a wariness by Javanese Islamic leaders concering the attachment of the Javanese to their ancient musical traditions which by extension means their ancient Hindu-Buddhist religious beliefs. Part of the problem for the *ulamas* (Islamic religious leaders) of Java is that the gamelan ensemble often accompanies dramas of Hindu origin and philosophy such as the Mahabharata and Ramayana. Beyond this, there is a suspicion, not clearly articulated, that the music itself runs counter to Islamic doctrine. I believe these critics are right; the music of the gamelan is deeply imbued with Hindu-Buddhist conceptions of reality, and foremost among these is the concept of time. Before analyzing the time organization

Figure 1. A Javanese *gamelan* ensemble.

of a particular gamelan composition, I would like to turn my attention to the general question of time organization within musical events.

Most of the world's cultures present more than a single idea of time in musical events. Some music may be strongly linear with a dynamic thrust forward that compels movement from the beginning to the end (the First Movement of Beethoven's Fifth Symphony). Other music may be basically cyclical, like the American blues form, with linear elements primarily in textual changes. With different musical genres or different styles of music, different temporal conceptions are displayed. If the philosophers are not of one mind concerning the nature of time, the musicians are not either. The nonverbal character of ideas of time as present in music allows, I think, a greater diversity of conceptual notions. I doubt whether one would ever be asked, which time concept is truer in Ravel's *Bolero*: the recurrent cycles of rhythm, harmony and melody, or the linear changes in dynamics and instrumentation. Not only is one forced not to choose, but the idea of evaluation itself seems absurd. "You cannot argue with a song."[2]

The everyday, mundane experience of time is highly dependent on context, what one is doing or what is being done around one. We use such expressions as "time passed quickly," or "the hours dragged by" to convey a sense of time as passage, as movement. Paradoxically, if many events or changes happen between one week and the next, we perceive time as having passed quickly while we may simultaneously feel that the preceding week was "a long time ago," i.e., many changes occurred. There is nothing homogeneous or consistent about our experience of time. No moment is the same as any other moment. Each has a unique feel to it, a special meaning. Another paradox in our experience of time as change lies in our relationship to what we see changing. We witness the change of seasons, our friends growing older, people being born and people dying, yet generally we do not feel ourselves changing (or feel it only mildly). We are surprised to see grey hairs in the mirror, or to notice the changing attitude toward us of students who are always the same age. We may experience a duration of time as rich and thick, or poor, thin and empty. Time can also "stand still" as in moments of intense concentration.

The various experiences of time in everyday life as described above are the kinds of time that are most often represented in musical events. A musical passage can metaphorically represent time passing quickly; slow time; rich, dense time; or sparse time. Music can also present teleological versus non-teleological time. Tonality is a perfect metaphor for teleological time with its inherent insistence on a return to the tonic. Non-teleological time can also be expressed tonally as the use of the electronic fade-out demonstrates. Cyclic time and linear time are represented musically and elements of both are found in all music systems. Repetition of one element while another element changes is a fairly universal way of structuring musical events. Western musical traditions prefer a cyclic metric structure with a superimposed linear melodic structure. A ballad has a strophic, cyclic metric, harmonic and melodic structure with a linear text structure. As far as I know, all cultures have a full range of musical possibilities for expressing many kinds of everyday, humanly experienced time. Tempo, rhythm, texture, tonal quality, dynamics, harmony, melody, etc., any one or all of which may be manipulated as a metaphorical statement of man's time experience as change.

There is another sense of time, also experienced, which is not dependent upon the perception of change, but rather its opposite, no change, the intimation of eternity, of Nirvana, the sense of being outside of any time framework at all. This is the time invoked by the music of many religious rituals, the "time out of time" one may experience in prayer, in meditation or in trance. The experience always seems to be accompanied by great joy and peacefulness, a feeling of sanctification and blessedness. Among the Australian aboriginees, eternity is the reenactment of original creation, a going backwards in time, or bringing primordial time into the present so that the individual may live his mortal life within a mind-set of eternity.[3] For the Christian or Moslem, eternity lies after the final judgment and must be anticipated through good deeds, ritual and prayer. For the Buddhist, timelessness is possible at any moment and comes with the attainment of enlightenment and the release from the timebound cycles of birth and death. The notion of timelessness, while not usually in the forefront of our minds, has, nevertheless, a universality that clearly allows us to identify it as one of man's common modes of perceiving time, if one can forgive the linguistic and logical violation involved in calling timelessness a kind of time.

Contradictory as it may seem in an art form usually described as timebound, musical means for expressing timelessness are found worldwide. Trance music is one example. Most trance music involves an unvarying musical cycle of fairly short length. However, not all trance music specifically is of this type. For some people, listening to music or performing any kind of music may induce a mild trance or meditation. Certain kinds of musical structures, such as unvarying cycles, or certain textures, such as the human voice, a fiddle-type instrument or an oboe-type instrument are considered more appropriate than other kinds of instruments and formal structures for inducing trance or for representing the eternal. Much ritual music, such as the long single lines of Gregorian chant, the long, slow periodic phrases of the voices of women who induce trance in Bali; the displays of overtone structure of an intoned syllable of Tibetan Buddhist chants, contain elements which apparently project one into a timeless consciousness, allowing one to be open to an understanding of the holy, the sacred, the timeless.

The central Javanese gamelan piece I will analyze here is a kind of timeless music, a relic from the pre-Islamic, Hindu-Buddhist era of Javanese history. The piece is called *Langen Bronto*, a Sanskrit title which means something like "passionately attached to Beauty." It is representative of a fairly sizeable genre of old-fashioned gamelan compositions, which display the same kinds of temporal and melodic characteristics. These kinds of pieces, while considered somewhat archaic, are still popular today as ritual pieces, as processional pieces, and as accompaniment for warrior dances.[4] *Langen Bronto* is generally repeated many times over with no pause between beginning and ending, and with a constant speed and dynamic level (Figure 2).

The endless cyclic repetitions, the unvarying tempos, the steady dynamics and textures are not, as some might suppose, a consequence of enfeebled imagination. Staticness is precisely the intent of this piece. I believe it represents a specific kind of meditation aid, a manifestation of a Hindu-Buddhist view of the nature of time, an aural *mandala*. Pieces such as *Langen Bronto* are anachronistic in one sense—many modern Javanese no longer subscribe to the epistemology in which this piece, and others like it, was conceived. Like Westerners, some modern Javanese, including some musicians, find these pieces "boring." More modern styles with incessant change and much melodic elaboration are preferred. Yet the fact that old pieces like *Langen Bronto* continue to be played, not as historical artifacts but as part of a viable tradition, demonstrates that the older ideas of time (and ultimately, older religious beliefs) may still evoke a response in modern Javanese listeners and musicians.

Large-scale conversion to Islam began in Java after the 14th century when many local princes embraced Islam. Before that, Java had been under the cultural influence (though not the dominion) of a series of powerful civilizations in India. Hinduism and Buddhism came to all Southeast Asia through Brahmin priests and Buddhist monks from roughly 500 A.D. to 1400 A.D. The two Indian faiths were not kept clearly separate in Java, nor were they always distinguished from authochthonous beliefs. In spite of the present widespread conversion to Islam and Christianity, Hindu-Buddhist beliefs can still be found in Java, particularly in artistic forms such as the shadow-puppet theater; in music; in dance, and in the doctrines of mystical meditation groups.

While the instruments and the music of the gamelan ensembles are indigenous, the development and elaboration of this indigenous music appears to be heavily influenced by Hindu-Buddhist concepts. Sanskrit words are common in the titles of pieces, in certain of the technical terminology and Hindu-Buddhist concepts are frequently encountered in the verbalized meanings and ethos of the gamelan. More than this, the musical structures themselves appear to be affected by Indian philosophical concepts.

The Hindu-Buddhist doctrine of time as infinitely recurring, immensely large cycles is a centerpin, an axis to which many other related doctrines, such as the idea of selfhood are linked. It is the intellectual edifice which underlies, supports, and

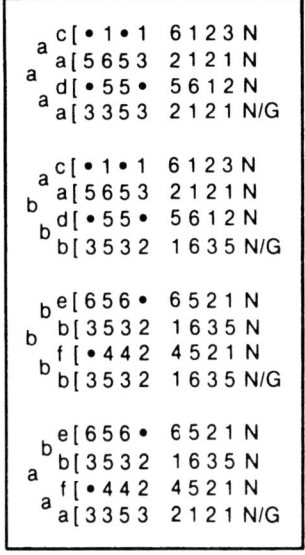

Langen Bronto in Javanese
Kepatihan notation.

Langen Bronto

Figure 2. I have analyzed the melodic structure of *Langen Bronto* according to the Javanese way of characterizing units by their endings.

The analysis consists of three hierarchic levels. The first level is what is called the gong unit (*gongan*) marked at its ending by a stroke on the largest gong and marked by G in the transcriptions. The gong unit consists of two lines in the western transcription, four lines in the Javanese cipher notation.

Each gong unit is divided into four kenong units (*kenongan*) marked at the end by a stroke on the kenong, marked N in the transcription.

The gong units marked *a* end with the kenong unit (in cipher notation) 3353 2121, or in western notation *eege dcdc*. The gong units marked *b* end with the kenong unit 3532 1635, or *eged caeg*.

The next hierarchic level of the analysis is half of each gong unit, one line of western notation or two lines of Javanese cipher notation. This unit is also characterized by its ending pattern. The *a* unit is, in cipher notation, either 5653 2121 or its variant 3353 2121, or, in western notation, *gage dcdc* or its varient *eege dcdc*. The *b* unit at the second level is 3532 1635 in cipher notation or in western notation *eged caeg*.

The third hierarchic level of the analysis consists of ¼ of each gong unit, 1 kenong unit, half a line of western notation, or 1 line of Javanese cipher notation.

The *c* unit is characterized by its ending, in cipher notation 6123, or in western notation *acde*, the *d* unit 5612 in cipher notation, or in westzrn notation *gacd*, the *e* unit in cipher notation 6521, in western notation *agdc*, the *f* unit in cipher notation 4521, in western notation *fgdc*.

sustains ideas about the nature of reality, the nature of man, and his relationship to all the universe.

Time is believed to be an illusion of the phenomenal world, a deception of our senses which we must overcome if we are to clearly understand our world. Because mankind continues to be seduced by the illusory phenomenal world, the illusion of time is perpetuated endlessly. A vivid metaphor is employed in Indian mythology to illustrate both the immensity of time cycles and their ultimate dissolution.

The god Brahma sits on a lotus, which emerges from the body of the sleeping god Vishnu. At Vishnu's feet, sits his wife Laksmi. Brahma, Vishnu, and Laksmi are all floating on a raft-like couch which is actually the body of a mythical snake, suspended in the middle of an endless cosmic ocean. As Laksmi strokes her husband's leg, he dreams; and from his dreaming emerges the lotus from which the god Brahma appears (Figure 3).

Every day (and night) of a Brahma lifetime (100 Brahma years), the god slowly opens and closes his eyes 1000 times. Each time they open, a universe appears, which dies away again as Brahma lowers his eyelids. The opening of Brahma's eyes creates a universe of four declining stages, after which the universe fades again into nothingness.[5]

The four stages of each universe are as follows:

1. The Krita Yuga: "a golden age of 4800 divine years in which the Cow of Virtue stands on all four legs and men are perfect in virtue" (one divine year = 360 human years.)
2. The Treta Yuga: "an age of 3600 divine years, diminished in virtue by one quarter, and in which the Cow of Virtue stands on three legs."

Figure 3. Vishnu, Laksmi and Brahma on the Cosmic Sea. Taken from J. Campbell, *The Mythic Image*, Plate 22, p. 140, from Edward Moor, *The Hindu Pantheon*, London: J. Johnson, 1890, Plate 7.

3. The Dvapara Yuga: "an age of 2400 divine years, virtue now diminished by one-half, with the Cow balanced on only two legs."
4. The Kali Yuga: "an age of only 1200 divine years, the Cow of Virtue tottering on one leg, a lawless age, declining toward catastrophe, our own age."

At the close of each Brahma lifetime (approximately 311,040,000,000,000 human years) Brahma and the lotus and the sequence of universes dissolve again into the body of the dreamer.[6]

Like all Hindu-Buddhist doctrines in Java, this one was reinterpreted and re-worked to include Javanese elements, i.e., naturalized and no longer associated with India. The extant Javanese documents which bear witness to the presence of these beliefs about the nature of time, although of 19th century provenance, are not well-known in Java, and are not among the historical/mythological documents studied in schools. They are available in museums—faded, curled, yellowed—but still treasured in certain literary circles.[7] But if the Javanese child does not ever read or have read to him the *Serat Manik Maya*, he may well hear the piece *Langen Bronto*.

Compared to most of the music which we hear, gamelan pieces are characterized by minimal tempo changes, rhythm changes, texture changes, and dynamic changes. The example, *Langen Bronto*, is an extreme representative of these characteristics. Each instrument plays at a constant rate of speed and dynamic level, continuing round and round the melodic/temporal cycle until the signal is given to end. Time does not go backwards nor forwards, no perceivable beginning again or feeling that the piece is finished. It is finished when it stops, only then. The musical cycle is not a background for something else which changes like the repetition of a strophic verse form. The unvarying cycle is all there is. By obliterating change and contrast, these pieces obliterate human time.

Within the all-encompassing cycles of gamelan music, other smaller aspects of time emerge which are also related to Hindu-Buddhist beliefs. In all gamelan music, one finds an overwhelming "four-ness." Musical periods, marked at their endings by a stroke of the largest gong, (marked with a capital G on Figures 2, 4 and 5a) consist of units of beats which number 4, or 8, or 16, or 32 (as in *Langen Bronto*), continuing on to 64, 128, 256, 512 and 1024 beats, all divisable by four. Most gong units are themselves divided into fours by a stroke on a smaller gong (kenong), as in *Langen Bronto*.

The *Yugas* are four in number; and the first and best is Krita Yuga, associated with the number four and signifying totality, complete and self-contained. In the Indian dice game, the dice-throw of four wins the jackpot. The Krita Yuga is the perfect yuga, conceived of as possessing all its four quarters and resting firmly, like a sacred cow, on all four legs. In this Yuga, "men and women are born virtuous. They devote their lives to the fulfillment of the duties and tasks divinely ordained by Dharma (Ideal Justice)."[8] Each of the succeeding yugas, Treta (three), Dvapara (two), and Kali (one) represent losses in virtue and harmony among men.

Mahayana Buddhism exhibits this same preference for four as the perfect number. On a temple ceiling in the holy city of Lhasa one finds a mandala, or yantra diagram, which could from the description serve as well as a spatial diagram of *Langen Bronto*.

". . . The personification in the center is the primal, eternal Adi-Buddha, or Vai-rochana. Radiating from him to the four quarters and the four points between are eight doubles or manifestations of his essence, differing in their special colors, gestures, and attributes. These denote the specific constituents going out from the immovable Absolute into the world. Illuminating and holding the universe, they are represented as contained within the heart of the cosmic flower. This, in turn, is set within the square sanctuary, and to each of the four quarters stands a meticulously pictured door. . . . Finally, the outermost rim of the lotus of the created universe is represented as a gigantic corolla of sixty-four varicolored petals."[9]

Mandalas, or yantra diagrams, in the Hindu and Buddhist faiths are meditation aids, symbols of wholeness, unity, completeness. Mandalas are perfectly symmetrical and balanced, circular spatial forms combined with square forms (Figure 4).

Langen Bronto acts as a aural mandala. The perfect balance of the melodic units and the large cyclic repetitions tend to focus concentration, to subdue distraction in

Figure 4. An abstract representation of *Langen Bronto* superimposed upon a mandala. Capital letters indicate strokes of gongs. Small letters indicate melodic content. G = largest hanging gong; N = large horizontal gong; P = large hanging gong.

the minds of players, dancers and listeners. By inhibiting the mind's tendency to follow and seek change, these pieces are inducive of meditative states. Ideally, at least, one leaves the realm of humanly experienced time and momentarily enters eternity. Many pieces displaying the mandala-type symmetry found in *Langen Bronto* are still associated with the central Javanese kingdom of Mataram, or present-day Yogyakarta. Another mandala, the 9th century Buddhist stupa Borobudur, stands within the hegemony of the former kingdom of Mataram (Figure 5).

Other aspects of Javanese and Balinese culture reiterate the idea of static time, or the immediacy of the present as against the demands of the past or future. In Bali, where Hindu-Buddhist religious traditions remain strong, the system of generational naming is cyclic and tends to minimize one's sense of personal uniqueness. There are generational terms applied to child, parent, and grandparent. After that the terminology repeats so that a great-grandparent has the same term as the child. At funeral rites, all younger relatives of the deceased are expected to pay homage to the spirit of the deceased except his great-grandchildren who are forbidden to do so. According to the Balinese, "they are the same age." Thus generational sequences do not mark the unique biological history of an individual life occurring in an unrepeating historical process, but rather illustrate a recurring cycle, longer than an individual lifetime, of familial descent.[10]

In language also, time as progression, sequence (as in before or after) is minimized. Balinese, Javanese and Indonesian are all tenseless languages. In any verbal form, no indication of time is inherent. Most Indo-European verbs have past, present, or future—often combined with aspectual markers—in every utterance. If the speaker of Javanese wishes to indicate temporality, he must add a separate time word such as "before," "now," "after," or an intention word such as "want" to indicate future action. The verbal form cannot be modified to indicate temporality. Temporality has to be consciously superimposed upon an utterance if the speaker feels it necessary.

Javanese syntax also discourages a sense of linear, progressive time. Narrative sequence, the idea that if two sentences are stated sequentially, the second occurred after the first, is not inherent in Javanese utterances. "I saw the kite" and "I blew the whistle" does not necessarily mean that the events occurred in that order. Temporality is always an added feature and is usually omitted altogether.

Chronicles of the reigns of kings in medieval and renaissance Java and accounts of current events in present-day Java are highly troublesome to translate into English with its inherent tense structure. Often in these manuscripts an event will be related several times over but each time from a different perspective. The unsuspecting translator often perceives these situations as sequential, separate events and links them with the English temporal conjunction "and then." The only way to convey in English the synchronicity of the events one finds in these texts is to interpolate explanatory passages.[11]

The coming of Islam to Java with the eschatological time of the ultimate judgment, plus the 400 years of Dutch occupation, has now made linear temporality an integral part of Javanese conceptual time. The older time sense remains strong, however, in indigenous calendrical systems and in art.[12]

The Hindu-Buddhist concept of time, as expressed in the ethos and music of the Javanese gamelan, is also a concept of self. The perfectly symmetrical, recurrent

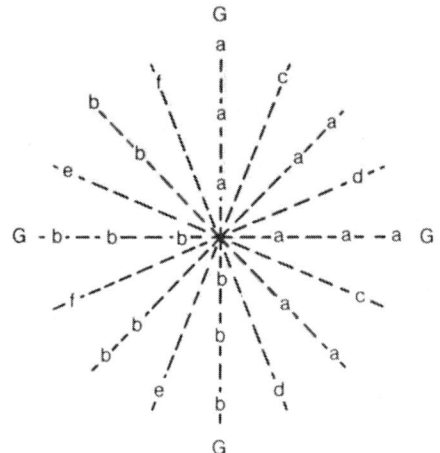

Abstracted melodic content of *Langen Bronto*.

Borobudur, Buddhist Stupa, 9th century, Java.

Figure 5. Aerial view of Borobudur, seen from southeast, Central Java. Taken from Claire Holt, *Art in Indonesia: Continuities and Change*, courtesy of Dinas Purbakala, Archaeological Service, Republic of Indonesia.

cycles of the music, the absence of dynamic changes, timbre changes, tempo changes, frustrate efforts to "express oneself." There is no place or space for personal display, creativity, originality. When playing or listening to gamelan pieces like *Langen Bronto*, a Westerner's reaction of the first few minutes is likely to be a feeling of impatience, a desire for some change. If one can relax the charged activity of the brain and settle into playing or listening to the piece, one gradually loses one's strong sense of personal uniqueness and comes to feel that the piece is playing itself. The particular identity of any particular musician, and especially, one's own identity, becomes irrelevant. One becomes a vehicle through which the music is played. It may be claimed that this is true of all good performances and that the musician taps resources of his and our minds which take us beyond personal ambitions. Of course that is true. But what does not apply to all performances is the fact that it is *only* by getting beyond the wish to be entertained that gamelan pieces like *Langen Bronto* can be appreciated at *all*. They demand a certain mental state from the performer and listener or they are intolerable. Since they also induce the necessary mental state, a kind of relaxed attentiveness, a surrender of the desire for stimulation, pieces like *Langen Bronto* can themselves create the context in which they may be appreciated.[13]

Indian religious theories, whether Hindu or Buddhist, reject the idea of a highly personal, unique selfhood as an illusion with dangerous consequences for the spiritual development of a person. It is only by seeing ourselves as a part of all other beings, by transcending the play of ego, that we may begin to hope for understanding. The Buddhists go so far as to reject entirely the idea of the personal self as being nothing more than an ephemeral collection of tendencies and inclinations which harden into habits. Stamped with ego's identifying insignias, these habits of thought, or feeling, of behavior, become the real, solidified "me." The kind of time of the dreaming Vishnu or the declining cycle of yugas unambiguously belittles the role of the individual. No human being affects, either for good or bad, the ultimate destiny of the universe. Universes rise and subside, are born, live and die without affect from anyone. Causation is far beyond the human realm. This stern view of the nature of the world and man's role in it is scarcely inspiring for one who sees the world as a stage on which to act out one's fantasies, fears, hopes and aspirations.

As the endlessly dreamed universes deny the significance of personality by their relentless cyclic recurrence, so does *Langen Bronto* discourage self-centeredness by the same devices. The emphasis on personal humility, on oneself as a vehicle for the continuance of the tradition, is a strong, verbalized element in the teaching of gamelan music. The aim of the ensemble is for no single musical line to stand out. Ideally, the members of the group become anonymous and essentially interchangeable. It is this aspect of gamelan music and music-making which often turns modern youth to other forms of music-making. What Javanese young people perceive as the quiescence of colorless old men does not, today, always inspire emulation. Within the epistemology of progressive, linear time sequences, gamelan·time and gamelan people may seem monotonous, nondynamic, noncreative and dull. On the other hand, to the old traditional musician, who is still within the mind-set of gamelan music, young people and their changing, linear, goal-oriented music seem restless, harassed, aggressive and unfocussed.

It may be that our ideas of what we are, as well as our ideas of time, are more or less constantly being nourished as we listen to music or perform it. Musical expression is a repository of culturally specific ideas which come to us subliminally, thus unselected, unfiltered and uncensored. Herein lies its power.

Notes

1. Lewis Rowell, "Time in the Musical Consciousness of Old High Civilizations—East and West," in *The Study of Time III*, eds. J.T. Fraser, N. Lawrence, and D. Park, (New York: Springer-Verlag, 1978), p. 578 ff.

2. Ernst Bloch, "Symbols, Song, Dance and Features of Articulation," *Archive of European Sociology*, 15 (1974): 71.

3. W.E.H. Stanner, "The Dreaming," in *Cultures of the Pacific*, eds. T.G. Harding and B.J. Wallace (New York: The Free Press, 1970), p. 304.

4. *Langen Bronto* can be heard in the repertoire of the gamelans which play at the annual *Sekaten* festival, as processional pieces for the entrance of *Bedhaya* dancers to the dance pavillon, and as accompaniment for the warrior dance *Lawung*. *Langen Bronto* appears on the commercial recording *Gamelan Garland*, Fontana, stereo 858614 FPY, recorded by Ernst Heins, printed in Holland.

5. Joseph Campbell, *The Mythic Image*, Bollingen Series C (Princeton: Princeton University Press,), pp. 140–144.

6. Ibid., p. 144.

7. Sartono Kartodirdjo, *Tjatatan Tentang Segi-Segi Messianistis Dalam Sedjarah Indonesia* (Yogyakarta: Penerbitan Lustrum ke-II, Universitas Gajah Mada, December, 1959). In this monograph, Sartono discusses, among others, the manuscripts *Serat Manik Maya* and *Serat Kanda*.

8. Heinrich Zimmer, *Myths and Symbols in Indian Art and Civilization*, J. Campbell, Bollingen Series 6 (New York: Pantheon Books, 1953), p. 13.

9. Ibid., p. 144.

10. C. Geertz, "Person, Time and Conduct in Bali," in *The Interpretation of Cultures*, (New York: Basic Books, Inc., 1973), pp. 374–375.

11. I owe my understanding of the time sense of Javanese grammar to my husband, A.L. Becker, Professor of Linguistics, University of Michigan.

12. Becker, J., "Time and Tune in Java," In *The Imagination of Reality: Essays in Southeast Asian Coherence Systems*, eds. A. Yengoyan and A.L. Becker (Norwood, N.J.: Albex Publishing Corp., 1979), pp. 197–210.

13. Gregory Bateson once made the remark that "all great works of art create their own context." Only now do I begin to understand what that means.

The Zones of Time in Music and Human Activity*

L. BIELAWSKI

The outstanding Polish logician and philosopher, Kazimierz Ajdukiewicz, believed that "time" has four different meanings: (1) "a moment, an exact date, a point of time"; (2) "a period of time, a span of time, a time interval"; (3) "duration, the length of a time period"; (4) "an all-embracing period of time." My analysis and interpretation of these meanings is seen in Figure 1.

The distinction between the second and third meanings is particularly important. In the 2nd meaning, a period is located within a temporal succession and is contiguous with neighboring periods on a temporal line between the infinite past and infinite future. Occurrence of separate phases within the context of temporal succession provides them with qualitative characteristics. The context can also be simultaneous, as in polyphonic music. Everything that exists in time exists in such a time context. All human acts; all musical, linguistic or artistic utterances, all lives; and all historical processes comprise the content of neighboring periods. Such a notion of time is well illustrated by Heraclitus, *panta rei*. All aspects of the time process can be seen in it, all syntagmatic relations are based on it. Such problems are common in the theory of art and music.

The subject of this study is time in Ajdukiewicz's third sense. Leading to a totally different set of problems, it deals with "duration, the length of a time period— different from the period itself." As Ajdukiewicz explains further, "two different time periods can have the same duration, just as two different segments of a line can have the same length."

Durations, however, can be not only equal but also different, *i.e.*, more or less similar to one another. Because of their duration they are somehow independent of their context. One can imagine a scale of all possible durations whose lengths would

*The Author and the Editors wish to express their appreciation to Prof. Garry L. Brodhead of Ithaca College, N.Y. for his valuable editorial assistance.

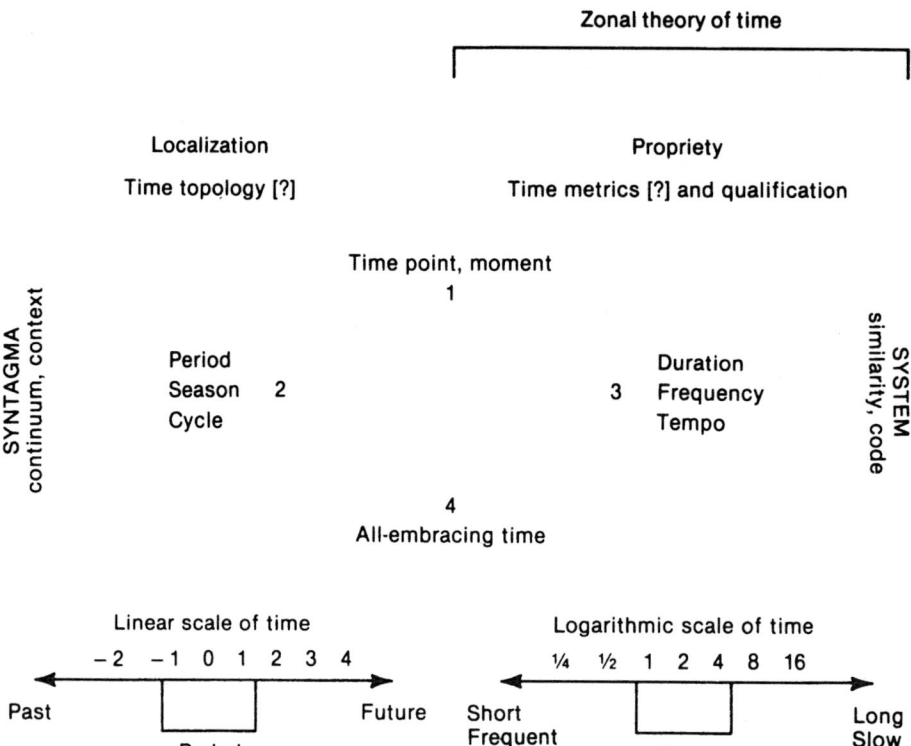

Figure 1. Four different meanings of time as distinguished by K. Adjukiewicz, seen from two opposite perspectives: the syntagmatic perspective, with its linear scale of time stretching between infinite past and infinite future, and the systemic perspective with its logarithmic scale of time extended between infinitely short and infinitely long time intervals (durations, frequencies, tempi). On the last scale, all temporal phenomena characteristic of music as well as all human activity and existence are associated with specific zones (ranges), which is the basis for the zonal theory of time.

range from infinitely short to infinitely long. Each real time interval is a choice of one value on this scale. At the same time it is an element on the scale of temporal sequence in the context of other time intervals. Time intervals, therefore, have dual links and can be interpreted in two ways. One is expressed through relationship to the code (in our case, to the scale of all possible durations), the other, through relationship to the context. Each of these links connects a time interval with a different set of intervals; in the former case, through alternation, in the latter, through alignment. A given time interval can be replaced by other intervals—fact reveals its meaning, while the contextual situation is defined through its relationship to other intervals in the same sequence, which also influences its meaning.

In his definition, Ajdukiewicz stresses duration. But we should not forget its counterpart, frequency or tempo. We perceive music in both of these categories and other time structures are not devoid of them either. We talk about the tempo of a theatrical performance, the tempo of historical changes, about varied tempi, etc. Obviously, the notion here is something other than the tempo of movement in space.

Musical tempo is abstracted from real space; it is a frequency, *i.e.*, a number of impulses or time intervals per unit of time. The natural scale which expresses a scale of similarities in a mathematical way is the logarithmic scale. Regardless of numerical size, proportion is marked by a constant unit on this scale; distance is therefore a natural measure of similarity. Application of this scale is further substantiated by the Weber-Fechner law—the basic law of perception, which ascertains that the quantity of impressions depends on the logarithm of an impulse. Although it is agreed that, in music, this scale applies to pitch and dynamics, it has not been realized that it also applies to such qualities as duration or frequency—that is, to time.

In understanding time, two different scales are applied, consciously or inadvertently: the linear scale of the continuum and the logarithmic scale used in evaluating the similarities between duration or frequency (Figure 2).

Thus, time has two dimensions for man. Analysis of all time's coursings and shapings can best be done on a plane with two coordinate axes: the axis of the temporal continuum, *i.e.*, the linear scale and the axis of similarities of duration and frequency, *i.e.*, the logarithmic scale. Here the multilayered quality of the time continuum reveals syntagmatic qualities of time, while a projection on the axis of interval similarity points to its systematic qualities. Time and space distinguish themselves by this dual nature. On the one hand they form a continuum (simultane-

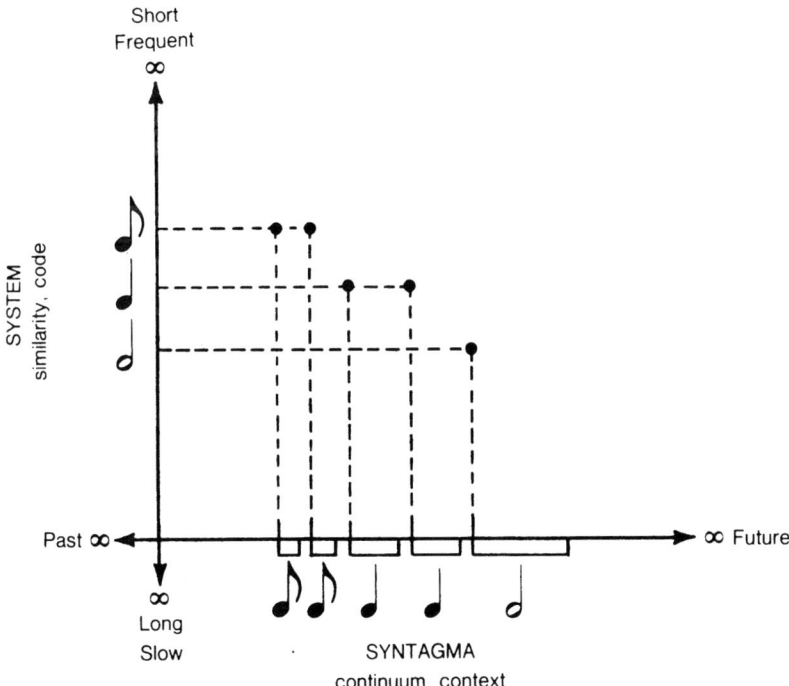

Figure 2. Two "dimensions" of time: time as succession of temporal intervals and time as duration or frequency. Each real temporal structure can be analyzed on a plane of two coordinate axes. The projection on the horizontal axis manifests the syntagmatic features of time, the projection on the vertical axis, the system of used time intervals and time zones.

ous and successive context) for all existing qualities, phenomena and structures. On the other hand they are some of the qualities of reality such as duration, frequencies, tempos, distances and proportions. Le Corbusier realized the qualitative scale of space when he defined his modulor, though he should not have limited it to the sequence of the golden proportion, thereby making it practically useless as a research tool.

The total scale of duration and frequency accessible to the physical sciences is comprised of 42 decades and is limited at one end by a time quantum—10^{-24} seconds, and at the other by the duration of the universe, approximately 20 billion years (Figure 3).

Homogenous and comparable phenomena fill specific zones or ranges on the scale. Seven primary zones deal with man and his activity, including artistic activity.

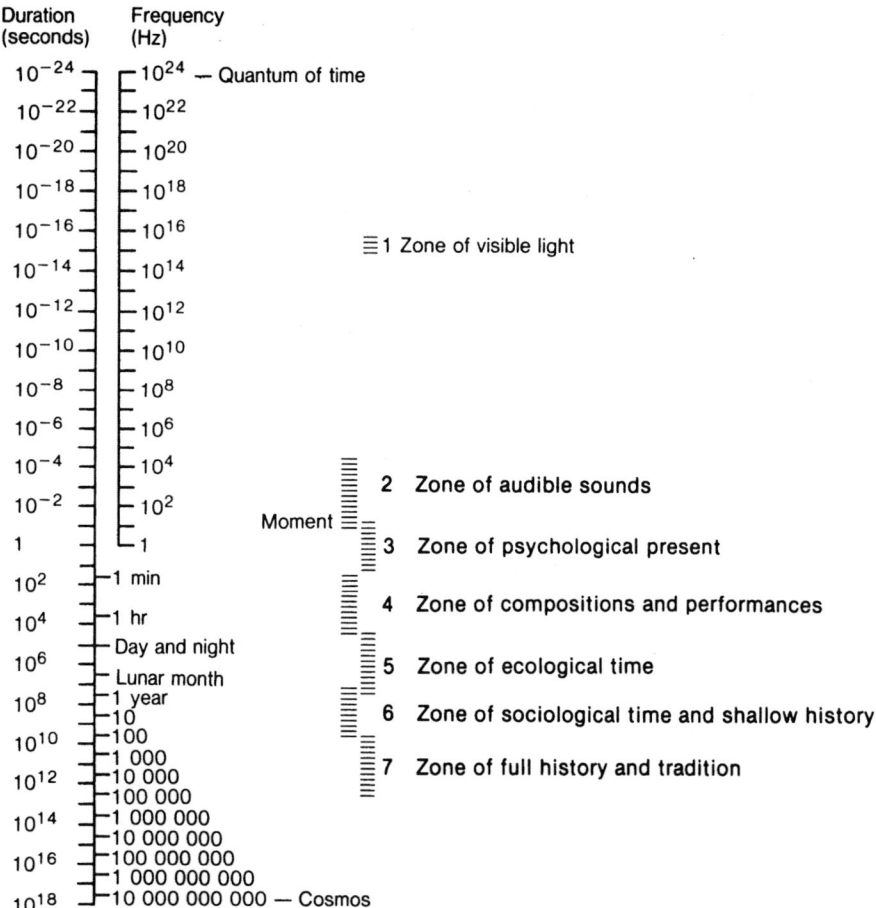

Figure 3. Seven principal temporal zones characterize human activity and existence. The first two zones are not perceived as time but as light or sound, though from the physical point of view they are temporal (frequencies of light waves and acoustic waves). The next four temporal zones are experienced as time. The last zone is certified only by tradition and history.

They are: (1) zone of visible light, (2) zone of audible sounds, (3) zone of psychological present, (4) zone of compositions and performances, (5) zone of ecological time, (6) zone of sociological time and shallow history, (7) zone of full history or cultural tradition.

Two of these zones have a temporal character found only within the realm of physics: the zone of light wave frequency which determines our ability to see colors—its range, small—and the auditory zone which provides for all auditory activities, particularly linguistic and musical ones. The remaining five zones are fully temporal in character.

The first of these, the zone of the psychological present, is the most important with regard to the time aspect of events; it determines the human perspective of time as it is the sphere of all activity (Figure 4).

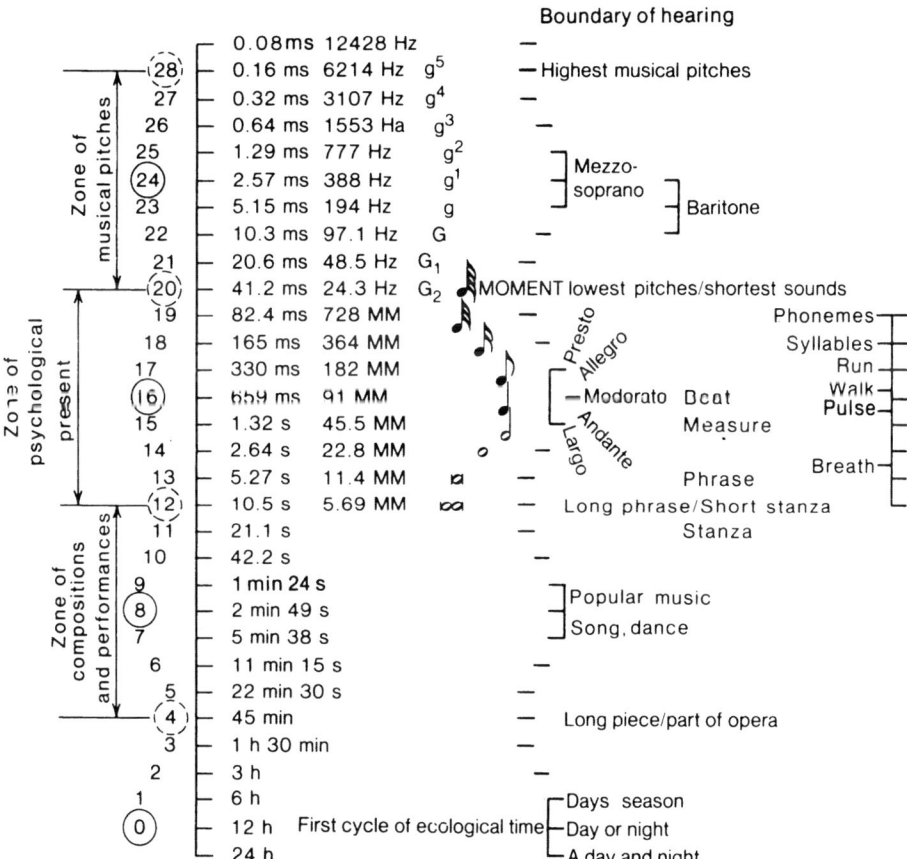

Figure 4. Three main zones of time are used in every musical activity: the zone of audible sounds (musical pitches), the zone of psychological present and the zone of compositions and performances. Each of these zones can be divided into sub-zones which fulfill a specified function in music and human activity. Localization on the logarithmic scale of time is shown in comparison to the day and night cycle, the first cycle of ecological time.

It is within its limits that we speak, play and listen (even though acoustic waves obviously have a different range of frequency). Within this zone we see (the frequency of light waves being altogether different), notice events, observe movement and move ourselves, use tools, paint, write, look at art works, react to impulses coming from the surrounding world and act upon this world. We can, of course, think of a different time but even this thinking will take place in the unchanging present which always accompanies our awareness of existence. The psychological present embraces a duration range varying from a human moment—approximately .50 millisecond, to approximately 10 seconds. In the middle of this zone is the medium musical tempo, moderato, approximately ⅔ second, or MM-90. This quantity can in fact be called a human second.

Within the large zone of the psychological present we can also distinguish sub-zones, or sub-ranges. They depend on the character of the phenomenon they deal with but, interestingly enough, certain ranges characterize many different phenomena which probably points to deeper and common determinants. In the psychological present, the sub-zone of language phonemes corresponds to the sub-zone of melodic ornamentation in music, which lacks distinct rhythmic function. The sub-zone of syllables has as its counterpart in music and dance the units of movement combined into basic metric units or beats. In experimental psychology it is the so-called "short-times." The central sub-zone is the most important. I identify it as the "immediate present." It contains basic logical units of the language: i.e., words, basic metric units and tempo units in music and dance, and in experimental psychology, so-called "long-times" or the times of a normal human reaction to outside impulses. Larger and more complex sub-zones are syntactic units: phrases and sentences in utterance, film sequences, segments of stage action.

All works and all human activity that develop in the psychological present have specific time dimensions which themselves belong to another zone—to the works and to the time in which they are performed, played, represented, danced, etc. It is hard to find a sufficiently general term for this zone. It ranges from approximately 10 seconds to a few hours.

The other time zones which are not a concern of this study are as follows: The fifth, also important in artistic activity, applies to ecological time, i.e., time set forth first of all by diurnal and annual rhythms and, to a lesser extent (at least in our culture), by lunar rhythms. The 6th zone is connected with recent history—that remembered by living generations. The 7th is the zone of complete history or cultural tradition.

So far, no argument has been given which would prove that musical time and its perception, or in fact human time as such, are conditioned by race. To the contrary, everything seems to affirm that this is not the case. Independent of race, people see within the same range of light waves, hear within the same range or acoustic frequency, and everything seems to speak in favor of their having the same range of psychological present. Also people have a common situation in the world, structured by the recurrence of days and nights, the succession of seasons, the cycles of human development and life.

Man has one constant perspective of time and space and he can have no more than an indirect notion of any other. By enlargement, for instance, we can see a world which is normally inaccessible to us; an electron microscope allows us to see the shape and structure of a cell. We see it as it is seen by another apparatus which has a different perspective and focuses on considerably smaller dimensions. Likewise, technology enables us to change our perspective on time. Through slow-motion photography we can see a runner's movements from a different perspective than the natural one; film projected at a quicker speed shows a bud becoming a flower. We know that the flower moves, but it does so too slowly for us to experience it; on the screen it appears real.

People can communicate because they are attuned to the same frequencies of light waves, acoustic waves and the same frequencies and durations of present time. Their perceptual organs are in accord, which makes it possible to send messages. Even a slight alteration can result in a lack of coherency within the informational systems. This can be observed when old movies, shot at 16 frames per second, are shown at the standard speed of 24 frames per second. The difference is not large but sufficient to make the scenes seem comic, artificial or somehow inhuman. More significant differences could result in a total inability to communicate. We thus have a constant perspective of time and space which establishes our communicative possibilities. Only within this framework can we choose, limit variety, get used to these limitations, and accept them as common to a certain culture or epoch. Cultural customs may change; special training may make us more sensitive to certain phenomena, yet it cannot change the basis of our existence in time and space and the way we perceive it, which is determined by our human condition.

Notes

K. Ajdukiewicz 1963. "Czas" ("Time"). In *Wielka encyclopedia powszechna PWN*, vol. 2, Warszawa, pp. 703−04.

L. Bielawski 1976. "Strefowa teoria czasu i jej znaczenie dla antropologii muzycznej" ("Zonal Theory of Time and its Significance for Musical Anthropology"). PWM ed. Krakow.

———— 1972; 1973. "Stefowy charakter czasu muzycznego" ("Zonal Character of Musical Time"). In *Muzyka. Kwartalnik poświecony historii i teorii muzyki*. IS PAN Warszawa, no. 4; nos. 1, 2.

Le Corbusier 1954. *The Modulor. A Harmonious Measure to the Human Scale Universally Applicable to Architecture and Mechanics*. Cambridge.

On Musical Continuity

D. Epstein

Introduction

In his paper on "The Creation of Audible Time" Lewis Rowell has developed with elegance some perspectives of musical time—perspectives of culture, of aesthetics, of structure.[1] This study will examine one of these perspectives in greater detail, probing the mechanisms by which musical time is structured in aspects of Western tonal music. Its concerns come naturally to those of us who spend our days making music, for the encounter with time structure is a continual one that faces us whenever we rehearse, perform, and indeed think about music. More often than not these issues lie at the heart of musical interpretation and, thereby, of musical preparation. For it is not the wrong pitch or the misplaced rhythm that worry musicians at work; they are easily heard and corrected. It is the ambiguities of nuance, of dynamics, of tempo, rubato, motion that call for clarification and decisions.

The bases for these decisions are rarely satisfactory. They tend to rest upon: 1) Personal authority—of teacher, coach, conductor ("It *must* go that way"); whether such authority is deserved is a matter in itself; 2) Tradition ("It's always done that way"), which brings to mind Gustav Mahler's famous dictum that "Tradition is *Schlamperei*," if it is not, as another musician put it, "the last bad performance"; 3) The text ("It says so in the score"), though just what the score says, and whether its indication is precise (a pitch, for example) or a suggestive symbol (a dynamic, an accent) may be debatable; 4) Intuition ("I feel it that way"). Intuition is the richest and deepest musical resource we have—in fact, the ultimate one. It works well, though it must be trained, of course. It does not work as well, however, when we come up against ambiguities that demand a cogent theorem or other rationale as a criterion for decisions. It is such a systematic view of temporal structure that we lack, in sufficient richness, in music. This study is a gesture in that direction.

On the face of it, beginnings and endings are reasonable demarcations of temporal experience. Is not the end in the beginning, as we have been told? Does not formal logic reveal that initial premises condition conclusions, much as initial circumstances affect the outcome of events? In making a duality of beginning and ending, however, we omit a third area of experience—that critical period which lies between the two extremes. That is, the passage of time itself—our experience of it and, in some cases, our structuring of this experience, which implies the powerful principle of continuity.

Continuity figures more significantly in some temporal passages than in others. It is of little interest in the seemingly "objective" measure of time, but more so in our evaluation of this measurement. Were an engineer, for example, to concern himself with a time/motion study based on the formula that Distance = Velocity × Time (involving, let us say, train travel between Paris and Amsterdam), Velocity and Time, the two temporal components in this formula, would be seen mainly as quantitative elements. There would be little concern about their qualitative nature. Were the same engineer to be a passenger on this train trip, however, his focus would incorporate qualitative concerns—the physical sense of speed; images of scenes as the train pursued its journey; fellow passengers met; conversations held enroute. Some of these events would be structured: the route; and the train's velocity over particular segments of track. Others might fall more to chance; some might be purely random (the order, for example, by which a group of passengers might line up in the corridor on leaving the train).

If our engineer was to change roles further, to become a dramatist and to write a play about this trip, the qualitative aspects of these experiences—and of time passed thereby—would become paramount: personal encounters, the interplay of personalities, the motivations, drives, feelings of the *dramatis personae*. Structure would immediately loom as a necessity in writing this play, for even the most powerful dramatic elements would not make for successful theatre unless they were controlled in their intensities, their energies directed, the flux and interplay among characters proportioned in terms of *dramatic time*, and all of these elements guided toward conclusions and resolutions. But all this implies more than structure *per se*. This is goal-directed structure—structure oriented in time and proportioned in time; structure which controls the flow of time through the molding and pacing of events as they move toward conclusions. It is, in other words, structure informed by a sense of continuity.

Our discussion touches the essence of music, for music is supreme among the time-oriented arts in achieving continuity through temporal control. One more stage in our train trip drama is necessary to bring home this point. Let us assume that this play has been written and is now upon the boards, its opening scene a discussion between two characters ensconced in their travel compartment, the lighting and dramatic ambiance of whatever mode you may wish. However effective the dialogue provided for these characters, the timings and pacing of its delivery would be open to considerable interpretation by the actors. Lines could be sped up, held back, drawn out, spaced by interpolated pauses unspecified in the script—all this by way of intensifying nuances, indeed even creating them, while none of it need be unfaithful to the text. The drama, in other words, while a major temporal art, and while structured with a control of temporal flow and of continuity, is nonetheless flexible in its delivery—in the all-important moment, that is, of performance.

The point is essential, and its reference to music now becomes clear. For music, too, is a temporal art in which flow, structure, and continuity are essential. However, no such flexibility in its performance could be taken by musicians, as was the case with drama, without wholesale distortion of the musical text. In this sense music is perhaps unique not only among the arts but in our experience of time itself. For music not only takes place in time; time is in fact a fundamental element of the art. There is no role here for time as an index of some other function, as was the case in our Distance/Velocity/Time formula. Music actually *structures* time—and the flow of time—in precise quanta and proportions, controlling this flow, its intensities, its direction, its speed, its goal orientation, to a degree unmatched in other domains of our temporal experience.

It would be well at this point to clarify the grounds within which this discussion moves. As with most matters involving human activity, we would be wise to eschew universal statements. Like language, art, custom and other areas of culture, music has its global aspects. They are so diverse, however, with many of them known in only approximate terms, that one speaks of universals in no more than general terms, and cautious ones at that. At this stage of the game, musical universals, though we may intuit them, make undependable yardsticks for precise studies.[2]

This study seeks a more limited model which, if it can be defined, may be extensible to other musical eras, possibly to other geographical musics. Our model is confined to Western art music largely of the German-Viennese tradition during the so-called classic-romantic period, as delimited by Haydn to Brahms (though it makes a brief excursion to French impressionism in one example). It is concerned, moreover, with instrumental music only; the problems of time structure in this literature are complex enough that it is well to avoid the further influences of language and prosody. The advantage of this model is its clarity; its forms, its elements (harmony, counterpoint, key structure, instrumentation, folk traditions) have been more studied and more codified than perhaps any other music.

Within this limited model ambiguities and confusions still abound, particularly concerning modes of temporal structure. Less is understood, in fact, of rhythm and meter, of the roles played by accent, dynamics, articulation, motion, tempo—all elements of temporal definition—than of any other aspect of this literature. Certain places in the repertoire are famous for the temporal ambiguities they pose. Where, for example, is the "true" opening downbeat in the G Minor Symphony (No. 40) of Mozart? Likewise, in the Second Symphony of Brahms? Why, in the music of Brahms, are the rhythmic strong points of melodies so often notated on weak parts of the measure—"written backwards," so to speak? In the Scherzo of the Beethoven Fifth Symphony, is the close of the first phrase (at the fermata) a downbeat or an upbeat bar? And so forth, through a large array of the most distinguished works. Indeed, to answer these questions some theory of time structure in tonal music is needed beyond what presently exists.

Toward a Theory of Musical Time[3]

Musical time is but a specific species of time in general. It would be helpful, therefore, in forming a theory of musical time, to take account of properties by which all manifestations of time, in all domains of experience, are ruled. Prime among these are: 1) the duality of time; 2) the intangibility of time; 3) hierarchic levels on which time can be structured; and 4) the need for—and the means of—temporal demarcation. From these properties there emerges motion, or flow through time, and it is this motion that demands and in turn leads to continuity.

Duality

Duality in time has often been referred to in philosophical writing in terms of objective/subjective time. The terms, in fact the concepts, seem inappropriate to music. For all time in music is subjective in the sense that it is experienced in a personal way. To speak of objective time, therefore, is confusing as it implies a time outside of music, time somehow not experienced. More meaningful distinctions for musical time would not involve objectivity or subjectivity but rather the orders, or varieties, of temporal experience created by music. These, too, are dual in nature and might be usefully characterized as "chronometric" and "integral."

Chronometric time, similar to "clock time," refers to that essentially mechanistic, evenly spaced and in large part evenly articulated time set up within a musical measure (and larger units as well). Its measurements and demarcations are in the main pragmatic and convenient periodizations.

Integral time, on the other hand, denotes the unique organizations of time intrinsic to an individual piece—time enriched and qualified by the particular experience within which it is framed. The mechanisms by which this temporal organization is established are likewise unique to a given work and must be studied anew in each case. Thus while the chronometric "beat" of two Mozart piano pieces of like tempos might be similar, the integral organization by phrases, section, motive, "breath" will be different in each work and cannot be usefully generalized. Given only the information that a particular Mozart piano piece is in $\frac{3}{4}$ meter, we could reasonably predict that this meter (part of its chronometric organization) probably extends to the end of the piece; we could say nothing however of its integral temporal character until we knew more of the music—its motives, themes, at the least.

Intangibility

The intangibility of time presents both contradictions and musical quandaries. Because time is intangible, it constitutes in our perceptual experience a dimension unique from the 3-dimensional world of space. That which is spatial, in music as in all else, is accessible to the senses; it is visible, audible, tangible. Further, it is quantifiable in terms integral to the senses, hence controllable: long, short; high, low; soft, loud.

The temporal by contrast is intangible, unavailable to the senses,[4] its perception partly a trained intellectual act. Yet any demarcation of temporal periods must rely upon tangible (and sometimes spatial) cues—the visual symbols on the clock face; the aural, sonic signals of the clock's tick, the metronome's click. Thus time in music poses a contradiction: in a world of phyical sound it is a basic element that is itself nonsonic, yet dependent for its projection and demarcation upon sound. More than any other musical dimension, time depends upon forces outside its own proper domain—that is, time depends upon sound.

Failure to specify this fact has resulted in much confusion in discussing time events in music. For while we think we are speaking of time, we are in fact speaking of sound events, not always making clear this distinction or this contradiction in terms. Furthermore, the emphases by which segments of time are marked off are of varied quality—beat and pulse, rhythmic, metric, agogic or harmonic means. Thus temporal statements are qualitatively different though not always qualitatively distinct. Even the criteria for discussing or perceiving temporal structures are neither fully clear nor consistent with the domain of time itself.

Hierarchization

Obviously time can be structured on differing levels of duration. It is only the means and the nomenclature that make this process different in music than in other fields. The dual aspects of musical time persist in this layering of duration, and can best be seen in the following chart.

MUSICAL TIME

Chronometric (Metric)	Integral (Rhythmic)
Beat	Pulse
Measure	Motive (or motive-group)
Hypermeasure	Phrase
Macrolevels (of hypermeasure groups)	Macrolevels (periods, sections, etc.)

Clearly, the metric/rhythmic duality provides complementary classifications on each successively broader level of duration.

In its smallest practical dimension the chronometric unit of musical time is the beat. In larger dimensions metric units group into measures, with their attendant arrangement of beats. In classic-romantic music, beat and meter were givens of a system that existed prior to any act of composition. Through convention and stylistic evolution they had been imbued with certain musical characteristics: duple and triple groups; preestablished distinctions between strong and weak beats, up- and downbeats; dance qualities associated with certain meters (and tempos). Emphasis within this domain is metric accent, largely mechanical and virtually automatic, associated mainly with those beats of a measure (or larger dimensional level) that are strong as determined by convention.

Beat and meter—the chronometric aspects of time—were more than purely mechanistic features of classic-romantic music. While they were givens of a preexistent system, at the same time they assumed life anew, set up, as it were by each work. Classical metricality not only established regularity of beat; it also established expectations of regularity within the perceiver. Regularity is one standard by which the passage of time within the work is understood.

Integral time, the other half of the temporal duality, is wholly bound up in the experience of a particular work, arising from the organization of time inherent in that work. This unique temporal organization serves, in fact, as a major premise intrinsic to the work.

The smallest unit of integral time, as contrasted with the chronometric beat, is pulse. The distinction between them is significant. For while we generally experience beat as precise and regular, stated with the blunt, immediate articulation of a click (foot, metronome), pulse is experienced with a far broader range of articulations and is understood, moreover, in intimate relation to bodily experience. Thus pulse may correspond to the gradual rise-emphasis-decay sensation of pulse in the circulatory system; to a similar cycle in breathing, where the tension-repose dimensions take longer to complete; to the precise, sudden jerk of a muscular spasm; to erotic movement, or to a host of other bodily sensations.

In all cases the experience of pulse has both physical and psychological connotations. Its nature and quality are intimately allied with articulation. Nor do all articulations necessarily fall into equally-spaced time patterns. Some will pull time or push it, stretching or compressing by small degrees the precise, even flow of chronometric time in response to the inner tensions and strains of a musical phrase. Thus a difference, and often a conflict, is experienced between the regularity of metrical beats and the slight deviations from regularity caused at times by the internal demands and the articulations of pulse.

Like chronometric time, integral time is also grouped in larger dimensions, as the chart indicates. Unlike meter, whose strong beats are in the main implanted precompositionally by means of a system already determined, the strong pulses of rhythmic groupings arise contextually from patterns intrinsic to a work. Because temporal phenomena cannot demarcate themselves, rhythmic strengths and weaknesses are affected by events in other domains, such as harmony—in its progression, tension and relaxation, stability and instability; by melodic contour; by cadence.

Thus time on all musical levels exists in a continual state of coordination and/or conflict, found in the relations of the metric to the rhythmic, with their attendant properties. Beat and pulse, the minimal units of time, often deviate slightly from each other, as we have seen. On larger levels, metric and rhythmic quanta are also dissynchronous at times, the pull and conflict between the two domains creating a continual balance of tensions until the ultimate resolution of final closure. At these points of dissynchrony it is the rhythmic, with its elements intrinsic to the piece, that prevails over the metric. By contrast the coordination, on middle levels of structure, of strong metric and rhythmic impulses can create structural downbeats—major points of simultaneous harmonic and rhythmic arrival so powerful that they turn what precedes them into their own upbeats.

This coordination/conflict relationship between the metric and rhythmic operates upon all levels. On the measure-motive level, motives or motive-groups generally lie within the confines of measure. The two domains are generally coordinate at this level; strong and weak pulses of motive usually agree with the metrically strong-weak beats of measures, the latter in fact determining the former. Beginnings (upbeats) and endings of motives may extend slightly beyond the measure or measures that frame them metrically.

Coordination/conflict relationships on the broader levels of hypermeasure-phrase are more complex. As with measure and motive, hypermeasures and phrases may or may not be congruent in their extreme limits; it is not unusual to find phrases with an anacrusis of almost measure length, thus creating different boundaries between the limits of phrase and those of the underlying hypermeasure.

These boundaries, whether congruent or otherwise, are not difficult to determine. More problematic is the determination of the emphasis properties that mark a phrase or hypermeasure—in brief, whether the phrase is downbeat or upbeat oriented, and whether this emphasis pattern is the same or different between the phrase and its complementary hypermeasure. Indeed this question is crucial, for it is downbeat and upbeat qualities at this level of structure that in large part engender musical motion—that ultimate result of temporal mechanisms.

Demarcation/Emphasis

We are led, then, to the fourth property of musical time essential to structure: emphasis, the means by which time is demarcated. "Emphasis" is used here as a generic term, encompassing all manner of intensification by which one particular point in time is distinguished from another and given greater prominence. Within this all-inclusive category there are three significant distinctions:

1) The difference between structural and ornamental emphasis;
2) the fact that both kinds of emphases are found in both metric and rhythmic domains; and
3) the need to delineate structural emphases on the various levels of structure.

With reference to Point 1: Ornamental emphases add foreground interest to a musical phrase (i.e., articulations, inflections of contour, dynamics, register, texture, weight, and the like). Hypothetically they could be removed from their context without changing the music in any deep-structure sense. The passage would be less lively, perhaps, but still basically recognizable. Structural emphasis (accent), by contrast, cannot simply be stated (nor can it be removed). A sudden loud noise, for example, could not create such an accent. For accents, being structural, are built or worked into the musical fabric.

Concerning Point 2: Metric time and rhythmic time each have their own variety of accent. In classic-romantic music, metric accents are largely determined by convention and style, as we have seen earlier. In the main they prevail on the level of measure. Essentially these are the familiar patterns of strong-weak beats found in duple and triple meter, implanted by usage over a long period of time, though set up anew with each piece. Hypermeasures also have conventionally determined strong-weak patterns. These are generally duple in nature, with an alternation between strong and weak bars; or, a four-bar pattern in which the initial bar is primary, the third measure of secondary strength—a correspondence in effect with duple pattern in a $\frac{4}{4}$ measure. The four-bar hypermeasure is a staple module of classical music, though the pattern is not continual; often asymmetric alterations result from metric-rhythmic conflicts that in turn generate powerful upbeat tensions.

As for Point 3: The principles that determine strong-weak pulses on the rhythmic levels of phrase, or on still wider spans of integral time, are complex. The most powerful element controlling these accents is harmony, or more specifically the degrees of relative instability-stability, tension-resolution or tension-repose that are inherent in harmonic progression. Harmonic stability is downbeat-oriented; instability, upbeat-oriented. Thus, consistent with the contradiction we have seen between time-points and the non-temporal means needed to articulate them, it is harmony that effects the most widespread articulations of downbeat, upbeat, and accent on broad integral-time levels. It is a control not only far-reaching but subtle. For harmony is capable of total stability (tonic chords in root position) and a spectrum of tensions, inflections and articulations ranging from quasi-stability to great instability, achieved by the syntax of chord progression and the establishment of temporary harmonic regions (so-called "key centers").

We have described at this point a structure that is dual in nature and multi-levelled. We have also delineated the means by which time-points in this structure are articulated, and we have examined the qualities of these articulations, or emphases. In effect, we have described the mechanisms used in this music for temporal control. But control of time is only part of the picture; it is the controlled *flow* through time, i.e., motion, that is the goal of this mechanism and the means as well by which continuity is achieved, as we have discussed earlier.

Musical Motion

Motion is indeed the ultimate desideratum of music, perhaps its most powerful affective quality. Without it music does not work; it cannot function. Without controlled and directed motion, continuity in classic-romantic music consists only of logical interrelations among themes and sections, lacking the "glue" of real-time flow that binds these parts into coherent movements, composed and heard from beginning to end. Furthermore, without fully controlled and directed motion a piece is bound to fail as a performed entity; it risks those drops in intensity that are the death-knell of musical projection, and the sign of weak composition. Indeed, many musicians can compose music with logically interrelated parts; only masters can further control the time flow of a work so that its peaks, valleys, and ultimate conclusions occur where desired.

A musical structure with such control of the flow of movement through time might be compared to the roller-coaster in an amusement park, unlikely as this may seem. Were a capable engineer to design such a roller-coaster, he would plan for the cars (in our case unpowered except for gravity) to start at a peak point "A" in the structure and to end at some later specified point "B". Between A and B would be the familiar hills and valleys of track, each carefully planned in its height, length, gradient of rise and descent, all of this calculated together with the mass of the cars, their coefficient of friction and the like. With these factors designed and counterbalanced, the cars could begin their ride and traverse the various slopes, their speed modifying with each hill at rates of acceleration and deceleration planned in advance. Ultimately the cars would come to their resting point, also pre-planned, when the kinetic energy which engendered their motion was expended.

The process of musical composition and performance is indeed similar. Energy is engendered when the performance of a piece is begun, yielding what we know as musical motion.[5] The time structure of the piece is likewise planned and designed by the composer, in terms outlined in this study, so that a system of checks and balances is created for motion, sometimes impelling the music forward, elsewhere holding it back, creating at certain places moments of (incomplete) rest, elsewhere of acceleration or deceleration, and, with final closure, creating the concluding point of total rest by the total expenditure of energy. It would not be an exaggeration to say that the ultimate purpose of all factors in the design of a work is motion—its creation, maintenance and closure.

This perspective seems to provide the most coherent and integrated view of a work, one in which all its parts are seen not only as a structure of logical design, but as a design intended to be unified through movement in time; a design which truly incorporates time as an essential dimension in the organization of the piece itself, rather than time used as a medium only, through which the piece must travel. A deep grasp of time as a structural dimension of music provides the composer with the ultimate mechanism for building a composition that performers will find workable— successful in performance. For whether they are consciously aware of time structure or not, performing musicians are gifted with a remarkably acute time sense—an intuition for the proper flow of time, for the molding of music within the dimension of time. Ultimately composers and performers inhabit the same time-bound musical universe, only their positions different, placed on one side or the other of the designing/performing axis that runs through this universe.

Every composer in the period under study had his own technique and disposition in the way he structured and designed the time flow and the temporal checks and balances within his music. We will examine some instances here, one of them, the second movement of the Brahms Second Symphony, serving also as an insight into the anomaly already noted in Brahms between the heard rythmic qualities of a phrase and its notated character. Whereas in many composers of the period these aspects of a passage would be congruent, in Brahms they are as a rule disparate. What emerges in this study is the fact that this disparity exists only in the small perspective; it is in fact part of a deliberate large-scale design, serving as one of the temporal checks and balances of that design.

The opening phrase of this movement illustrates this common "anomaly," for the melody in the high cellos (F# down to B#), if it is heard with no reference to the score, sounds as if it begins on the downbeat of the measure. This is not the case; it is displaced from the downbeat by one beat. An overview of the entire movement from the perspective we have developed shows the reason for this: *The movement is designed in such a way that no clear and unimpeded downbeat exists anywhere until the final chord.* The entire piece is thus on an upbeat footing—a gigantic anacrusis to the one true thesis and point of rest—its final note. The rhythmic-metric anomaly of the opening phrase is but one of the many devices used as temporal checks to regulate and control degrees of tension, or of partial resolution, at all points throughout the movement.

What is achieved by this design is motion. Because of its continual upbeat footing, because waypoints along the path to the final cadence are only partial points of rest—moments of only partial rhythmic-metric congruence, or of partial harmonic resolution, the music is kept in a continual though varied state of tension. Energy remains unresolved, driving the music toward closure.

This is a remarkable structure, shaped and guided by a composer who had perhaps one of the deepest insights into the nature of muscial motion, and who developed a technique to assure the continuity of that motion precisely as he wished it. This is not the case in all music. For with even the finest of earlier composers there occur moments in their works when the forward motion of a structure may be in danger of bogging down. This arises not from flaws of craft (indeed!), but from the fact that the means for achieving motion are different and, perhaps, more vulnerable in the hands of poor performers. Thus there are moments, for example, when the harmony of a section may have reached a point of stasis, however temporary, persisting on this level for bars at a time. Movement in such passages inheres in other musical aspects—the by-play of line, perhaps, or the lessening of energy that accompanies seeming resolution. Beautiful as such moments are, they run the risk of losing intensity in performance; a lesser player may fail to understand in which dimension activity prevails, thereby responding to the stasis of the harmonic dimension and "dropping the line" as a result.

One suspects that Brahms, with his background as a pianist and his friendship with Joachim and other performers, must have developed a canny awareness of this performance danger, constructing a technique which in its constant counter-pitting of structural elements made such a thing impossible.[6] For if there is one element that cannot be destroyed—in this symphonic movement and in most of Brahms—it is that of motion. While the music can be mis-performed in many ways (alas, all music can), its forward motion cannot be impeded, for this motion is built into the structure; it does not depend upon the perceptions and sensitivities of the performer.

A glance at the excerpts from the movement (Example 1) will show the diverse means by which Brahms controls this flow of motion. The opening two phrases (measures 1–2 with upbeats), as noted, pit the natural rhythmic emphases of the melody, in which a downbeat of some degree is felt at the first note, against a metric grid from which the melody has been displaced by a beat. Following this, the obvious downbeat feeling of the melody on the first beat of measure 3 is counterbalanced by a non-tonic harmony. The subsequent phrases in measures 4 and 5 are left harmonically unresolved; the resolution of harmony in measure 6 is ambiguous at beat 1 and is deceptive when the chord on beat 2 is sounded. A deceptive resolution is also used in measure 10 (second beat); and the resolution of the entire phrase in measure 12 takes place on the weak third beat in such a way that it is unclear whether the harmony is to be heard as a major chord on D (♮III) or on F#(V). More subtle

Brahms - Symphony No. 2
Second Movement

Example 1.

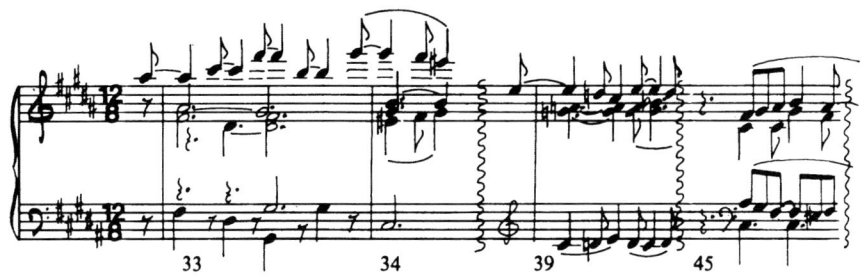

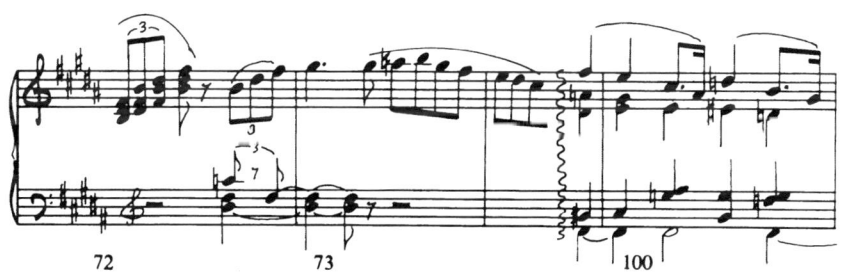

Example 1. *(con't.)*

ambiguity marks the closure of the motive at measure 17: Though a fully resolved tonic harmony sounds as this bar opens, it is unclear whether the downbeat is to be heard melodically as strong—that is, as a downbeat-oriented close of the phrase, or whether it is weak, ending a phrase whose strong pulse is its initial one (measure 16, first beat).

Subtlety of a different mode marks the passage beginning at measure 33, the second thematic group of the sonata-form movement. The quality of the motion is more relaxed here, more at rest. Only the mild anticipation of each melodic note keeps the music off balance in any obvious sense. Though the passage is calm in this surface aspect, it is a deceptive calm, for in another structural domain, that of harmony, tension prevails. The passage lies in the dominant region, so that any lesser tension that is felt locally in the disposition of melody and rhythm is counterbalanced in the long-range perspective by a tonal region at dissonance with the tonic key of B major.

Interested readers can pursue this work from score, as well as from the remaining excerpted examples. They will find variations of the means discussed above as well as further devices by which some degree of incongruity among rhythm, meter, harmony, melodic contour, tonality are all used to maintain various levels of tension and irresolution until the final cadence. Like the motion of the roller-coaster in our analogy, the motion of this movement is designed so that its pacing, its affect, its character will emerge and will proceed by plan, controlled until a specified point of resolution.

If this movement by Brahms is unusual, even anomalous, the Menuetto from Mozart's "Jupiter" Symphony is even more so in the sense that there is *no* clear and unequivocal downbeat emphasis in the entire movement. At all points where downbeats occur in some aspect of the music, they are offset by some counter-element which diminishes their force. The quality of motion of the piece is conditioned by this fact: it is continually of an "up" or "quasi-floating" character (words are futile for expressing these musical qualities), never fully punctuated by a "downward" force.

With the exception of measures 25-27, the entire movement is constructed of 4-bar hypermeasures, occasionally subdivided into two-measure subgroups. (See Example 2). Invariably the music is so disposed that the metric downbeats that occur at the head of each hypermeasure are at variance with the melodic accents of the phrases contained within these hypermeasures. This can be seen at the start of the movement, where the tonic harmony on the first beat is only quasi-stable (I⁶₄), the fully stable root position occurring in the middle of the phrase. At measure 25, when the C major tonic is heard, its harmonic definition is withheld until the third beat, thus leaving the downbeat of this bar harmonically ambiguous, though metrically powerful, its power reinforced by the loud dynamic. By measure 30, where a root-position tonic occurs, an extended passage has been underway for two bars, thus blunting the force of this moment. Moreover, the downbeat of measure 30 is an elision of two phrases. This further weakens its downbeat force, as its character at this point is unclear: Is it to be heard as the ending of one phrase or the beginning of the next, or both? The elision at the opening of measure 40 has a similar effect on this later phrase.

Mozart - Symphony No. 41
Third Movement

Menuetto

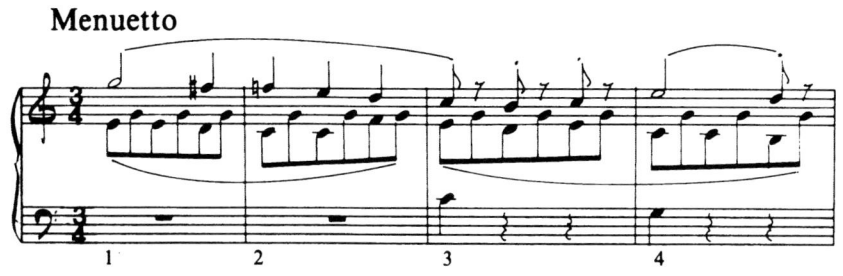

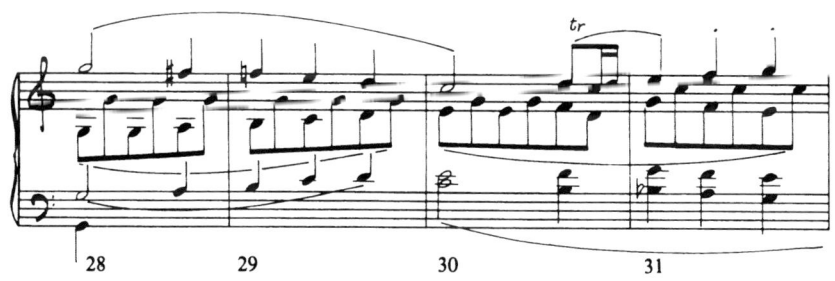

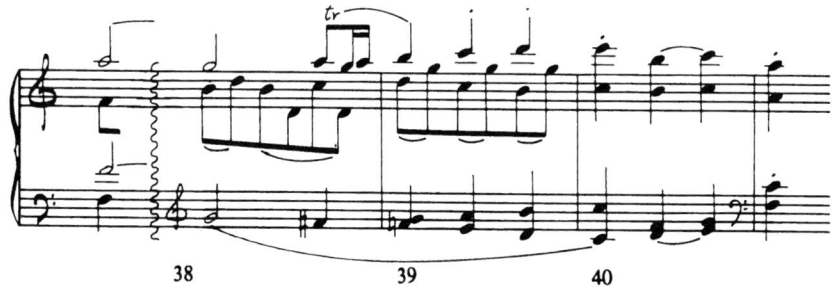

Example 2.

Trio

Example 2 *(con't.)*

Measure 52 poses a more potent downbeat in its return to root-position tonic harmony, intensified by the *forte* dynamic. Again, however, the force of this moment is to some extent diminished, since the note may serve as a cadential close to the previous phrase. This is an ambiguous moment, an ambiguity reinforced by orchestration. For only the timpany and basses play the brief quarter-note C, the root of this seemingly resolute harmony; the balance of the orchestra, with its considerable force, plays parts of lesser stability.

This sense of a clouded downbeat, diminished by countervailing factors, persists throughout the Trio. The few passages that do present a stable tonic chord, a model of which is seen in measures 60–61, do so by a weak cadence, where the resolution to stability occurs on the second bar of the phrase (measure 61 in this example), at odds with the strength of metrical articulation in the initial bar (here at measure 60).

Our concern so far has been with rhythmic-metric structure as the mechanism for motion; a mechanism which in engendering motion also creates continuity. The next example studies a different aspect of motion—its qualitative nature, or affect. The piece is the first of Debussy's Three Nocturnes for Orchestra, "Nuages." (Example 3)

The opening bars of this work are enigmatic in terms of performance technique and interpretation. The meter is indicated as $\frac{6}{4}$, which conventionally suggests a duple arrangement of the bar, essentially two beats with each having three subpulses. The first question is whether the music can be conducted in two at the

Debussy - Nocturnes
I - Nuages

Modéré

Example 3.

indicated *Modéré* tempo. It is quite slow, entailing a very broad beat that runs the danger of being vague, thus of little help in holding together a large orchestra. Yet this must be conducted in two, for the alternative of 6 beats per measure (or even of two broad beats with each sub-divided by three lesser strokes) so chops up the music as to destroy its quiet legato character.

There are further ambiguities: Is the opening passage really to be felt in two, or does it by its melodic structure parse as in three? It has, in fact, both aspects, as Example 3 shows. The upper melodic line by its contour inflects as a series of 3, 2-note groups, this pattern reinforced by the oboe rhythm in measure 3 (though in ambiguous notation). As the upper is the most easily perceived of the 2 lines, we tend to hear 3 as the dominant metric. The lower line, however, is less clear; it can be heard as two groups of 3 notes each (i.e., in 2), possibly as 3, 2-note groups (thus in 3), or even in some sort of asymmetrical plan, as indicated in the example. Its contours and their potential inflections allow any of these possible perceptions.

The situation, on the face of it, is totally confused: two lines with opposing metric character, to be played at a moderate tempo of such broad pulse that the conductor's beat must describe a wide and slow arc that verges on the unreadable. Moreover the lines must be played totally legato—devoid of articulation other than the inflections of the contours themselves. Nor are these inflections any guide to a metric plan; in fact they suggest contradictory meters.

The ambiguity of time plan is not only deliberate; it is the very element that creates the ambiance and the affect of the music, resulting in a mode of musical motion as remarkable as it is unique, uncanny in its intrinsic relation to its impressionistic aesthetic. For if the music is conducted in two, with the necessarily vague beat this entails, and if the woodwind lines are played as indicated, totally legato and without articulation, a subtle cross-tension emerges between the upper wind line, which tends to inflect in three, and the duple beat of the conductor. The "three-ness" of the upper line also conflicts with the ambiguous metric of the lower line in the winds. The effect is almost to cancel out any sense of meter within the bar, resulting in a sense of "hovering" motion. The music seems to go nowhere, almost to "float"—still, imbued with some sort of movement, yet movement unclear in its direction or goal. How close this is, say, to the movement of clouds on a windless day, clouds that seem to hang motionless, suspended.

We cannot be sure this was Debussy's intended impression. We can, however, make clear distinctions between the uses to which musical language can and cannot be put. One cannot "paint" clouds with tones; one can, however, engender musical motion that suggests the movement of clouds. In this case art truly copies life, evoking an ambiance closely resembling experience.

Epilogue

These studies suggest that the time dimension of tonal music—its metric-rhythmic structure, its directed motion—creates more than the continuity and closure any composition needs for performance. The affective, or expressive, or "emotional" content of music also inheres within this unique qualitative character, created and controlled by the forces that guide the flow of movement through time.

The perspectives of this discussion suggest grounds for exploring musical motion that lie beyond music itself. If time in performance is controlled through tempo and flow, for example, clearly it is a biological, physical system that is entrained and coordinated to make possible such control. Our biological clocks, in other words, function in musical performance as they do in other aspects of life. But how do these mechanisms function—to what degrees of control and within what small dimensions of time? It seems essential that we understand these processes and their implications. For a musical performance obviously is not the expression of a composition alone; it communicates human thought, feeling, symbolization as well. To gloss over these dimensions of music, or to accept them simply as givens, is to lessen the richness of the experience.

By the same token what is controlled physically, mechanically, in performance via biological entrainment has psychological implications. Musicians are keenly aware of this. As they control tempos, contours, intensities, they cannot but sense the affective power of these nuances. They are aware as well, however consciously, of the potential psychological response they may evoke as they shape such moments. The physical, in music as in the rest of life, lies intimately interlinked with the psychological, as does the psychological with the affective. Here, too, is rich ground to be tilled. In studying musical affect in terms of motion and temporal structure, we may have a way out of the subjective fantasizing that marked musical aesthetics a generation or two ago—writing that sought the "heroic" in Beethoven's Third Symphony or unrequited love in "Tristan" through adjectives, metaphor, and personal fancy. Our aversion to this fantasizing has for some decades skewed our focus away from even considering what music may have to "say," other than in its demonstrable properties of structure.

What is suggested here is a circular musical world, structure resting upon physiology which evokes psychology and in turn affect. These are not the parameters of a closed circle, however, fencing in our views of music. It is more a spiral, each perspective by its interconnections strengthening a process that continually carries us onto deeper levels of musical meaning. The key to it seems to lie in the structuring of time.

Notes

1. Rowell, Lewis, "The Creation of Audible Time", paper delivered at the fourth congress of the International Society for the Study of Time, Alpbach, Austria, July 1979 (in this volume).

2. The one universal we can rely upon with certainty is human biological structure itself—a structure whose limitations must in turn constrain our physical actions and our perceptions, as they pertain to all music.

3. The concepts in this section emanate from the studies of rhythm and meter in my book, *Beyond Orpheus: Studies in Musical Structure* (Cambridge, Mass.: M.I.T. Press, 1979). I have taken the liberty here of quoting and/or summarizing some of these ideas. Interested readers may want to refer to the more detailed discussions of these matters in the book itself.

4. I.e., we do not "see" time, feel, touch, smell or hear it, though in some less precise or definable way we may "sense" its passage.

5. Of what this motion consists is unclear. Certainly all of us sense movement when hearing music in performance. Performers, in fact, must move in order to play—the conductor's beat, the violinist's bow arm, etc. The performer's motion, however, is outside the music proper; it is more mechanical motion, required to produce music. And certainly there is no physical motion in space—from point A to point B, when music is played. Yet we experience in music physical sensations akin to specific kinds of physical motions—swaying, martial steps, dancing, erotic movement, tension, gesture, etc. Is this sensation a distilled "essence" of physical motion without that motion itself? Or can all this be explained on another level, as the human energy that underlies physical gesture? The problem is beyond the scope of this paper. I use the terms motion, movement, gesture and energy somewhat freely (and equally) here, recognizing that they all pertain to the same phenomenon, one which to date has not been adequately explained.

6. To be sure, this technique also complemented the ambiguities of harmony, tonality, and structural prolongation that were hallmarks of Brahms' musical personality.

The Creation of Audible Time

L. ROWELL

I. Introduction

John Cage once remarked about David Tudor, "His music has no beginning, no middle, and no end."[1] What Cage meant by this cryptic comment was that Tudor's music studiously avoids the conventional rhetoric—the various opening and closing gambits, tactics, and behaviors that audiences have come to expect in different genres and styles of music. This "conventional rhetoric" is the subject of the present paper. Our method is an exploration of how musical compositions begin: what it takes to transport the listener from the external world of clock time to the internal time which a piece of music creates and which is shared by composer, performer, and listener. And how—by means of specific actions, energy, duration, speed of pulsation, patterning, and a variety of other clues—we are brought under the control of an audible, hierarchical time that is more palpable, more insistent, more clearly articulated, and more flexible than the world of everyday time to which we eventually return.

Although textbooks on music are full of examples of musical beginnings, only a few authors have addressed our main problem in any depth. We obviously take musical beginnings for granted! Two short but provocative discussions deserve to be mentioned here: Ernst Toch's *The Shaping Forces in Music* and Edward T. Cone's *Musical Form and Musical Performance*.[2] The approach taken in this paper is consonant with theirs, but the discussion goes considerably beyond the scope of the works cited.

One could easily write a history of music based on beginning mannerisms and gambits. In fact the analogy to the game of chess is a very good one: just as in chess there are certain combinations of opening moves (gambits) that implement the strategic principles of the opening game, namely (1) the rapid development of pieces and (2) control of the center of the board, in music there are various gestures which are designed to achieve some important strategic objectives of what we might call the "opening game" of music. Such a history of music would be fun to write, but the goal of the present paper is a more modest one—to set forth some general principles that govern beginnings of musical compositions and thus establish a framework of ideas within which various musical beginnings may be studied and compared.

This paper deals with the *commonplace* in music rather than with brilliant, individual solutions. The pianist Artur Schnabel has been often quoted as saying that "genius begins to be apparent only at about the fifth measure," by which time the composer has delivered himself of one of his assortment of opening gambits and now faces a far more critical question—"What next?" The beginning of a piece of music accomplishes simultaneously an amazing number of objectives, but—like the various chess openings—it can be mastered by any first-year student. So, although most of our examples are taken from great masterpieces of the Western symphonic repertoire, they should be regarded as typical rather than atypical. If a bias toward the Austro-Germanic "mainstream" of 18th- and 19th-century composers is detected, let it be attributed to the spell cast by the lovely Tyrolean hillsides on all those attending the Conference.

I should like to emphasize how different are the situations that the composer confronts at the beginning and at the ending of a piece: nearing the end, he has deployed an intricate array of forces, an accumulation of energy, momentum, probabilities and expectations, tonal tendencies that demand resolution—internal forces that have been developed within the compostion, entailing certain musical consequences. Certainly there are conventional ending gestures, and certainly musical endings display some of the same rigidity of cadence structure that we find in poetry, but a successful resolution of a composition requires a much more sophisticated solution than simply tacking on one's favorite fanfare or chord progression. But at the beginning *nothing* already exists—no accumulation, no momentum, no tendencies—only silence and inertia to overcome. Musical endings require a selection from among the potential consequences of the musical accumulation; musical beginnings are selected from all the possible choices in the world. To narrow the range of choices from this overwhelming array, composers have resorted to specific *opening behaviors*.

We make certain exclusions: most types of functional musics in which the beginning is directed by an extra-musical idea—descriptive music, music for the theatre and films, church music,[3] marches, dance music, wake-up music, and background music (which should ideally have *no* beginning). The storm music that opens Verdi's opera *Otello* and the pistol shot that serves as the sole overture to Wolf-Ferrari's *I Gioielli della Madonna* are strikingly effective as openings but do not represent an intrinsically-musical solution. They do, perhaps, serve as microcosmic equivalents to the "Big Bang!"

II. Principia

There is a lot to be accomplished at the beginning of a piece of music, and most of it happens within the first few seconds. The immediate, tactical objectives include these: first, the translation from external time to the internal time of the composition;

second, the overcoming of the inertia of the surrounding silence, the zone of atemporality that serves as a frame for the music—music takes energy to set it in motion, and this energy must be felt (and sometimes *seen*) by the auditors. A third objective is the demarcation of the tonal field: laying out the boundaries within which the game is to be played, setting temporary high and low "edges" for the pitch spectrum, and establishing a tonal focus or perspective therein. Fourth, the beginning must also give a forecast of the scope of the whole composition, its accentual weight, tonal quality, and energy level; the listener is thus ushered into an appropriate scale of reference along which he may judge correctly the proportions of the Gestalt. Fifth, the beginning initiates the listener's train of expectations, predictions, and retrodictions, moving from ambiguity to certainty; and sixth, a feeling of motion is established. There are other, more practical objectives too: warming-up the performers and audience, enlisting a sense of community and joint participation in the musical process, pitch-giving, getting the attention of the listener and directing it in an appropriate way; and perhaps even an announcement of the composition, a kind of signal or commercial message: "Here beginneth the first movement of the Fifth Symphony of the prophet Beethoven!"

A few comments on some of these objectives: many pieces seem to announce themselves, *i.e.*, to begin with an encapsulated statement of seminal musical substance, the primary motive from which the piece develops. This may be one of a number of curious vestigial survivals that one occasionally finds in music, possibly the equivalent of the liturgical incipit that typically opens a motet or Mass section from the Middle Ages or Renaissance. This type of beginning was a standard procedure in the musical Baroque (roughly 1600–1750): the initial motive stated—in capsule form—both the musical and affective substance of the composition, and Baroque arias frequently began with a title announcement of the text (known as the *Devise*). In another sense, this opening announcement may also serve as an artist's signature or benchmark: "Hoc fecit Mozart." The fanfare is a popular type of musical announcement borrowed from court life and the theatre. And, in many ethnic musics, an announcement of the piece's title and/or composer's name is considered an integral part of the music.

Just as the onset of a piece of music must establish all the properties of its temporal dimension, so must it lay out the *musical space*—the coordinate dimension of pitch. I return to the analogy of a playing field on which coordinate boundaries, proportions, lines, and paths are clearly marked. The great theorist Heinrich Schenker called attention to what he called "space-opening motives"—opening motives that cover a relatively wide zone of pitch, usually ascending, that say in effect: "This space is now available for use."[4] It also happens very often that provisional high and low points are established early in a piece as a kind of pitch frame for the musical events to follow. The principle of tonality serves not only to coordinate the pitch boundaries of a composition but also to lay out *acceptable paths* within the pitch spectrum, to be followed by means of tonal probabilities and the listener's expectation. In short, the deployment of pitches is directly analogous to the rapid deployment of chess pieces and serves both a tactical and a strategic function.

Beginning tactics are greatly affected by the way in which one conceives of a piece of music. In the Western philosophy of art music, we have been led to think of a musical masterpiece as a kind of sound monument, a "thing" and an object of permanence.[5] But music is also very much *process*, and in certain genres of music the processive principle is more prominent than the sense of "thingness" and the design that the composition ultimately assumes—fugues and variations, in particular, fall into the processive category. And in many world music traditions music is clearly more process than product. Products and objects seem to generate more impressive, more rhetorical beginnings than do processes; a process need not be as clearly separated from other activities, and it need not begin all at once. I do not mean to suggest that pieces can be easily divided into products and processes—the two conceptions are intermingled, but I do suggest that a piece tends to have its edges emphasized the more it is conceived as object.

And there are other interesting conceptual possibilities, now under exploration by composers: music can also be viewed and experienced as a field, a state, or even a situation. Much recent music suggests to me that the composition is conceived as a field within which the placement of individual events is unpredictable; the analogy of a "Brownian motion" has, in fact, been applied by Leonard B. Meyer to such compositions.[6] On the other hand, recent compositions that feature much improvisation and that depend upon the lively projection of the performer's personality can well be seen as a kind of situation—a happening or encounter that is based on a minimal script. And finally, Fraser's musical "ecstasies" (as described in his *Time as Conflict*) are excellent illustrations of musical states of timelessness—stabilized processes that suggest the illusion of stasis and suggest *being* more than *becoming*.[7] Whether one thinks of a piece as object, process, field, situation, state, or something else, that conception entails certain consequences for the piece's beginning (as well as its ending).[8]

I think it is useful to identify four stages in the inception process: first, the framing silence, a buffer zone of timelessness between external and internal time; second, the attack proper;[9] third, time creation—the establishment of a metrical framework or some other hierarchy of controlling periodicities; and fourth, formal processing—the articulation of the first structural units. Of the four, we will have the most to say about the third stage: time creation, the representation of time in audible form.

It is very important to consider the timeless zone of silence that precedes the opening attack, even though the ritualistic silence of today's concert hall is not typical of all musical societies. It seems to me that this period of no-time is extremely useful as a preparation for the new time which is to be audibly created and sustained. It is a highly artificial, tensed silence—a way of erasing our previous consciousness of time and of external events, a period of intense focus, concentration, and pure expectation during which we are poised on the brink of time, as it were, and made ready to process the rapid succession of temporal clues we are about to receive.

We pass over the attack itself, although its length, weight, sharpness, and other properties carry enormous consequences for our perception of the rest of the piece. But the actual creation of music's internal time is probably the most remarkable achievement of all, especially when it is accomplished in so few strokes. Musical time is, I believe, mainly the result of perceived patterns of accentual weight, although accent itself (using the term broadly) is the result of a complex combination of stresses, pitches, and patterned durations. A framework of more-or-less regular pulsation is perceived at some level, and upon this is superimposed a matrix of strong and weak (thesis and arsis), or some kind of alternation that produces in us the psychophysical response of tension and release. Musical time is a very physical thing, an insistent phenomenon, and its communication by gesture and our response to that gesture are vital elements in the musical process. Musical time is a matter of weight and energy, and extremely minute variations in perceived weight are enough to establish a metrical framework to our complete satisfaction. Musical time is always a hierarchy,[10] although our immediate apprehension of this hierarchy is limited to a few adjacent levels—perhaps no more than two or three in the initial stages. And finally, this newly-created time is maintained as a grained, textured surface or background against which the themes, events, and phrases can be projected.

What seems to me so remarkable about this phenomenon is how quickly it can happen and how the various reinforcing temporal clues render it virtually impervious to error and impossible to misjudge. By means of a few simple sound events we are entrained into the scale of proportions, periodicities, and accentual structure within which the actual "game" is to be played.

We remarked earlier that musical beginnings are dominated by *behavior*: private and public rituals, practical necessities such as tuning, and various other activities of transition between external and internal time. Most of these may be considered part of the "frame" in a social context—special buildings, special clothes, special times, tickets, lights dimmed, ritual homage, fidgeting, tuning, social contact, sense of community, ear-cleansing silence, focus of attention, announcement, and preparation for the actual attack. Sociologists have not neglected these behaviors, but what has received little comment is that these behaviors often spill over into the music itself.

We have already noted one such example above, the tendency to begin a composition with an opening "announcement." And it is easy to cite other "framing behaviors" that are often found at beginnings of pieces: introductions, exploratory tonal figurations which suggest tuning procedures, attention-getting and attention-focusing devices, and beginning with an extended musical duration (which suggests a continuation of the framing zone of atemporality). Musical beginnings since 1600 have tended to become increasingly rhetorical, especially in concerted (as opposed to *solo*) music, and much of this larger-than-life opening rhetoric is behaviorally consistent with the honorific gestures and charismatic personality-projection that have come to mark public concert life.

A prime characteristic of musical beginnings is a tendency to begin with relatively pure, uncomplicated sounds: sustained triads or other simple sonorities, clear textures, and well-delineated tonality. The significance of this palate-cleansing material may well be to mark the transition from noise to music, from nonperiodic to periodic vibration structure. And even the preparation for the first attack (the conductor's upbeat of preparation, the raising of the violinist's bow, and other such gestures) is often paralleled within the piece by means of complicated *anacrusis* or "upbeat" beginning motives.[11] In sum, musical beginnings rely heavily on framing behaviors, often presenting them not singly but in clusters—multiple frames.

Patterns of energy and perceived metric weight are vital in the communication of musical structure; on each level of the musical hierarchy (beat, measure, phrase, *et al.*), structure is dependent upon the alternation of *arsis* and *thesis*, the feelings of lift and descent. On the largest structural level, most Western symphonic compositions of the last few hundred years may be viewed as a single large arsis and thesis, a preparation and resolution. Second in importance to the final thesis (and on a lower structural level) is the beginning thesis, but this may be preceded by an extremely long and complex arsis—such as the slow introduction to a symphonic first movement. Western music displays an astonishing assortment of anacrusis beginnings, but one almost never finds an "afterbeat" beginning such as is common in Indian and Japanese music. To me this is strong evidence that traditional European art music is beginning-accented, as opposed to the clear preference for end accent that I see in the art musics of South and East Asia. So a proper Western beginning involves a feeling of weight and stress that is not equaled until it is topped by the final thesis of the piece; but there is virtually no limit to the amount of upbeat material that can successfully precede this first thesis.

III. An Annotated Anthology of Musical Beginning[12]

The compositions cited below illustrate (among other things) four favorite opening gambits of Western composers: (1) *assertion*, beginning the composition with a dramatic gesture, (2) *entrainment*, in which the listener is drawn into the musical time fabric without ceremony, (3) *emergence*, a kind of creation out of nothingness and the musical equivalent of time as *becoming*, and (4) *pure duration*, beginning with an extended, unarticulated musical duration.

Richard Strauss. *Don Juan*. The example *par excellence* of the dramatic, assertive type of opening popular since 1700; this highly rhetorical opening procedure was especially favored by Beethoven and later 19th-century composers.

Johannes Brahms: *Symphony No. 3*. The opening progression of three chords illustrates a typical *space-opening motive* as described above.

Johannes Brahms: *Symphony No. 4*. The work launches immediately into the lyrical main theme, illustrating the process of entrainment. Other famous examples include Mendelssohn's *Violin Concerto* and Mozart's *Symphony in G Minor, K. 550*. It is interesting to see the traces of Brahms' indecision in the autograph score: at a relatively late stage in the composition, he apparently decided to preface the opening theme with four measures of sustained chordal background but then thought better of it and crossed out the inserted measures.[13]

Anton Bruckner: *Symphony No. 9*. This work, like most of Bruckner's other symphonies, features the gradual emergence of the opening theme out of a hushed string tremolo. The first and most celebrated piece to begin this way is probably Beethoven's *Symphony No. 9*, and Gustav Mahler's *Symphony No. 1* continues the process for more than five minutes of the opening movement. I would like to suggest that both the emerging and assertive openings display what I have referred to as "framing behavior," the former emphasizing the surrounding silence and (in a sense) incorporating it in the music, the latter placing a clear, definite "edge" at the beginning of the composition.

Ludwig van Beethoven: *Symphony No. 1*, the slow introduction to the fourth movement (see Figure 1.). This example deserves careful study, since it incorporates a number of interesting features. The movement begins with an unmeasured duration, the opening *fermata* ("hold"), and continues with a gradual "wind-up" in the violins that leads ultimately to the upbeat to the fast portion of the movement. The latter procedure is unique, but many Classical symphonies and concerti begin with long durations—often unmeasured and long sustained, perhaps to as much as three times the value of the note. This extremely significant opening continues the step-by-step introduction of the listener from external time, to atemporal silence, to atemporal duration, and finally to structured internal musical time. Such durational beginnings are usually unpunctuated or unsupported by any rhythmic activity. And, since we have difficulty in judging long periodicities, even pieces which begin with a series of very slow durations may serve the same function.

Ludwig van Beethoven: *Quartet Op. 59, No. 2*, illustrating other types of framing behavior: two sharp opening chords are followed by a pause, then by two isolated "motto" statements before the movement swings into regular, measured time.

Felix Mendelssohn: Overture to *A Midsummer Night's Dream*. The overture begins with the same progression of long-held chords that Mendelssohn later used to underscore Puck's epilogue. Here again there is no clear projection of a sense of ongoing musical time.

African Festival Music from Chad, illustrating music *qua* process with the performers gradually joining in to build up a complicated rhythmic background for the improvised solo of the master drummer.[14]

Igor Stravinsky: *Le sacre du printemps*, another beginning that suggests music as process; although the music is strictly notated, the time gives the illusion of being under minimal control and the instruments not fully synchronized with each other.

Hector Berlioz: Overture to *Benvenuto Cellini*. Beginnings can be deceptive, tricking the listener's expectations for continuity!

IV. The Free/Strict Archetype

Thus far we have been concerned with the first few vital moments of a musical composition; for the last part of this paper I propose to examine what I believe to be a universal and instinctive beginning strategy for music, as embodied in a formal archetype that pervades much of the world's music. In simplest form, this archetype

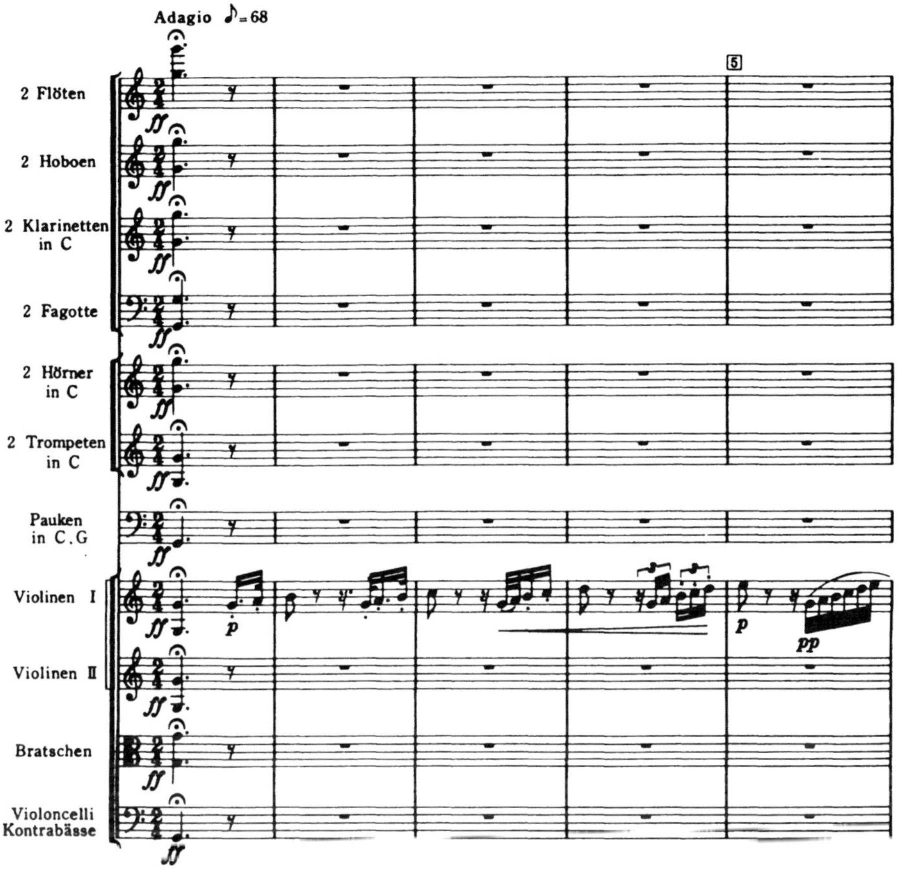
Adagio ♪=68
5
2 Flöten
2 Hoboen
2 Klarinetten in C
2 Fagotte
2 Hörner in C
2 Trompeten in C
Pauken in C.G
Violinen I
Violinen II
Bratschen
Violoncelli Kontrabässe
ff
p
pp

(Exposition)
Allegro molto e vivace 𝅗𝅥=88
10
I
Vln.
II
Brn.
Vlc.
Kb.
p

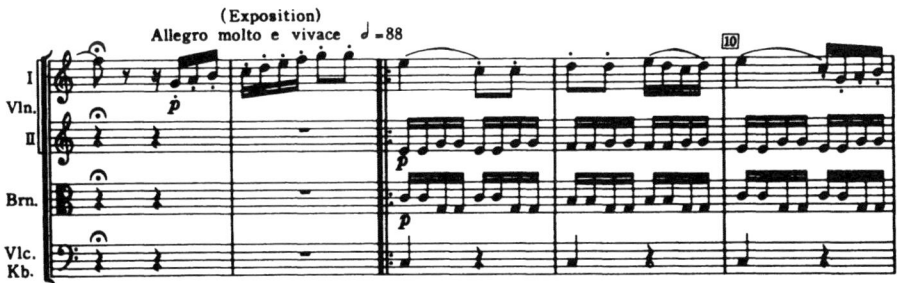

Figure 1.

consists of two elements: a slow, exploratory piece with a loose temporal organization, followed by a faster, more strictly-organized piece. I consider this to be a very
deep-seated human response to the idea of time as first *becoming*, then *being*—a
cosmological statement in tonal form. Musical compositions and musical processes
move generally from ambiguity to certainty, toward more strictly-defined pitch and
time organization—very rarely the other way around.

In the music of Japan the formal pattern known as *Jo - Ha - Kyū* has been the
most significant and pervasive musical design since Zeami set forth the structural
principles of the Noh theatre in the 15th century.[15] The names of the three sections
are usually translated as "introduction," "scattering," and "rushing to conclusion." I
do not wish to introduce confusion by discussing a tripartite form—our archetype
appears in various world musics with various kinds of endings; what I consider
significant is the special relationship between the first two sections, in this case the *Jo*
and the *Ha*.

When we look closely at the free/strict archetype, it is obvious that the musical
details are subject to local variation and carry distinctive cultural meanings. The
Japanese variants are these: *Jo* and *Ha* are recognized on many different hierarchical
levels, from a single phrase of music to an entire day's program. Thus a larger *Jo*
section will contain its own (sub) *Jo*, *Ha*, and *Kyū*. Thus Japanese music is consistent
with the whole of Japanese culture in that beginnings overlap with the endings of
other events, are subject to multiple interpretations and ambiguity, and are often
imperceptible; a musical event often begins in concept before it becomes audible or
visibly communicated by gesture. The time of *Jo* is the smooth, continuous time of
breath and of gesture and is symbolized in the music of the Noh drama by the
kakagoe, the sustained calls of the Noh drummers. Japanese beginnings are considered important and auspicious, and the moment of preparation (the intake of breath,
the wind-up for a drum stroke, the lift of the foot before a step) has the force of
musical accent; the resultant sound is *Ha*, a consequence.

In the calendar *Jo* is located at the coldest time of the year when the seeds of
spring are about to sprout. The character for *Jo* has the radicle "roof," perhaps the
roof of the theatre, and the inside character signifies "preparation," a symbol of
internal energy and potential, heightened by suppression and constraint—physical
feelings that are enormously important in the Japanese concept of beginning and
clearly evident in the tonal quality of Japanese singing.

The apparent "loose" rhythm of a musical *Jo* section is deceptive and should not
be interpreted as "freedom." Enormous importance is placed upon exact timing in all
the arts of Japan (from the musical theatre to sumo wrestling) as well as in Japanese
conversation. The concept of *Ma* signifies the regulated timing that separates events,
in music as well as life—a feeling of tensed silence controlled by breath and timed
physical gesture.

The time of *Ha* is closer to what we in the West regard as regular, metrical
rhythm and is experienced in Japanese music as a steady, well-articulated, and
highly-organized temporal flow. Thus, on hearing a typical Japanese statement of the
archetype, the listener will notice that the time audibly firms up until the beat
structure resembles the familiar Western $\frac{4}{4}$ metric pattern.[16]

We find an equally compelling interpretation of the free/strict archetype in the music of India, once again with distinctive local color. In a typical performance the opening *ālāp* is an improvisation for solo voice (or instrument) that establishes the pitch framework of the scale, the *rāg*. This section has no specific time framework other than the constraints of the singer's breath, his sense of pattern length, and the overlapping imitations of the accompanying violin that bridge the gap between vocal phrases. The sense of continuity is further reinforced by the drone that is sustained throughout the performance on the tambura. The *ālāp* is followed by a faster and more rhythmic section in one of the strict, cyclical rhythms known as *tāl*; this section takes the form of a series of variations over the tonal and temporal matrices that have been established, underscored by brilliant drum patterns which further vary the rhythmic cycle. Indeed we may describe the basic process of Indian music as one of improvised variations over pre-established pitch and time frameworks, which are introduced successively in the two opening sections.[17]

Our formal archetype has important cosmological significance in Indian culture: the constant drone symbolizes the process of continuous creation and the underlying unity of all matter. The threefold world process of creation, preservation, and destruction (presided over respectively by Brahma, Vishnu, and Shiva) is reflected in the Indian concept of music—the opening improvisation represents matter undergoing differentiation and emerging as structure, a process of pure *becoming*. The change to measured rhythm and strict time symbolizes the universe after it has been set in motion, and the cyclic organization of the underlying *tāl* is a microcosmic parallel to the macrocosmic cycles within which Indian time unfolds. Becoming and being are not emboxed, as in Japanese music, on many hierarchical levels but are considered as two clearly separate stages of the musical process. A beginning in Indian music has no specific locus, no precise moment—it is pure emergence and arises from a continuous and undivided substratum of internal sound. The model for Indian music is the human voice: internal, breath time has no divisions nor structure; it is absolute *flow*, and it is this sense of flow which the *ālāp* seeks to communicate. Only when sound leaves this innermost sanctum and is manifested as articulate sound does it become subject to beats, organized time, and rhythmic repetition. Thus, the free/strict archetype in the music of India retraces symbolically the emergence of sound from its internal source. Struck sounds are obviously inconsistent with this model of vocal sound, so Indian attacks are not as sharp as in other world musics based on instrumental models. And we may add that accent in Indian music is a property of endings—not of beginnings.[18]

It should be stressed that the above examples are not merely isolated ethnic curiosities but local variations on a universal theme. We find several versions of the same archetype in genres of Western music, of which we will mention just three:

The most important instrumental version has been popular in keyboard music since the Renaissance: the combination of a quasi-improvisatory prelude or toccata and a strict fugue. The first element is almost always very loosely organized, is flexible in time, exploratory in its tonal design, texture, and figuration, and is frequently studded with short cadenzas (melodic flourishes in unmeasured rhythm). Its successor, the fugue, is one of the most highly-organized of all musical genres; its hallmarks are rhythmic regularity and continuity, unity of thematic content, and a strong sense of linear development. It is hard to imagine a better Western response to the sequence of becoming and being. And we may add that the prelude or toccata frequently suggests tuning behavior and often proceeds by means of a systematic opening up of musical space through the different registers of the instrument.

In vocal music the archetype appears as the recitative and aria, in this case contrasting the time of speech and the time of music. Recitative, as the name implies, arose around 1600 as a style of dramatic musical speech and rapidly found a permanent home in opera; subsequently the characteristics of recitative were imitated in instrumental music (by Beethoven and others) where there was no longer any need to preserve the rhythms and accents of a text.[19] The recitative embodies all the characteristics of the "free" stage in the archetype: no large controlling organization other than the text phrases with their rhythmic and accentual structure, a loose tonal plan, and great flexibility in performance. I will venture to speculate on what this means in Western musical culture: the time of speech has, from classical antiquity, served as the first model for the time of music—at least for "art" music. Early polyphonic compositions generally began with liturgical incipits, text forms were important models for early musical forms, and rhetorical devices served as models for musical structure. The transition (in the recitative and aria) from speechlike time to songlike time seems to me to be a kind of *recapitulation*, a symbolic reenactment of our achievement of intrinsic musical time. The beginnings of instrumental music may feature tuning, tonal orientation, and other such behaviors, but the symbolic beginning of vocal music is speech.

In more recent Western music we find the archetype in the combination of a slow introduction and a faster, more organized piece—the latter typically in sonata-allegro form. This describes the role of many symphonic introductions in the 17th and 18th centuries, the standard concert overture, and the more recent pairing of introduction and allegro; the same pattern is also found in many pieces of descriptive music, generally for extra-musical reasons. We can more easily trace the spread of the free/strict archetype if we agree to interpret the concepts of becoming and being more broadly. For example, in the famous slow introduction to Mozart's "Dissonant" Quartet, the freedom is more tonal than temporal.[20] Nevertheless the contrast between two modes of musical organization is striking, and the introduction serves as an enormous, slow upbeat to the main part of the movement.

I am obliged to remind readers not to overestimate the importance of the free/strict archetype in Western music; it is but one, relatively-limited class among the array of Western beginnings. Far more frequent are pieces that introduce tonal and rhythmic certainty and a high degree of organization at the start. Accent is a property of Western musical beginnings, and I believe this explains why we take exploratory musical beginnings as extended upbeats, interpreting the start of the strict section as the initial thesis.

Our discussion would be incomplete without some mention of the range of beginning tactics found in various ethnic musical traditions: because many world musics are, as ethnomusicologists constantly remind us, more process than product, many musics begin with some kind of signal or first entry with the rest of the performers gradually joining in the ongoing process. Some other musics (as in Korean kayageum playing) blur the line between tuning and the composition proper—all at once the piece is in full swing without anyone noticing. And the

American Negro song sermon, which *is* conceived as a piece of music, gradually crosses over the line between impassioned speech and pure song with no clear indication of where the transition occurs.

How long is a beginning? The question is proper but impossible to answer precisely with reference to music. We have taken as our target the transition from external time to the internal time of a composition, whether achieved in a few quick strokes or by means of an elaborate opening section. Sometimes the transition requires a neutral zone of atemporality, sometimes a gradual and imperceptible emerging, and at other times an enforced expenditure of extra energy. We have dwelt at some length on questions of "framing," and we have asserted that the frame is sometimes within, sometimes without, the musical composition. Local beginning gambits are patterned after familiar cultural models, are often ritualistic and highly mannered, and are communicated by means of specific behaviors. The evidence clearly suggests that musical beginnings reflect (albeit in a very naive way) the more general concept of beginning as held by the parent culture.

Notes

1. In a television interview.

2. Ernst Toch, *The Shaping Forces in Music* (New York: Criterion Music Corp., 1958), pp. 217–227; Edward T. Cone, *Musical Form and Musical Performance* (New York: W. W. Norton, 1968), pp. 11–31.

3. Cone, *loc. cit.*, suggests with tongue in cheek that one might compose an organ prelude without a beginning and a postlude without an ending!

4. Schenker's term is *Anstieg*. For a good discussion in English see Allen Forte, "Schenker's Conception of Musical Structure," in *Readings in Schenker Analysis and Other Approaches*, ed. Maury Yeston (New Haven: Yale University Press, 1977), pp. 20–23.

5. Barney Childs, "Time and Music: a Composer's View," *Perspectives of New Music*, vol. 15, no. 2 (1977): pp. 194–219. This provocative article sets forth what he describes as an "alternative position" to traditional Western European aesthetic assumptions. High on his list of "invalid" notions are these: the "masterpiece" idea and permanence as an aesthetic value.

6. In a lecture at the University of Rochester.

7. J. T. Fraser, *Time as Conflict* (Basel: Birkhäuser Verlag, 1978), pp. 298–300.

8. To make this discussion more concrete, I suggest the following typical examples:

   an object - Beethoven's *Ninth Symphony*
   a process - Ravel's *Bolero*, a jazz improvisation, or see Note 14
   a field - Witold Lutoslawski's *Symphony No. 2*, typical of "sound mass" pieces
   a situation - John Cage's *4'33"* (of silence!)
   a state - Terry Riley's *A Rainbow in Curved Air*

9. This as well as other words for "beginning" in various language traditions conveys clearly the implicit message that "beginnings" imply a certain amount of aggression!

10. Here is a typical set of hierarchical levels:

                        composition
                        movement
                        section

> phrase-group
> phrase
> measure/grouping of beats
> beat
> beat divisions

11. For a superb example see the introduction to the finale of Beethoven's *Symphony No. 1*, quoted in the previous pages.

12. The following recorded illustrations were played during the original lecture. Space limitations preclude the use of musical examples, but care was taken to select readily-available pieces. And beginnings have the additional advantage of being easy to locate on disc recordings!

13. Johannes Brahms, *Symphony No. 4 in E Minor, Op. 98*, facsimile edition of the autograph (Adliswil-Zürich: Edition Eulenberg GMBH, 1974).

14. Ocora OCR 36, side A, a typical example of many African musics.

15. For a comprehensive description see William P. Malm, *Japanese Music and Musical Instruments* (Rutland, Vermont: Tuttle, 1959), pp. 110–112.

16. The example chosen was an entrance song from a Noh play, recorded on Lyrichord LLST 7137, side one, band one.

17. The example chosen was taken from a recording by the celebrated South Indian singer, K. V. Narayanaswamy, Nonesuch H-72018, side two, band three.

18. Lewis Rowell, "Time in the Musical Consciousness of Old High Civilizations—East and West," in *The Study of Time, Vol. III.*, J. T. Fraser, N. Lawrence, & D. Park (New York: Springer-Verlag, 1978), pp. 578–611, especially p. 606.

19. A typical Beethoven example may be seen in the first movement of the *Piano Sonata in D Minor, Op. 31, No. 2.*

20. K. 465.

Miscellaneous Contributions

The Branch System Hypothesis: A Critique*

S. C. Schwarze

The relation of entropy to temporal direction and temporal anisotropy has been a persistent, nagging problem to both scientists and philosophers. Since the advent of statistical mechanics, which revealed the temporal symmetry of the microphenomena thought to underlie the temporally asymmetric macrophenomena of the Second Law of Thermodynamics, there have been a number of puzzles, including whether or not entropy, increasing or decreasing,[1] tells us anything at all about time. Adolf Grünbaum has been one of the foremost proponents of a connection between entropy and time, claiming that his branch system hypothesis describes a temporally asymmetric entropy increase which confers anisotropy on (at least) the present temporal epoch.[2]

The branch system hypothesis is an attempt to recapture the temporal asymmetry of thermodynamics which is apparently lost when thermodynamics is reduced to statistical mechanics. Thermodynamic processes have long been in the forefront in the discussion of temporal anisotropies. The reason for this is the well-known consequence of the Second Law of Thermodynamics; namely, that the future direction of time is always indicated by the direction of increasing entropy in a closed system. It would appear that from the Second Law one has a temporal anisotropy that needs little or no further discussion or justification. The advent of statistical mechanics, however, seriously weakened the credibility of the Second Law. For the laws of statistical mechanics, which govern the microphysical processes underlying the macrophysical thermodynamic phenomena, are temporally reversible.

That is, the Second Law of Thermodynamics, a law about observable phenomena, rests on the observed generalization that heat never spontaneously leaves a cold body for a hotter one. Rather the spontaneous transfer of heat is always from the hotter body to the colder one. For example, consider the thermodynamic system

---

*I am indebted to Paul Fitzgerald and Michael Friedman for earlier comments on this argument.

of a glass of water into which seconds before an ice cube has been dropped. The ice cube will melt. The water will cool. The temperatures of the two bodies will tend to equalize, causing the entropy (a state function describing heat transfer) of the system as a whole to increase. In no case is the Second Law observed to fail. One never sees ice cubes grow colder and the water warmer, or the entropy of the system decreasing. Yet with the advent of statistical mechanics the phenomenon of a decreasing entropic system was shown to be merely improbable, not impossible.

Statistical mechanics concerns the behavior of the microparticles which make up the macrophenomena of thermodynamics. The original formulation of statistical mechanics was made by Ludwig Boltzmann. Using the kinetic theory of gases, Boltzmann was able to relate the behavior of the microparticles of a system to the entropy of the system. The behavior of the microparticles of the system, however, is governed by the principles of Newtonian mechanics. These classical mechanical principles are reversible. It is just as probable that a molecule have the reverse or negative of its actual velocity. This means that entropy decreases in a closed thermodynamic system should be just as probable over time as entropy increases. Hence, the entropy curve of a closed thermodynamic system, when viewed from the point of view of statistical mechanics, is temporally symmetric and does not yield a temporal anisotropy.

A suggestion was made by both Erwin Schrödinger and Hans Reichenbach—later developed more fully by Grünbaum—that the phenomenological irreversibility of the Second Law of Thermodynamics (which is lost when thermodynamics is reformulated in terms of statistical mechanics) might be regained if thermodynamics is reformulated in terms of subsystems or *branch systems* instead of permanently closed systems. They hoped in this way to avoid the temporal symmetry of statistical mechanics as it applies to closed or isolated systems.[3]

A branch system, according to Reichenbach, is a thermodynamic subsystem which branches off from the comprehensive system, remains isolated or quasi-isolated for a length of time and then merges or reunites with the larger system. The general idea of Reichenbach's account is that the temporal symmetry of statistical mechanics is avoided because individual branch systems do not exist for long enough periods of time to exhibit the temporal symmetry of closed thermodynamic systems. Rather they separate from the larger, more comprehensive thermodynamic system (which may be temporally symmetric) usually in a state of low entropy (a likely outcome of the separating process); and since high entropy states are more probable than low entropy states, they increase in entropy over time. They rarely decrease in entropy since low entropy states are improbable and branch systems do not exist long enough for this improbable (but not impossible) situation to occur.

Familiar examples of branch systems are a glass of water into which an ice cube has just been dropped, a cup of coffee into which cream has just been poured, an open bottle of perfume left in a closed bathroom, etc. Unfortunately, much more needs to be said about the nature of a branch system. It has only been vaguely characterized by its proponents, Reichenbach and Grünbaum, who have tended to rely on our intuitive notions about local, familiar thermodynamic systems.

For present purposes, this vague intuitive notion is sufficient. Hence, this paper will speak of branch systems and space ensembles of branch systems, the latter being a group or collection of branch systems (at any entropy) existing at an arbitrary temporal moment. With these concepts in mind it is now possible to turn to a consideration of Grünbaum's thesis.

The branch system hypothesis of Grünbaum has been attacked on a number of different grounds by a variety of authors. John Earman, a persistent critic of Grünbaum, rejects Grünbaum's claims for the hypothesis on the ground that a de facto temporal asymmetry cannot render time anisotropic; that is, he rejects the Grünbaum view on the basis of the definition of *anisotropy* as it applies to time.[4] Lawrence Sklar also finds fault with the branch system hypothesis. On Sklar's account, one ought to be suspicious of the thesis because it rests on Boltzmannian statistical mechanics rather than on the more sophisticated Gibbsian mechanics.[5] On the other hand, P.C.W. Davies in his book *The Physics of Time Asymmetry* is not concerned with these more philosophical issues and he uncritically presents and espouses the branch system hypothesis as revealing an important temporal asymmetry, although he relies on the earlier version of Reichenbach rather than that of Grünbaum.[6]

Curiously, none of these authors looks closely at Grünbaum's actual thesis; namely, that on the branch system hypothesis time is rendered anisotropic; that is, it provides a structural distinction between the two opposite time senses. One must ask: is the branch system hypothesis an account of an irreversible process such that the two directions of time are distinguishable by it? I will argue in what follows that Grünbaum has not accomplished the task he set out to do. By his own definition of *temporal anisotropy* the branch system hypothesis fails to satisfy the relevant criteria. Although the conceptual and reductionist issues raised by Earman and Sklar are most interesting in their own right, they need not be persuasive. (Arguments from definition are rarely persuasive.) It is important to see that the weakness of the branch system hypothesis lies within, not without.

Grünbaum does not explicitly define his use of *anisotropic*. He comes closest to a definition in the following words:

> . . . the existence of irreversible processes structurally distinguishes the two opposite time senses as follows: there are certain kinds of sequences of states of systems specified in the order of decreasing time coordinates such that these same kinds of sequences do not likewise obtain in the order of increasing time coordinates. Accordingly, if there are irreversible kinds of processes, then time is *anisotropic*.[7]

That is, according to Grünbaum, if the two directions of time are determined by serial ordering, then time is anisotropic if there exists a structural difference between them. A structural difference occurs when some sequence of states of a physical system characterizes events in one temporal direction but not the other. A well-noted feature of this definition is that this structural difference may be nomological or de facto in origin.[8] Although this characterization of the anisotropy of time is problematic, it is not the focus of our attention here. Clearly, Grünbam's intention is that the two temporal directions are distinguishable in virtue of some property of sequences of states of physical systems. Hence, this characterization will be accepted as the basis for the critical assessment of the branch system hypothesis.

Grünbaum's branch system hypothesis holds the following: phenomenologically we observe the Second Law of Thermodynamics to obtain. That is, in the thermodynamic systems we observe, entropy either increases over time or remains at equilibrium. We never observe entropy decreases in quasi-isolated systems (when properly described) with increasing time. According to Grünbaum, we can understand how this occurs despite the temporal reversibility of statistical mechanics, if we note that these observed systems are characterized by two de facto conditions. The conditions are (1) these systems are branch systems, and thus they have not been and will not remain permanently isolated [hereafter referred to as the duration condition] and (2) the microstates which make up the initial macrostates of each branch system are random samples of all the microstates which can constitute a macrostate of that entropy [hereafter referred to as the randomness condition][9] According to Grünbaum, these two conditions when taken as boundary conditions on the laws of statistical mechanics yield the observed asymmetry of increased entropy because they generate the following two statistical regularities:

> (1) With regard to space ensembles of branch systems whose branch systems are initially in states of low entropy, a majority of space ensembles of such systems will have a majority of their branch systems in a state of higher entropy after a time, T. There are two reasons for this. First, the duration condition prevents these branch systems from reproducing the temporal symmetry of an isolated system. That is, these branch systems cannot exhibit the temporal symmetry of permanently isolated thermodynamic systems since they have not existed prior to their initial low entropy states nor will they—if T is not too long—exist long enough to decrease spontaneously in entropy as an isolated system eventually would. Second, because of the randomness condition these branch systems will have the following property of permanently isolated systems: since large entropic decreases are far less probable than moderate ones, most submaximum entropic states of a permanently closed system are on or very near the upgrades of entropic curves. Hence, because of the randomness condition most branch systems of the space ensemble having initial low entropy states will have low entropy states located on or very near the bottom of their entropic curves and are likely to be found on the upgrade of their entropic curves after being quasi-isolated for a limited period of time.[10]
>
> (2) Space ensembles of branch systems each system of which is initially in relatively high entropy will continue to have a vast majority of their branch systems in relatively high entropy after a finite time T. For the randomness condition guarantees that most of these systems fall well within the plateau of the one-system entropy curve, not near its extremities, and hence are not likely to decrease in entropy in the near future. Or to put it another way, since high entropy states are more probable, most high entropy states endure for long periods.[11]

Since all space ensembles of branch systems have branch systems which are initially in states of relatively low entropy or relatively high entropy, they fall into one of these two categories and exhibit the regularities described by them. On the basis of these two regularities Grünbaum concludes:

> . . . it is . . . apparent that the statistical distribution of these entropy values on the time axis is such that the vast majority of branch systems have the *same direction of entropy increase* and hence also the same opposite direction of entropy decrease. Thus, the statistics of entropy increase among branch systems assure that *in most space ensembles the vast majority* of branch systems will increase their entropy in one of the two opposite time directions and decrease it in the other: in contradistinction to the entropic time-symmetry of a single, permanently closed system, the probability within the space-ensemble that a low entropy state *s* at some given instant be followed by a higher entropy state S at some later given instant is much greater than the probability that *s* be preceded by S. In this way *the entropic behavior of branch systems confers the same statistical anisotropy on the vast majority of all those cosmic epochs of time during which the universe exhibits the requisite disequilibrium and contains branch systems satisfying initial conditions of "randomness".*[12]

Grünbaum's account here raises many questions, such as the meanings of *vast majority* and *cosmic epoch*, the length of duration of the systems under consideration, etc. Rather than deal with these questions, however, what I want to do here is to assume a general consensus on the nature of branch systems and to ask the more meaningful question: whether or not Grünbaum has in fact described a genuinely anisotropic time—as he himself has characterized anisotropy and the branch system hypothesis. That is, what one wants to know is whether or not the laws of statistical mechanics taken together with the two de facto boundary conditions of duration and randomness are sufficient conditions for a statistical temporal anisotropy as Grünbaum claims them to be. A closer examination, I think, reveals that they are not. The reason for this is Grünbaum's failure to describe the behavior of branch systems in a non-question-begging way. Rather, his description of the nature of branch systems is temporally loaded such that we cannot be sure if a genuine temporal anisotropy exists or one which follows merely from the temporal terms themselves.

To see this, consider Grünbaum's characterization of the relevant temporally asymmetric branch systems: they are those whose *initial* conditions satisfy his postulate of randomness. Such temporally asymmetric branch systems, however, do not render time anisotropic—in the Grünbaum sense—just in case there also exists with the same relative frequency other systems (call them "merge" systems)[13] whose "final conditions" satisfy Grünbaum's randomness postulate. That is, the Grünbaum thesis as stated does not speak to, deny, or rule out the possibility of quasi-isolated thermodynamic systems which are similar to branch systems except for the fact that they obey the randomness condition in their final states and, presumably, the non-randomness condition in their initial states. Hence, if we consider all the thermodynamic subsystems of the more comprehensive system which satisfy the randomness condition at either one of their terminals—and surely we must do so if we are not to beg the question—it may turn out that we would not find any temporal anisotropy, statistical, or otherwise with regard to entropy increase in branch systems.

Grünbaum is aware of the temporal loadedness of his description of the branch system hypothesis but claims that the statistical anisotropy of entropy increase in branch systems does not depend on either (1) "assigning the lower of two time coordinates t_1 and t_2 to the lower of two entropy states of a branch system" or (2) "designating the lower of these two states as the 'initial' state rather than as the 'final' state."[14] If, says Grünbaum, we correctly specify the boundary conditions at either of the two extremities of the careers of branch systems, we will have a statistical anisotropy.[15]

Using our usual time coordination, at time t_1 a typical space ensemble of branch systems will have its members falling into either of two classes: Class A systems are in relatively low entropy and Class B systems are in relative equilibrium. Grünbaum now claims that if the microstates of these branch systems have the de facto property of randomness, then as the discussion above showed, at t_2 (where $t_2 > t_1$) the Class A systems will be in relative equilibrium. More importantly, according to Grünbaum, at time t_2 the Class A systems will lie temporally near or at the extremity (i.e., near the t_1-extremity) of the one-system entropy curve.[16] Hence, the microstates which correspond to the high entropy states of Class A systems are not random samples of the totality of microstates which correspond to macrostates of that high entropy. On the other hand, Class B systems at time t_2 continue to lie well within the plateau of the one-system entropy curve. Hence the microstates underlying these equilibrium macrostates continue to be random samples of high entropy microstates. Grünbaum believes that if we specify the conditions of Class A and Class B at time t_2, we will find that A systems decrease in entropy in the t_1 direction while Class B systems continue in relatively high entropy in that direction. Hence, it does not matter which time, t_1 or t_2, we consider the initial time. Nor does it matter if we replace the usual time coordination with a transformation $t \rightarrow -t$. We will still get a statistical anisotropy of entropy increase in one temporal direction and entropy decrease in the other.

Perhaps a few pictures will help clarify Grünbaum's argument. Consider the following two entropy curves of a closed system:

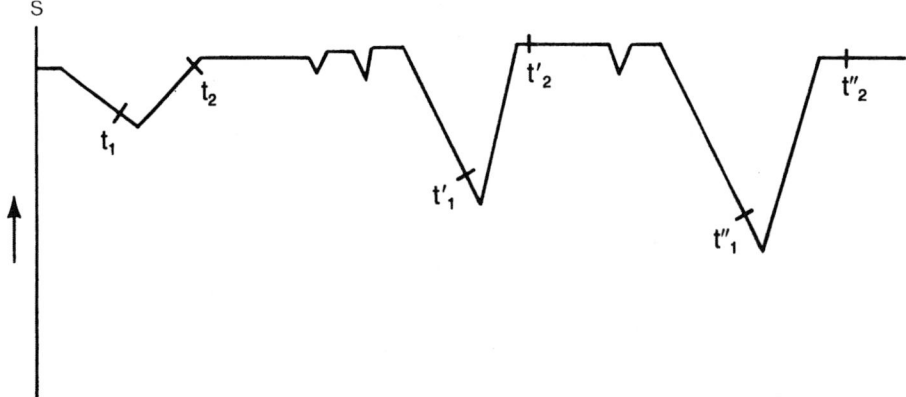

Figure 1. Entropy curve exhibiting Grünbaum's Class A thermodynamic systems.

According to Grünbaum, a Class A branch system with a random microstate at t_1 or t_1', etc. will have a "non-random" microstate at t_2 or t_2', (respectively) etc. in the sense that, although in a high entropy state, it is near or on a dip. Hence, that *same* system—if t_2 is considered its initial state—will exhibit a decrease in the t_1 direction.

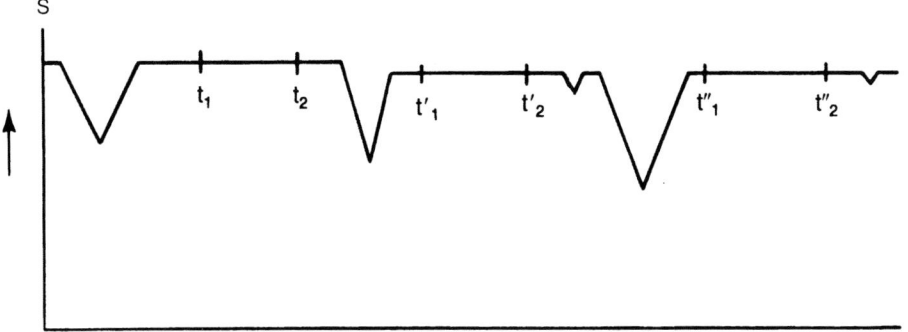

Figure 2. Entropy curve exhibiting Grünbaum's Class B thermodynamic systems.

A Class B branch system with a random microstate at t_2 or t_1', etc. will continue to have a random microstate at t_2 or t_2' (respectively), etc. Hence, that same system, if t_2 is considered its initial state, will exhibit entropic equilibrium in the t_1 direction.

Notice, however, that in Grünbaum's account there are two different specifications of the relevant branch systems. To pick out the relevant branch systems at t_2, one must use a different specification; namely, for the Class A systems we now pick out those systems which do not have microstates which are random samples of the totality of microstates corresponding to macrostates of that entropy. (Of course, for Class B systems the same specification may be used.) What this means, says Grünbaum, is that we can pick out the same statistical anistropy whether we do so at t_1 or at t_2.

However, his conclusion is unjustified. The temporally loaded character of Grünbaum's argument is revealed by the fact that one must know which terminal of a space ensemble of branch systems—the initial or final—one is talking about to know which boundary conditions to specify. That is, it does make a difference which end of a branch system we call its initial state and which we call its final state, not because we are to call the lower state the initial state (which is the objection Grünbaum is trying to avoid), but because we are told to consider only initial states satisfying the randomness condition (i.e., that the microstates which are random samples of all the microstates which can constitute a macrostate at that entropy) and final states (of Class A systems) which do not.

The question then arises as to whether or not the above-mentioned merge systems exist, that is, Class A systems with non-random initial states and random final states. Nothing in Grünbaum's account rules out their existence or would suggest that their existence is any less frequent than that of the relevant branch systems. Should the latter be the case—i.e., should merge systems be just as frequent as branch systems—there would, of course, be no justification on the branch system hypothesis for claiming that time is anisotropic. To put the issue another way, if we consider *all* the thermodynamic subsystems of the more comprehensive system which satisfy the randomness condition at either one of their terminals—and surely we must—it may turn out to be the case that we would not find any temporal anisotropy, statistical or otherwise with regard to entropy increase in branch systems. Entropy might tend to increase in more or less equal numbers of branch sytems in each temporal direction—with those branch systems satisfying the randomness condition in their initial states increasing in the positive direction and those satisfying the randomness condition in their final states increasing in entropy in the negative temporal direction. Perhaps a diagram will clarify what is involved here.

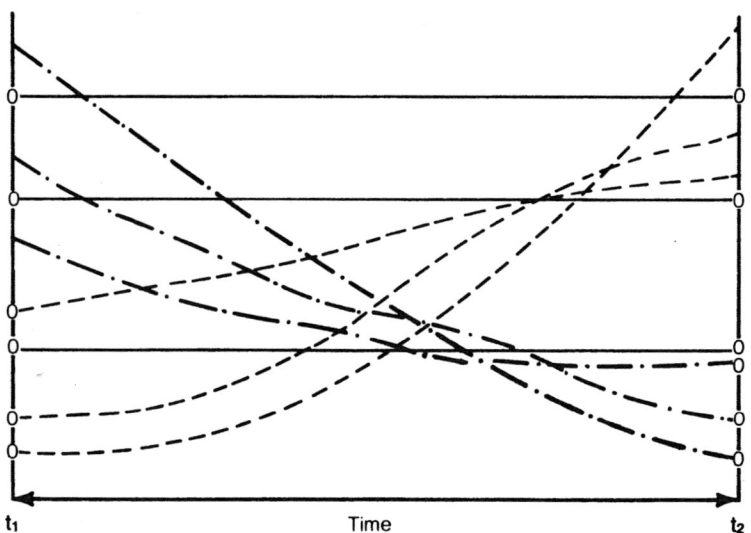

Figure 3. A space ensemble of branch systems over time.

In Figure 3:

———— designates Class B systems, systems which maintain equilibrium throughout their existence and which are virtually irrelevant here.

– – – – designates Class A systems which satisfy Grünbaum's randomness condition at t_1.

– · – · – · designates "merge" systems, systems which satisfy Grünbaum's randomness condition at time t_2.

   0   indicates point at which randomness condition is satisfied.

At this point, it seems to me, that Grünbaum's only argument would be to claim that the "final states" of merge systems do not truly satisfy his randomness postulate.

For given that such states would either be low entropy states or high entropy states, if high entropy states, they are simply the above-mentioned Class B systems and are irrelevant for temporal asymmetry. If low entropy states at t_2, however, they must have non-random microstates since they most likely existed as high entropy states at t_1; therefore, for the most part, they are representative of the highly improbable microstates which underlie the transition of a system from a high entropy macrostate to a low entropy macrostate.

But surely this counter argument would be question-begging. For it clearly depends on the assignment of the lower of two time coordinates to the initial states of branch systems, thereby rendering them able to satisfy the randomness postulate. Under the transformation of $t \mapsto -t$, however, the "final states" of merge systems would also be capable of satisfying the randomness postulate.

What one wants here and what is missing from Grünbaum's argument is some independent, temporally neutral account of randomness and why it might characterize one end of a branch system and not the other. Perhaps there is some fact about the universe or about cosmic epochs. If so, then the anisotropy of time may rest there and not with thermodynamics. Perhaps the justification might be something like that of Reichenbach who justified his branch system hypothesis on the basis of observation.[17] Although there is something to be said for this kind of justification, it is obviously not without problems. For, although we are very familiar with branch systems of increasing entropy such as cream mixing with coffee and ice cubes cooling a glass of soda, we are also quite familiar with thermodynamic subsystems which become more complex and hence might be said to decrease in entropy over time. The growth of a biological organism is a case in point. Moreover, we need not turn to biology for possible counter-examples. The growth of chemical crystals and even the gradual erosion of the beach by the sea might be considered possible counter-examples. On Grünbaum's account, one faces the further difficulty that observation would have to be in terms of the satisfaction of the randomness postulate and not simply the observation of the predominance of quasi-isolated systems of increasing entropy.

In conclusion, Grünbaum's branch system hypothesis does not do what he claims for it; it does not show that certain sequences of states of affairs exist in increasing serially ordered time which do not obtain in decreasing serial order. Hence, it does not render time anisotropic, according to Grünbaum's own criteria of temporal anisotropy. For on Grünbaum's account the temporal directedness of entropy increase in branch systems does not come from temporally neutral boundary conditions together with the laws of statistical mechanics, but rather from the temporal directedness of the application of the boundary conditions. One must assume a temporal direction and the temporal asymmetry Grünbaum describes is then generated by this assumption.[18] What we do have here is an explanation of why the Second Law of Thermodynamics appears to hold. That is, if certain boundary conditions obtain on one end of the temporal interval of the space ensemble and not the other. Grünbaum's claim, however, is stronger than this. He claims to have distinguished structurally the two opposite time senses. The branch system hypothesis, as we have seen, fails to meet this claim.

Notes

1. Karl Popper, "Time's Arrow and Entropy," *Nature* 207 (July 1965):233—34. For the suggestion that temporal direction might be connected with entropy *decreases*.

2. Adolf Grünbaum, *Philosophical Problems of Space and Time* (New York: Alfred A. Knopf, 1963), pp. 209—81.

3. Erwin Schrödinger, "Irreversibility," *Proceedings of The Royal Irish Academy*, LIII, Sec. A (1950): p. 189; idem, "The Spirit of Science," *Science and Nature*, ed. J. Campbell (New York: Pantheon Books, 1954), pp. 337—41; both cited by Grünbaum, *Philosophical Problems of Space and Time*, p. 245; Hans Reichenbach, *The Direction Of Time*, ed. Maria Reichenbach (Berkeley: University of California Press, 1956), pp. 117—193.

4. John Earman, "The Anisotropy of Time," *Journal of Philosophy* 47 (December 1969): 273—95; idem, "An Attempt to Add a Little Direction to 'The Problem of the Direction of Time'," *Philosophy of Science* 41 (March 1974):32.

5. Lawrence Sklar, *Space, Time and Spacetime* (Berkeley: University of California Press, 1957), pp. 379—94; "Review: Adolf Grünbaum: *Philosophical Problems of Space and Time*," *Journal of Philosophy* 74 (August 1977):497.

6. P.C.W. Davies, *The Physics of Time Asymmetry* (Berkeley: University of California Press, 1974), pp. 68—74.

7. Adolf Grünbaum, "The Status of Temporal Becoming," *The Philosophy of Time*, ed. Richard Gale (Garden City: Doubleday & Co., Anchor Books, 1967), p. 327.

8. Henryk Mehlberg "Physical Laws and Time's Arrow," in *Current Issues in the Philosophy of Science*, ed. Herbert Feigl and Grover Maxwell (New York: Holt, Rinehart and Winston, 1961) is the first to take exception to Grünbaum's de facto account.

9. Adolf Grünbaum, *Philosophical Problems of Space and Time*, pp. 255—57.

10. Ibid., pp. 257—58.

11. Ibid., p. 258. A branch system need not initially be in a low entropy state in the Grünbaum account.

12. Ibid., p. 259. [Italics Grünbaum's.]

13. A term first used in discussion by P. Fitzgerald.

14. Adolf Grünbaum, "The Anisotropy of Time," *The Nature of Time*, ed. Thomas Gold (Ithaca, N.Y.: Cornell University Press, 1967), p. 164.

15. Ibid.

16. This rather large assumption on Grünbaum's part depends for its truth on the relation between the duration of the branch system's existence and the length of the dip in the entropic curve. If the system endures too long for the depth of the entropy decrease, its end-point will obviously be well within the plateau of the one-system entropic curve. A similar assumption is also made concerning the duration of Class B systems.

17. Reichenbach, *The Direction of Time*, pp. 117—193.

18. J. M. Blatt in his article "An Alternative to the Ergodic Problem," *Progress of Theoretical Physics* 22 (December 1959): 745—756, cited by Davies and others, makes a similar mistake. He tries to show that the apparent temporal asymmetry of coarse-grained thermodynamics is well-founded in fine-grained thermodynamics, properly considered. Unfortunately, the temporal asymmetry of Blatt's fine-grained thermodynamics rests on an asymmetric application of boundary conditions.

Time Proverbs and Social Change in Belgrade, Yugoslavia

M. SPANGLER

In contemporary Belgrade the social idea of time has been changing in two ways. First, calendrical time no longer reflects Serbian Orthodox religious custom to a significant degree. The Church or Julian calendar has been gradually relegated over the past sixty years to a minor function of religious ritual; and Serbs, comprising over 80% of the population of the city (Bogavac 1974:665), generally calculate even the dates of many religious observances by the Gregorian calendar, civilly accepted in 1918 (Mišković 1966:106). Just as importantly, church holidays are no longer recognized by the postwar socialist government and are being gradually supplanted in their general observance by civil holidays (cf. Spangler 1979: Chapter Two).

Second, contemporary Belgradians are increasingly relying on the clock for daily time reckoning. However, since clock time is a late 19th century loan into the urban model of time, other components of the model, not yet being adjusted to clock time, do not facilitate a more precise and widespread use of the clock.[1] For example, on the basis of informal counts on the major streets of Belgrade, I found that only one out of four individuals wears a watch. Since there are not many public clocks, the majority of the urbanites estimates the time of day from previously timed activities or asks the time of someone with a watch. A personal watch is considered to be expensive or, more significantly, unnecessary according to general social opinion.

There are a number of consequences which are partly the result of indirect clock time reckoning.[2] The most important consequence is a less demanding punctuality in many situations and an even more demanding one in cases where an activity is viewed as crucial to the individual. The 'isolate' of the time system, the smallest practical unit of time, is 15 minutes rather than one to five minutes as in the United States (Hall 1959:130−1). Time is not divided further because of difficulties in more precise time reckoning.

As I have implied above, there is also stability in the transformation of the institutionalization of time in Belgrade. Certain evaluative and emotional features of

19th century rural temporal concepts and attitudes have persisted in the city. Primary causes for this persistence lie in the relatively recent industrialization of the Serbian republic and extensive rural-urban migration over the past three decades. While time is rigidly scheduled in some contexts, such as factories and clerical offices, in many other social and economic situations time contines to be considered as an 'elastic' category. Timed activities are more freely lengthened or shortened since it is believed that time can be reallocated to necessary activities. Rules of time utilization seem to reflect a devaluation of certain scheduled work activities. Labor is generally not a value in itself in Belgrade, just as it was not in certain areas of 19th-century Serbia. There are a number of factors accounting for work values which become variable and complex with social and occupational stratification in Belgrade. I have suggested elsewhere that the cultural value of work is related to the elastic conception of time[3] (Spangler 1979). Thus the social definition of time is in a transition caused by large-scale processes of industrialization, urbanization, and the formation of a socialist nation-state. However, the transformation is ongoing in an expansive temporal and cultural system which retains some stability.

Methodology

A list of proverbs, compiled from a large corpus, constitutes the basis for an analysis of certain changes and continuities in cognitive, emotive, and evaluative concepts of time, which have accompanied the institutionalization of clock time and the civil Gregorian calendar in Belgrade.

Many social scientists have turned to the study of proverbs for a greater knowledge of shared perceptions of behavior, in particular, of social and political activity (i.e., Arewa and Dundes 1964; Firth 1926; Herskovits 1930; Herzog 1936; Messenger 1959; Giovannini 1978). Yet, with the exceptions of Dundes (1969) and Algar (1971), who have considered oral literature in complex societies, most of the studies mentioned above emphasize the conventionalized interpretation of proverbs. Leonard Doob (1971) exemplifies this approach in his programmatic treatment of time proverbs. According to Doob, proverbs serve three functions:

1) Proverbs convey information about traditional rules of behavior.
2) Proverbs are cited, in teaching and conversation, to enforce and justify rules of behavior.
3) Proverbs demonstrate that their user is conforming to the rules or that he is at least aware of what conformity means.

In the present study I shall argue that proverbs not only reflect collective 'folk wisdom' (traditional rules of behavior), but also change in meaning as people apply them to new situations. Hence proverbs may reflect new rules of behavior and be employed by individuals who may not yet be sure of what social conformity to the rules means. Social changes in temporal institutions in Belgrade provide us with experimental situations for the study of variability in temporal models and attitudes which I have typified above.

A sample of 200 Serbian informants, stratified by age, sex, and occupation, were asked to explain the meaning of eleven proverbs, illustrating their answers with examples which they have observed in everyday life. The content of the elicited responses was analyzed in light of a general hypothesis: The interpretive meaning of time proverbs varies with not only social definitions of time but also social and economic experience. More specifically, different social, age, and occupational groups evince a differential tendency to utilize more traditional concepts tied to religious conceptions and sun-time reckoning as opposed to contemporary concepts and attitudes based on civil and clock time.

The Proverbs

Eleven proverbs were selected from over six thousand proverbs in the Karadžić collection (1965) in order to elicit interpretations. Each proverb mentioned time (*vreme*) or was concerned with a significant temporal aspect. It soon became apparent from responses that only middle-aged and elderly informants or informants with a relatively recent rural origin used proverbs frequently. Of course, almost all of my informants professed familiarity with the proverbs. The original questionnaire list of proverbs was reduced to four proverbs in the analysis here because of limitations in scope. More importantly, the four proverbs proved more useful than the remaining in illustrating cultural attitudes toward time and require a detailed examination in themselves. We shall not be concerned with interrelationships between the proverbs although this is also a valuable problem for study (cf. Maranda and Maranda 1971; Fischer and Yoshida 1968).

Proverb 1: A clock is a lock.

This proverb derives from a fairly modern slang term. The usage is identified as a proverb (*poslovica*) by informants who are generally under the age of 25. The usage has been somewhat liberally translated from the Serbian. In actuality, the expression is more specific: a clock is a padlock. The slang term, padlock (katanac), arose in Belgrade in the early twentieth century and at first referred only to a pocketwatch (džepni sat) which did not work. Since the watch could not serve its purpose and was on a chain, it was derisively referred to as a padlock. The repair of watches was probably not adequate in Belgrade as well as being expensive.[4]

Questionnaire responses determined that few people over 65 and almost no one under 65 could recall the original slang expression. However, library research and interview data demonstrate that the slang was well-known among male upper-middle class circles in early 20th century Belgrade. Dragoslav Andrić derives the usage from the "underground," professional criminals and delinquents in Belgrade (1976). This assertion seems to be incorrect since so many members of the middle class (bureaucratic and professional occupations) used the expression without embarrassment or reference to its alleged origins in a group of criminals or 'rowdies' (mangupi).

The contemporary interpretation of the usage revolves around at least three ideas:

1) time is limited;
2) time obligates and dictates to people;
3) clock time is in opposition to human needs and measures.

These meanings contrast strikingly with the original meaning of the expression in early 20th century Belgrade and indicate the greater degree of familiarity with clock time in modern Belgrade, if not a wider acceptance of clock time. It is significant to note that the negative evaluation of clock time was not as common among Yugoslav informants as American ones to whom the same questionnaire was also administered. It seems that clock time is not as influential in Belgrade social life as in American life and does not warrant as many objections to its artificiality or institutionalization. Serbian peasants in the early 19th century also used proverbs which belittle the use or value of the clock:

> Ko nema spahije, neka kupi sahat (pa će potrošiti ono što bi plaćao spahiji) (Karadžić 1965:158).
> Let him who has no landlord buy a clock (so that he will spend what he would have paid to the landlord).

> Trbuh je najbolji sat. (Trbuh najbolje pokazuje kad je vrijeme jesti.) (Karažić 1965:287)
> The stomach is the best clock. (The stomach best shows when it is time to eat.)

Peasants simply had not adjusted to the artificial measurer of social life, the clock; much less have they been affected by it.

Proverb 2: Time is money.

It is doubtful that this proverb is an independent Serbian or Slavic invention. The proverb has been attributed to Benjamin Franklin's Advice to a Young Tradesman (1748), "Remember that time is money" (II:234). In 1572 Wilson wrote in A Discourse upon Usury, "They saye tyme is precious" (1925:228). Bacon (1612:162) also mentions the topic. Hence the proverb is more literary than folk in origin. It was probably recognized largely in the merchant classes of 17th and 18th century Europe and America.

Knežević would have us believe that the expression is Serbian since he lists it as a traditional Serbian proverb, "Time is the most precious money" (1957:89). However, his collection is too modern to insure the veracity of the citation. Dera (1894:16), a 19th century cleric, states unequivocally that the usage is English and not Serbian. If the Knežević proverb is authentic, it may not be common to all South Slav regions and probably referred to the shortness of the seasons and pečalba (migratory labor during the summer) rather than to production rates and contract deadlines.

Daniel Bell explicates the proverb in the following passage:

> First, it was an injunction against idleness (and waste); second, it was a view of time as something methodical, a set of divisions into hours and minutes whose very measure could regulate a calculus of utility and the allocation of energies.

However, the American sociologist's interpretation of the proverb is itself culturally bound or biased.

In my Yugoslav sample there were no significant differences in interpretations according to age group. Evidentally a stable cross-generational consensus underpins conceptions and attitudes toward time. Three attitudes readily emerge from the interpretations of the above proverb:

1) time is precious if one is paid for it;
2) time is precious because it is irreversible. When lost, time cannot be recovered;
3) one should plan and work effectively.

Two informants also believed that the proverb was foreign to South Slav culture.

Roughly ½ of the sample expressed sentiments decrying the desire for money and the concern for work. These attitudes support the weak work motive and oppose the concept of timed work and the pervasive monetary standard. However, ¹/₅ of the sample believed that "those with money could make their time valuable" (Informant 17). Everyone else, the same informant felt, had too much time and too little money. The final ³/₁₀'s of the sample, generally from professional occupations, argued for a general recognition of the value of time, but usually in a way counter to the work sentiment of the proverb as traditionally expressed in English sources. I have found in many of the responses a slightly critical, if not strong bias against the 'time is money' attitude, which is perceived as becoming dominant in modern Serbian society.

Proverb 3: Suffering does not take your soul, rather the judgment day does.

Kulišić et al. (1970:279) give the above proverb from 'folk verse' without tracing its source. The proverb is clearly Serbian and old since it can be found in Karadžić's 1836 list (1965:184). The phrase, 'judgment day,' is more aptly expressed as 'destined or appointed day.' The Serbian equivalent, *sudjeni dan*, is distinct from Judaic and Christian concepts of the judgment day although interrelated with them. I have translated the phrase as judgment day here because of the religious syncretism which will be explained below.

The proverb is directly related to the concept of fate. Grimm (1875:362) and, before him, the Roman historian Procopius of Caesarea have written that the ancient Slavs did not believe in Fate or its influence among people. While it may be true that the ancient Slavic beliefs were not exactly coincident with those of classical Greek beliefs, it is beyond doubt that Fate was an integral part of the ancient Danubic Slav religions as well as of the traditional Serb's conception of temporality. Hence we have the proverb

> Nema smrti bez sudjena dana (Milićević 1894:52). There is no death without the appointed day.

Moreover, when someone was sick, peasants usually quoted the proverb

> Ako ima veka, biće i lika (Milićević 1894:53).
> If there is time, there will be a cure.

Both proverbs stress the belief in an unavoidable fate. Many folktales and legends support the belief in the Sudjenice or three Sisters of Fate, who determined how long a person will live, in addition to other events in his life (cf. Spangler 1979: Section 3.6). The judgment day or sudjeni dan is thus the day on which the Fates decreed one was to die. Suffering could not kill a man, but only the decision of the Fates. Of course, the Fates were subordinate to God in the 19th century and thus the judgment day was a divinely appointed one. While the day did not refer to the Judgment of the Dead at the Second Coming of Christ, it did signify a personal day of Fate on which one left the world of the living.

Half of our contemporary sample believed that the proverb referred to the power of Fate. Older and rural informants especially recognized the association of the proverb with Fate. However, most informants denied the personal meaning of the fatalist doctrine in their lives. Moreover, approximately two thirds of the informants below the age of 25 felt that the proverb referred to the lack of ultimate importance to certain problems in their lives: problems can be overcome, if not avoided. Hence fatalistic doctrines of folk Christianity have been eroded in the urban environment.

Proverb 4: Save money for a rainy day.

This proverb has been liberally translated into English from the Serbian. The exact Serbian translation is: save white money for black days. White money traditionally referred to gold and silver, and a black day to a day of misfortune. The English equivalent does not distort this meaning. The English proverb at least goes back to the 16th century and is probably much older since it can be found in classical Greek and Latin sources.

Karadžić cites the Serbian proverb in his 1836 collection (1965:310). It can also be found in Muškatirović (1807:129). The proverb refers to the necessity of providing for one's financial security in the future.

The majority of responses in my sample suggested that the proverb referred to a specific time when a person had more money than he needed and thus could save for the future. Otherwise, the uncertainty of life and the shortcomings of plans in the face of the future seem to be more obvious cultural rules of time conceptualization than one of successful planning into the future.

Interestingly, a significantly greater number of male than female informants denied any application of the proverb in their lives.

Among their responses, I cite the following:

"To enjoy life one must start today."
"Money is not to be saved, for it is of little importance when death is so close."
"To be a real man is to show little concern for money."
"My wife will live longer than I shall. Let her receive her pension."

While most people in Belgrade save money, many men will not admit to it in certain settings, such as in a *kafana* (cafe-bar), because it is not only against the masculine standard of living in the present but also an admission of domination by their wives. Time perspective consequently becomes a way of defining sex and marriage roles to some extent. As Doob puts it, "The modal temporal perspective of a society reflects and affects a modal philosophy of values pertaining to other behavior" (Doob 1971:56).

The present time orientation is a traditional time perspective, reflected in many old Serbian proverbs:

Danas jesmo, a sutra ni jesmo.
Today we live, but tomorrow we do not.

Danas čoek. Sutra crna zemlja.
Today a man. Tomorrow black earth.

Danas sutra, dok i smrt za vrat.
Today tomorrow, while death is at your back.

(Karadžić 1965:85−7)

However, it would be unwise to conclude that Serbs act as if they will die tomorrow. The proverbs and similar arguments are simply given in certain social contexts as justifications for certain actions.

When I tested Belgrade college students on the proverb, many male students found the question difficult or pointless because they felt that they could not "know the future" and that "it is best to live from today until tomorrow." These reactions point to a subtle change from more traditional time attitudes, which emphasized religion, powerlessness, and masculinity. Tradition is not inviolably preserved in the present. Even if the rules for behavior are socially reinforced and culturally utilized, the original justifications for the rules change.[5]

Conclusion

Each of the four proverbs analyzed above offers us a glimpse into the complex changes in temporal concepts and attitudes in Belgrade. The first two proverbs demonstrate the emerging significance of clock time and of 'inelastic time,' the time of work periods and deadlines, qualified by work attitudes and socio-economic circumstances. The third proverb concerned with fate and the fourth with future time

perspective, however, indicate to some extent the stability of emotional and evaluative components of time inherent in 19th century and rural models of time, as well as
the urban model.

A more complete treatment of temporal concepts and attitudes would supplement
the data presented here with participant-observation and extended interview data. I
have attempted to defend two propositions:

1) Proverbs represent changing rules of thinking and acting.

The interpretation of proverbs from one social group to another reflects varying
cultural rules as well as values and attitudes.

2) Belgradians are evolving new temporal attitudes: at once an ending and a
   beginning in the constant social evolution of the idea of time.

The change currently revolves around the clock and the acceptance of rigidly
scheduled industrial time as opposed to more elastic agricultural time. Yet the notion
of inelastic time, inflexibly timed periods of human activity, has not yet been
accepted and perhaps will not be because of emotional and evaluational attitudes
which the proverbs have partially revealed.

Joel M. Halpern suggested that "forms of organization are increasingly evident
which doubtless derive from shared values about the instrumental utility of science
and derivative technology for the maintenance of contemporary society" (Halpern
1964:1). Of course, the commitment to a universal technological subculture is
variable within a society as well as cross-culturally. Postwar Belgrade presents a case
against general convergence on a universalist technological culture. While people
have accepted the automobile, airplane, and clock, past cultural development still
dominates the use of technology. I have tried to show that even the clock is utilized
in a different manner from that in America. Older values and attitudes persist in the
South Slav state, despite social and economic changes.

The social and economic changes to which I have alluded in the preceding
paragraph have been glossed in several studies as *modernization*. Essentially, modernization presents a now outdated model of social convergence much as Westernization
and assimilation theory did before it (cf. Bendix 1967). At least three processes have
been unified under the rubric of 'modernity':

1) industrialization,
2) urbanization,
3) the transformation of traditional religious and social beliefs.

As a society undergoes the processes listed above, it becomes 'modern'. However,
such an approach is primarily institutional rather than social psychological. Material
culture and technological efficiency are seen as the main indicators of modernity.
Interestingly, Western Europe and American technology and industrial methods
usually provide the specific standards for evaluating modernity, even when a social

psychological approach is adopted. Hence Inkeles and Smith write that "modern men would be oriented to the present or the future rather than to the past . . . he would more readily accept fixed schedules as something appropriate, or possibly desirable" (1974:22). Significantly, such attitudes toward time permit "the factory to operate efficiently and effectively" (1974:19). Inkeles and Smith also consider "it more modern to be punctual" (1974:22).

Will industrialization necessarily lead to a uniform system of punctuality and fixed scheduling in Belgrade? Clearly, industrial workers in many countries must conform to such standards in their work situations. Diverse cultural factors must also support the standards outside of the labor situation if they are to be widely used in social life. It seems that Belgradians will not be modern in the Inkeles and Smith sense at least for some time since they believe that such standards are not justified or responsive to their needs (cf. Spangler 1979: chap. 5). We may even be witnessing a move away from more rigid time utilization in American and Western Europe as the work ethic declines and health and welfare dangers are perceived in such work and time attitudes. It is this shift in temporal notions which may well herald a greater change in the idea of time in the developed nations rather than in the underdeveloped nations of the world. As energy and natural resources decline, and inflation and health risks increase with modern technology, time itself may be redefined in terms of more elastic work and leisure conceptions which are no less appropriate and valid than the ones which preceded them.

Notes

The views expressed in this essay are those of the author and not necessarily those of the U.S. Department of State. IREX and the Fulbright-Hays Doctoral Program supported library and field research by the author in Yugoslavia during 1976–77.

1. Few Balkan towns had public clocks up to 1850. Privately owned clocks and pocketwatches were also rare up to the end of the 19th century.

2. Time scheduling and utilization are not explained by reference to time reckoning alone but also to affective attitudes and time orientation in general. Since social values and attitudes allow a lack of punctuality in certain situations, time reckoning has not become more exact. Nevertheless, the method of reckoning has placed some limitations on social attitudes and time orientation and scheduling. The system is an interrelated one.

3. Rézsoházy (1972:449–60) and Herskovits (1961:114–38) have generally argued that peasant and technologically primitive societies do not recognize precise deadlines or inflexible daily time requirements. In actuality, agricultural activities in such societies demand an extremely fine temporal organization during planting and harvesting times.

4. Austrian Serbs, Germans, and Jews dominated the clock-making trade in Belgrade and helped to establish Central European Time in Serbia although another time zone would have been more appropriate.

5. See Spangler and Petrovich (1978) for a consideration of future time perspective among college students in Belgrade and an American city.

References

Agar, Michael, 1971. Folklore of the Heroin Addict: Two Examples. *Journal of American Folklore* 84:175—85.

Andrić, Dragoslav, 1976. *Rečnik žargona* (Dictionary of Slang). Belgrade: Beogradski Izdavačko-Grafički Zavod.

Arewa, E. Ojo, and Alan Dundes. 1964. Proverbs and the Ethnography of Speaking Folklore. *American Anthropologist* 66:70—85.

Bacon, Francis. 1612. Essayes or Counsels, 1625 ed. in J. Spedding *The Works of Francis Bacon.* Eds. R. L. Ellis and D. D. Heath. 1625 ed. Vol. 12, London: Longmans, 1864.

Bell, Daniel. The Clock Watchers: Americans at Work. *Time* September 8, 1975:55—7. (Parenthetical addition is my own.)

Bendix, Reinhard, 1967. Tradition and Modernity Reconsidered. *Comparative Studies in Society and History* 9:292—346.

Bogavac, Tomislav, 1974. Demografske promene u Beogradu 1941—1971 (Demographic changes in Belgrade 1941—1971). In Vasa Čubrilović, Ed. *Istorija Beograda.* vol. 3. Belgrade: Prosveta. 649—84.

Dera, Djordje. 1894. *Radiše svega više, stediše joře više* (They worked as much as possible, they saved even more). Knjiga za Narod 40. Novi Sad: Manstirska Štamparija.

Doob, Leonard W. 1971. *Patterning of Time.* New Haven: Yale University Press.

Dundes, Alan. 1969. Thinking ahead: a folkloristic reflection of the future orientation in American Worldview. *Anthropological Quarterly* 42:53—72.

Firth, Raymond. 1926. Proverbs in Native Life, With Special Reference to Those of the Maori. *Folk-lore* 37:134—53; 245—70.

Fischer, John L., and Teigo Yoshida. 1968. The Nature of Speech According to Japanese Proverbs. *Journal of American Folklore* 81:34—43.

Franklin, Benjamin, 1748. Advice to a Young Tradesman. In The Works of Benjamin Franklin, ed. J. Bigelow. vol 2. New York: G.P. Putnam's Sons. 1904. (Page reference to the 1904 ed.)

Giovannini, Maureen J. 1978. A structural analysis of proverbs in a Sicilian Village. *American Ethnologist* 5:322—33.

Grimm, J.L.K. 1875. *Deutsche Mythologie.* vols. 1—2. Berlin.

Hall, Edward T. 1959. *The Silent Language.* Garden City, New York: Doubleday.

Halpern, Joel M. 1964. *Peasant Culture and Urbanization in Yugoslavia.* Waltham, Mass., Brandeis University. Department of Anthropology.

Herskovits, Melville J. 1930. Kru Proverbs. *Journal of American Folklore* 43:225—39. 1961. Economic change and cultural dynamics. In *Tradition, Values, and Socio-Economic Development,* ed. J. J. Spengler. Durham, N.C.: Duke University Press. pp. 114—38.

Herzog, George. 1936. *Jabo Proverbs From Liberia.* London: Oxford University Press.

Inkeles, Alex, and D. H. Smith. 1974. *Becoming Modern.* London: Heinemann.

Karadžić, Vuk S. 1965. Srpske Narodne Poslovice (Serbian Folk Proverbs). Volume 9 of *Sabrana Dela Vuka Karadžića.* Belgrade: Prosveta.

Knežević, M.V., ed. 1957. *Antologija govornih narodnih umotvorina* (Anthology of oral folk works). volume 1. Novi Sad: Zmaj.

Kulišić, Š., P.Ž. Petrović, and N. Pantelić, 1970. *Srpski Mitološki Rečnik* (Serbian Mythological Dictionary). Belgrade: Nolit.

Maranda, Elli Kongas, and Pierre Maranda. 1971. *Structural Models in Folklore and Transformational Essays.* The Hague: Mouton.

Messenger, John C., Jr. 1959. The Role of Proverbs in a Nigerian Judicial System. *Southwestern Journal of Anthropology* 15:64–73.

Milićević, M. Dj. 1894. Život Srba Seljaka (Life of serbian Peasants). *Srpski Etnografski Zbornik* 1. Belgrade: Srpska Kraljevska Akademija.

Mišković, B.B. 1966. Prilog ujednačenju gradjanskog kalendara (A contribution of the standardization of the civil calendar). *Glas* 263:28:93–147.

Muškatirović, J. 1807. *Pricte iliti po prostomu poslovice tjemza sentencije iliti rječenija* (Stories or simply proverbs also sentences or sayings). Buda: Royal Hungarian University.

Rézsoházy, R. 1972. The methodological aspects of a study about the social notion of time in relation to economic development. *The Use of Time*, ed. A. Szalai. The Hague: Mouton. pp. 449–60.

Spangler, Michael. 1979. Time and Social Change in a Yugoslav City. Unpublished Ph.D. Dissertation. University of Wisconsin Library, Madison, Wisconsin.

Spangler, Michael, and Olivera Petrovich. 1978. Future time perspective and feeling tone: a study in the perception of the days of the week by Yugoslav and American students. *Journal of Social Psychology* 105:189–93.

Stoianovich, Traian. 1967. *A Study in Balkan Civilization*. New York: Knopf.

Wilson, T. 1925. *A Discourse upon Usury*, ed. R.H. Tawney. New York: Harcourt, Brace, and Company.

Appendix

A Report on the Literature of Time, 1900–1980

J. T. Fraser

The three sections of this Appendix are intended to serve as a map to the literature of time in the Twentieth Century.

1. A Selected Bibliography of Books

This section contains over 800 citations of books, with no more than a handful articles in books. The entries were selected by the standard of whether they do, in some ways, extend the concerns of the papers included in the four volumes of *The Study of Time.*

Part A covers the period since the founding of the Society, that is, 1966–1980. Entries have been classified under twelve headings according to a division of time-related material that is employed in another publication, *The Voices of Time *.* The well-known problem of single subject classification is quite evident. Each book had to be entered under a single subject only, even if it qualified under several subjects. For this reason the classification is sometimes imperfect but, hopefully, always defensible.

Part B covers the period 1900–1965 and is arranged alphabetically by author or, in some cases, by title. To avoid unnecessary duplication, some 300 books published before 1966 and already cited in *The Voices of Time* have not been listed here.

The bibliography was not intended to include works simply because they manipulate the idea, or depict the experience of time, as a literary device.

---

*J. T. Fraser, ed. *The Voices of Time,* Second edition (Amherst: University of Massachusetts Press, 1981).

Each entry was pared down to the minimum information necessary for identification. A detailed, annotated and analytic bibliography of books pertinent to the study of time, including works before 1900, may very well be a useful reference book, but it could hardly have been made a part of the introduction to *The Study of Time IV*.

Part A. Works of Possible Interest, 1966–1980

1. Contributions Toward an Integrated Understanding of Time

Čapek, Milič, ed., *The Concepts of Space and Time: Their Structure and Their Development*. Dordrecht and Boston: Reidel, 1976. 570 p.

Doob, Leonard W. *Patterning of Time*. New Haven: Yale University Press, 1971. 472 p.

Elton, L. R. B., and Messel, H. *Time and Man*. New York: Pergamon Press, 1978. 112 p.

Fischer, Roland, ed. "Interdisciplinary perspectives of time." *Annals, New York Academy of Sciences* 138 (1967): 367–915.

Fraser, J. T., *Of Time, Passion, and Knowledge: Reflections on the Strategy of Existence*. New York: G. Braziller, 1975. 529 p.

———. *Time as Conflict: A Scientific and a Humanistic Study*. Basel and Boston: Birkhäuser Verlag, 1978. 356 p.

Fraser, J. T., ed. *The Voices of Time: A Cooperative Survey of Man's Views of Time as Expressed by the Sciences and the Humanities*. 2d ed. Amherst: University of Massachusetts Press, 1981.

Fraser, J. T. et al., eds. *The Study of Time*. Papers from the Conferences of the International Society for the Study of Time. Vols. 1– . Berlin, Heidelberg, and New York: Springer-Verlag, 1972–

Freeman, Eugene, and Sellars, Wilfrid, eds. *Basic Issues in the Philosophy of Time*. La Salle, Ill.: Open Court, 1971. 241 p.

Gale, Richard M., ed. *The Philosophy of Time: A Collection of Essays*. London: Macmillan, 1968. 514 p.

Gardet, L. et al., eds. *Cultures and Time*. 3 vols. Paris: Unesco Press, 1976–1979.

Grant, John, ed. *The Book of Time*. North Pomfret, Vt.: David & Charles, 1980. 320 p.

Pacault, Adolphe et al. *A chacun son temps*. Paris: Flammarion, 1975. 294 p.

Patrides, C. A., ed. *Aspects of Time*. Manchester: Manchester University Press; Buffalo, N.Y.: University of Toronto Press, 1976. 270 p.

Pratt, Sally, ed. *Man vs. Time*. Minneapolis: Graduate School Research Center, University of Minnesota, 1966. 209 p.

Sherover, Charles M. *The Human Experience of Time: The Development of its Philosophic Meaning*. New York: New York University Press, 1975. 603 p.

Wendorff, Rudolf. *Zeit und Kultur: Geschichte des Zeitbewusstseins in Europa*. Wiesbaden: Westdeutcher-Verlag, 1980. 720 p.

Whitrow, Gerald J. *The Natural Philosophy of Time*. 2d ed. Oxford: Clarendon Press, 1980. 399 p.

———. *The Nature of Time*. New York: Holt, Rinehart and Winston, 1973, 191 p.

Zeman, Jiri, ed. *Time in Science and Philosophy: An International Study of Some Current Problems*. Amsterdam and New York: Elsevier, 1971. 305 p.

2. Philosophy

Ahmad, Aziz. *Change, Time, and Causality: With Special Reference to Muslim Thought*. Lahore: Pakistan Philosophical Congress, 1974. 122 p.

Alexander, Samuel. *Space, Time, and Deity*. 2 vols. New York: Dover Pub. 1966.

Alther, E. *Das Absolute als Zeit-Raum-Verhältnis und Vorgang: beziehungsweise, das Wesen und Gesetz der den Erscheinungen im gesamten zu Grunde liegenden Ursache oder Kraft: dargelegt für Denkende, Wissenschaftler und Forscher*. Zürich: Kreis-Verlag, 1979. 84 p.

Andrade, Almir de. *As duas faces do tempo*. Rio de Janeiro: Livraria J. Olympio Editôra, 1971. 650 p.

Anscombe, Gertrude Elizabeth Margaret. *Times, Beginnings, and Causes.* London: Oxford University Press, 1975. 20 p.

"Basic Issues in the Philosophy of Time." *The Monist* **53,** no. 3 (1969): 325–518.

Bieri, Peter. *Zeit und Zeiterfahrung. Expositione. Problembereichs.* Frankfurt, a.M. Suhrkamp, 1972. 235 p.

Bîrsan, Gheorghe. *Timpul în ştiinţă şi filosofie. Analiză filosoică.* Bucuresti: Editura ştiinţifică, 1973. 248 p.

Böhme, Gernot. *Zeit und Zahl: Studien z. Zeittheorie bei Platon, Aristotles, Leibniz u. Kant.* Frankfurt am Main: K. Klostermann, 1974. 281 p.

Boschke, Friedrich L. *Und 1000 Jahre sind wie ein Tag. Die Zeit, das unverstandene Phänomen.* München: Bertelsmann, 1979, 224 p.

Brentano, Franz Clemens. *Philosophische Untersuchungen zu Raum, Zeit und Kontinuum.* Hrsg. u. eingel. von Stephan Körner u. Roderick Chisholm. Hamburg: Meiner, 1976. 242 p.

Bulhof, Ilse Nina. *Apollos Wiederkehr. Eine Untersuchung der Rolle des Kreises in Nietzsches Denken über Geschichte und Zeit.* Den Haag: Martinus Nijhoff, 1969. 169 p.

Callahan, John Francis. *Four Views of Time in Ancient Philosophy.* New York: Greenwood Press, 1968. 209 p.

Castelli, Enrico. *Il Tempo esaurito.* Padova: CEDAM, 1968. 256 p.

———. *Il tempo inqualificabile: contributi all'ermeneutica della secolarizzazione.* Padova: CEDAM: 1975. 155 p.

Castelli, Enrico, ed. *Il Tempo.* Padova: CEDAM, 1968. 248 p.

———. *Temporalità e alienazione.* Padova: CEDAM, 1975. 493 p.

Chahine, Osman E. *La durée chez Bergson.* Paris: H. Boucher, 1970. 104 p.

Cleugh, Mary Frances. *Time and Its Importance in Modern Thought.* New York: Russell & Russell, 1970. 308 p.

Convegno nazionale dei docenti di filosofia nelle facoltà, seminari e studentati religiosi d'Italia, 3d, Ariccia, Italy, 1970. Tempo e storicità dell'uomo. (Roma) Ariccia, 28–31 dicembre 1970. Padova: Gregoriana, 1971. 235 p.

Cvekl, Jiří. *Čas lidského života.* Praha: Svoboda, 1967. 82 p.

Decloux, Simon. *Temps, Dieu, liberté dans les "Commentaires aristotéliciens" de saint Thomas d'Aquin, essai sur la pensée grecque et la pensée chrétienne.* Paris: Desclée, De Brouwer, 1967. 262 p.

Delacre, Georges. *El tiempo en perspectiva: introducción a una filosofia del tiempo.* Río Piedras: Editorial Universitaria, Universidad de Puerto Rico, 1975. 171 p.

Erunvo, B. A., redaktor. *Filosofskie aspekty problemy vremeni: sbornik nauchnykh trudov.* Leningrad: Leningradskiĭ gos. pedogog. in-t im. A. I. Gertsena, 1978. 119 p.

Feldman, Gregorio. *Una unidad en tres: ser, tiempo, verbo.* Buenos Aires, 1975. 150 p.

Fraser, J. T. "Time as a Hierarchy of Creative Conflicts," *Studium Generale* **23** (1970): 597–689.

Gale, Richard M. *The Language of Time.* New York: Humanities Press, 1968. 247 p.

Ganduglia Pirovano, Mirtha. *Hacia la salud integral del hombre: tiempo, existencia y dios.* Buenos Aires: A.S.E.S. Ediciones, 1975. 111 p.

García Astrada, Arturo. *Tiempo y eternidad.* Madrid: Gredos, 1971. 120 p.

Gardies, Jean Louis. *La logique du temps.* Paris: Presses universitaires de France, 1975. 160 p.

Giacomini, Ugo. *Spazio e tempo nel pensiero contemporanea.* Genova: Tilgher, 1975. 246 p.

Giroux, Laurent. *Durée pure et temporalité: Bergson et Heidegger*. Paris, Tournai: Desclée; Montréal: Ballarmin, 1971. 136 p.

Goldschmidt, Victor. *Le système stoïcien et l'idée de temps*. 3. éd. rev. et augm. Paris: J. Vrin, 1977. 265 p.

Gonseth, Ferdinand, *Time and Method: An Essay on the Methodology of Research*. Translated by Eva H. Guggenheimer. Springfield, Ill.: Thomas, 1972. 453 p.

Granel, Gérard. *Le Sens du temps et de la perception chez E. Husserl*. Paris: Gallimard, 1968. 281 p.

Guitton, Jean. *Man in Time*. Translated by Adrienne Foulke. Notre Dame, Ind.: University of Notre Dame Press, 1966. 139 p.

Gurméndez, Carlos. *El tiempo y la dialéctica*. Madrid: Siglo Veintiuno de España Editores, 1971. 288 p.

Heidegger, Martin. *On Time and Being*. Translated by Joan Stambaugh. New York: Harper & Row, 1972. 84 p.

Heinrichs, Jürgen. *Das Problem der Zeit in der praktischen Philosophie Kants*. Bonn: Bouvier, 1968. 123 p.

Herrmann, Friedrich-Wilhelm von. *Bewusstsein, Zeit und Weltverständnis*. Frankfurt a.M.: Klostermann, 1971. 400 p.

Hinckfuss. Ian. *The Existence of Space and Time*. Oxford: Clarendon Press, 1975. 153 p.

Hintikka, Kaarlo Jaakko Juhani. *Time & Necessity: Studies in Aristotle's Theory of Modality*. Oxford: Clarendon Press, 1973. 225 p.

Hörz, Herbert. *Philosophie und Naturwissenschaft. Neue Aspekte im Verhältnis von Naturwissenschaft und marxistisch-leninistischer Philosophie, erläutert am Raum-Zeit-Problem*. Berlin: Dietz, 1968. 109 p.

Jaeglé, Pierre. *Essai sur l'espace et le temps: ou, Propos sur la dialectique de la nature*. Paris: Éditions sociales, 1976. 125 p.

Jordan, Pascual et al. *Zeit und Ewigkeit*. Karlsruhe: Badenia Verlag, 1970. 51 p.

Kosseleck, Reinhart. *Vergangene Zukunft. Zur Semantik geschichtlicher Zeiten*. Frankfurt, 1979.

Mandal, Kumar Kishore. *A Comparative Study of the Concepts of Space and Time in Indian Thought*. Varanasi: Chowkhamba Sanskrit Series Office, 1968. 223 p.

Maula, Erkka. *On the Semantics of Time in Plato's Timaeus*. Abo, Finland: Åbo Akademi, 1970. 37 p.

Mazzantini, Carlo. *Il tempo e quattro saggi su Heidegger*. Parma: Studium Parmense, 1969. 231 p.

Polato, Franco. *Louis Lavelle. L'essere e il tempo*. Ravenna: A. Longo. 1972. 568 p.

Prior, Arthur N. *Past, Present and Future*. Oxford: Clarendon Press, 1967. 217 p.

Pucelle, Jean. *Le contrepoint du temps. (Méthodologie de la liberté)*. Louvain: Éditions Nauwelaerts; Paris, Béatrice-Nauwelaerts, 1967. 347 p.

———. *Le Temps*. 5. éd. Paris: Presses universitaries de France, 1972. 112 p.

Pucciarelli, Eugenio, ed. *El Tiempo en la filosofia Francesa deal siglio XX*. Cuadernos de Filosofia, no. 13. Buenos Aires, 1970. 239 p.

Reenpää, Yrjö. *Über die Zeit: Darstellung und Kommentar einiger Interpretationen des Zeitlichen in der Philosophie. Über die Zeit in den Naturwissenschaften*. Helsinki: Distribuit Akateeminen Kirjakauppa, 1966. 83 p.

Reitmeister, Louis Aaron. *A Philosophy of Time*. Westport, Conn.: Greenwood Press, 1974. 452 p.

Salmon, Charles Ray. *The Book of Purpose: An Introduction to the Fourth Dimension*. Santa Maria, Calif.: Cronus College Press, 1972. 104 p.

Sambursky, Samuel, and Pines, S., comps. *The Concept of Time in Late Neoplatonism*. Jerusalem: Israel Academy of Sciences and Humanities, Section of Humanities, 1971. 118 p.

Schwarz, Gerhard. *Raum und Zeit als naturphilosophisches Problem*. Wien, Freiburg, Basel: Herder, 1972. 207 p.

Sherover, Charles. *Heidegger, Kant and Time*. Bloomington, Ind.: Indiana University Press, 1971. 322 p.

Silva Garland, Alejandro. *Los temas fundamentales: ciencia y filosofia*. Santiago, Chile: Editores Arancibia Hnos., 1970. 169 p.

Simon, Josef. *Sprache und Raum: philosophische Untersuchungen zum Verhältnis zwischen Wahreheit und Bestimmtheit von Sätzen*. Berlin: De Gruyter, 1969. 327 p.

Simonis, Walter. *Zeit und Existenz: Grundzüge der Metaphysik und Ethik*. Kevelaer: Butzon & Bercker, 1972. 170 p.

Spisani, Franco. *Significato e struttura del tempo*. Testo bilingue. Bologna: Azzoguidi, 1972. 161 p.

Stirn, François. *La soif d'éternité*. Paris: Hatier, 1979. 78 p.

Le Temps et la mort dans la philosophie contemporaine d'Amérique latine. Ouvrage collectif de l'équipe de recherche associée au C.N.R.S. no 80 (sur la philosophie de langues espagnole et portugaise). Toulouse: [Association des publications de Toulouse-Le Mirail], 1971. 211 p.

Le Temps et la mort dans la philosophie espagnole contemporaine. Toulouse: É. Privat, 1968, 239 p.

Terry, Bruce. *The Theory of Time*. New York: Exposition Press, 1974. 64 p.

"Time and Temporality." *Philosophy East and West* 24, no. 2 (1974): 119–225.

Van Fraassen, Bastiaan C. *An Introduction to the Philosophy of Time and Space*. New York: Random House, 1970. 224 p.

Weizsäcker, Carl Friedrich, Freiherr von. *Das Problem der Zeit als philosophisches Problem*. Berlin: Wichern-Verlag, 1967. 35 p.

Wiggins, David. *Identity and Spatio-Temporal Continuity*. Oxford: Blackwell, 1967. 83 p.

Wright, Georg Henrik von. *Time, Change and Contradiction*. London: Cambridge University Press, 1969.

Yamamoto, Makoto. *The Possibility of Metaphysics*. In Japanese. Tokyo: Tokyo University Press, 1977. 269 p.

Zalewski, Sylwester. *Czas i istnienie: z methodologii i filozofii klasycznej koncepcji czasu*. Warszawa: Pax, 1971. 162 p.

Zwart, P. *About Time: A Philosophical Inquiry into the Origin and Nature of Time*. New York: American Elsevier Pub. c., 1975. 266 p.

3. Religion

Efros, Israel Isaac. *The Problem of Space in Jewish Mediaeval Philosophy*. New York: AMS Press, 1966. 125 p.

Institut der Görres-Gesellschaft für die Begegnung von Naturwissenschaft und Theologie. *Weisen der Zeitlichkeit. Vortr. u. Diskussionen anlässl. d. Arbeitstagung d. Inst. d. Görres-Ges. f.d. Begegnung von Naturviss. u. Theologie.* Freiburg, München: Alber, 1970. 243 p.

Leenhouwers, P. *Het uur van de mens, het uur van de wereld: gedachten over een mogelijk perspectief.* Averbode: Abdij, 1977. 110 p.

Little, H. Ganse. *Decision and Responsibility: A Wrinkle in Time.* Tallahassee, Fla.: American Academy of Religion, 1974. 68 p.

Manipulierte Zeit? Wilhelm Stählin: *Christus und die Zeit.* Walther Bühler: Schöpfungsrhythmus und Weltkalender Stuttgart Evangelisches Verlagswerk, 1968.

Nakayama, Nobuji. *Bukky ō ni okeru toki no kenky ū.* Rev. ed. Kyoto: Hyakka-en, 1969. 398 p.

Picola, Thérèse. *Esprit, espace, temps, trinité créatrice, Dieu au carrefour de la science et de la foi.* Paris: le Courrier du livre, 1968. 127 p.

Rissi, Mathias. *Time and History: A Study on the Revelation.* Translated by Gordon C. Winsor. Richmond: John Knox Press, 1966. 147 p.

Tarthang Tulku. *Time, Space and Knowledge: A New Vision of Reality.* Emeryville, Calif.: Dharma Pub., 1977. 306 p.

Teilhard de Chardin, Pierre. *Réflexions et prières dans l'espace-temps.* Textes assemblés et annotés par Édouard et Suzane Bret. Paris: Éditions du Seuil, 1972. 158 p.

Torrance, Thomas Forsyth. *Space, Time and Incarnation.* London and New York: Oxford University Press, 1969. 92 p.

Watts, Alan Wilson. *Time.* Photographs by Joseph McHugh. Millbrae, Calif.: Celestial Arts, 1975. 63 p.

Wilch, John R. *Time and Event: An Exegetical Study of the Use of cēth in the Old Testament in Comparison to other Temporal Expressions in Clarification of the Concept of Time.* Leiden: E.J. Brill, 1969. 180 p.

4. History, Law, Society, Time Allocation

Antoine, Pierre, et Jeannière, Abel. *Espace mobile et temps incertains: nouveau cadre de vie, nouveau mileau humain.* Paris: Aubier Montaigne, 1970. 157 p.

Berliner, Paul. *The Soul of the Mbira.* Berkeley: University of California Press, 1978.

Bradley, Michael Anderson. *The Cronos Complex I: An Enquiry into the Temporal Origins of Human Culture and Psychology.* Toronto: Nelson, Foster & Scott, 1973. 165 p.

Burke, Peter. *The Renaissance Sense of the Past.* New York: St. Martin's Press, 1970. 154 p.

Demandt, Alexander. *Metaphern für Geschichte. Sprachbilder und Gleichnisse im historisch-politischen Denken.* München, 1978.

Engelhardt, Dietrich von. *Historisches Bewusstsein in der Naturwissenschaft.* Freiburg, München, 1979.

Finnegan, Ruth. *Oral Literature in Africa.* Oxford: Clarendon Press, 1970.

Georgescu-Roegen, Nicholas. *The Entropy Law and the Economic Process.* Cambridge, Mass.: Harvard University Press, 1971. 457 p.

Givens, Douglas R. *An Analysis of Navajo Temporality.* Washington: University Press of America, 1977.

Glasser, Richard. *Time in French Life and Thought.* Totowa, N. J.: Rowman and Littlefield, 1972. 306 p.

Grant, George Parkin. *Time as History.* Toronto: Canadian Broadcasting Corp., 1969. 52 p.

Gunnell, John G. *Political Philosophy and Time.* Middletown, Conn.: Wesleyan University Press, 1968. 314 p.

Haber, Francis C. *The Age of the World: Moses to Darwin.* Westport, Conn.: Greenwood Press, 1978. 303 p.

Hall, Edward Titchell. *Beyond Culture*. Garden City, N.Y.: Doubleday, 1976.

―――. *The Hidden Dimension*. Garden City, N.Y.: Doubleday, 1966. 201 p.

Harris, Wilson. *Fossil and Psyche*. Austin: African and Afro-American Studies and Research Center, University of Texas at Austin. 1974. 12 p.

Hersch, Jeanne, ed. *Entretiens sur le temps*. Paris, la Haye: Mouton et Cie, 1967. 352 p.

"History and the Concept of Time: Studies in the Philosophy of History." *History and Theory*, Beiheft No. 6, 1966.

Hughes, Peter. *Spots of Time*. Toronto: Canadian Broadcasting Corporation, 1969. 72 p.

Ibérico, Mariano. *La aparición histórica: ensayos y notas sobre los temas de la historia y el tiempo*. Lima: Universidad Nacional Mayor de San Marcos, 1971. 208 p.

Jackson, Michael. *The Kuranko: Dimensions of Social Reality in a West African Society*. New York: St. Martin's Press, 1977.

Kolaja, Jiris Thomas. *Social System and Time and Space: Introduction to the Theory of Recurrent Behavior*. Pittsburgh, Pa.: Duquesne University Press, 1969. 113 p.

Küng, Emil. *Freizeit in der nachindustriellen Gesellschaft*. Tübingen, 1971.

Le Goff, Jacques. *Time, Work, and Culture in the Middle Ages*. Chicago: University of Chicago Press, 1980. 400 p.

Leon-Portailla, Miguel. *Time and Reality in the Thought of the Maya*. Boston: Beacon Press, 1973. 176 p.

Lindgren, Kevin E. *Time in the Performance of Contracts: Especially for the Sale of Land*. Melbourne; Woburn, Mass.: Butterworths, 1976. 167 p.

Luhmann, Niklas. *Gesellschaftsstruktur und Semantik. Studien zur Wissenssoziologie der modernen Gesellschaft*. Frankfurt, 1980.

Michelson, William, ed. *Public Policy in Temporal Perspective: Report on the Workshop on the Application of Time-Budgets Research to Policy Questions in Urban and Regional Setting*. The Hague: Mouton, 1978, 210 p.

Moulton, Donalee. *The Time Budget: A Methodological Bibliography*. Monticello, Ill.: Vance Bibliographies, 1979. 5 p.

Ortiz, Alfonso. *The Tewa World. Space, Time, Being, and Becoming in a Pueblo Society*. Chicago: University of Chicago Press, 1972.

Pappas, P. *Time Allocation Study*. Athens: Athens Center of Ekistics, 1968. 191 p.

Piekarczyk, Stanilaw. *Historia, kultura, poznanie: ksiazka propozycji*. Warszawa: Państwowe Wydawn. Naukowe, 1972. 394 p.

Pütz, Karl. *Zeitbudgetforschung in der Sowjetunion: zur empirischen Sozialforschung in der USSR*. Meisenheim am Glan: A. Hain, 1970. 102 p.

Schöps, Martina. *Zeit unde Gesellschaft*. Stuttgart: Ferdinand Enke, 1980. 242 p.

Skvortšov, Lev Vladimirovich. *Vremiâ i neobkodimost' v istorii*. Moskva: Znanie, 1974. 63 p.

Thornton, Robert J. *Space, Time, and Culture Among the Iraqw of Tanzania*. New York: Academic Press, 1980. 275 p.

Volbers, Wilfried. *Fristen, Termine und Zustellungen*. Bad Godesberg: Asgard-Verlag, 1967. 128 p.

Vom Umgang mit der Zeit. Bad Salzuflen: MBK-Verlag, 1968. 16 p.

Webber, Ross A. *Time and Management*. New York: Van Nostrand Reinhold Co., 1972. 167 p.

Weiss, Leo. *Zeit Zeitlichkeit und Recht*. Zürich, 1968. 141 p.

White, Anthony G. *Time Management: A Brief Bibliography*. Monticello, Ill: Vance Bibliographies, 1979. 4 p.

Wiberg, Lars. *Personlig planering. Handbok om hur man hanterar siu tid*. Stockholm: Norstedt, 1973. 88 p.

Wollweber, Hans-Werner. *Die Fristen im Kündigungsschutzgesetz*. Köln, 1972. 125 p.

Wright, Lawrence. *Clockwork Man: The Story of Time, Its Origins, Its Uses, Its Tyranny*. New York: Horizon Press, 1969. 260 p.

242

242 J. T. Fraser

5. Art, Language, Literature

Alkon, Paul Kent. *Defoe and Fictional Time.* Athens, Ga.: University of Georgia Press, 1979. 275 p.

Barahona, Maria Alzira. *Para um estudo da expressão do tempo no romance português contemporâneo.* Lisboa, 1968. 207 p.

Bielawski, Ludwik. *The Zonal Theory of Time and Its Significance for Musical Anthropology.* In Polish. Krakow: Polskie Wydawnictwo Muzyczne, 1976. 228 p.

Brall, Artur. *Vegangenheit und Vergänglichkeit: zur Zeiterfahrung und Zeitdeutung im Werk Annettes von Droste-Hülshoff.* Marburg: N. G. Elwert, 1975. 324 p.

Brøndegaard, V. J. *Tid og tidsmåling.* Billedred: Anita Amundsen. København: Munksgaard, 1969. 72 p.

Buckley, Jerome Hamilton. *The Triumph of Time: A Study of the Victorian Concepts of Time, History, Progress and Decadence.* Cambridge, Mass.: Belknap Press of Harvard University Press, 1966. 187 p.

Castelli, Enrico, ed. *Il Simbolismo del tempo, studi di filosofia dell'arte.* Padova: CEDAM, 1973. 185 p.

Ejder, Bertil. *Dagens tider och måltider.* Lund: Gleerup, 1969. 503 p.

Émery, Éric. *Temps et musique.* Lausanne: Éditions L'Age d'homme, 1975. 696 p.

Engstrom, Alfred Garwin. *Darkness and Light: Lectures on Baudelaire, Flaubert, Nerval, Huysmans, Racine, and Time and Its Images in Literature: Being the First Series of Humanities Lectures for the Liberal Arts Forum of Elon College, 1967—1969.* University, Miss: Romance Monographs, 1975. 151 p.

Frank, M. *Das Problem "Zeit" in deutschen Romantik.* München, 1972.

Harmon, William. *Time in Ezra Pound's Work.* Hill: University of North Carolina Press, 1977. 165 p.

Hein, Birgit et al. *Film als Film, 1910 bis heute.* Stuttgart, 1977.

Higdon, David Leon. *Time and English Fiction.* London: Macmillan, 1977. 168 p.

Holl, Oskar. *Der Roman als Funktion und Überwindung der Zeit und Gleichzeitigkeit im deutschen Roman des 20 Jahrhunderts.* Bonn: H. Bouvier, 1968. 248 p.

Kermode, John Frank. *The Sense of an Ending: Studies in the Theory of Fiction.* New York: Oxford University Press, 1967. 187 p.

Kläui, Elisabeth. *Gestaltung und Formen der Zeit im Werk Adalbert Stifters.* Bern: H. Lang, 1969. 110 p.

Lenz, Bernd. *The time has been: die Vergangenheitsdimension Shakespeares Dramen.* Frankfurt am Main: Akademische Verlagsgesellschaft, 1974.

Macey, Samuel L. *Clocks and the Cosmos: Time in Western Life and Thought.* Hamden, Conn.: Archon Books, 1980. 260 p.

Masciandaro, Franco. *La problematica del tempo nella Commedia.* Ravenna: Longo, 1976. 149 p.

Morris, Jill. *Time and Timelessness in Virginia Woolf.* Hicksville, N.Y.: Exposition Press, 1977. 120 p.

Murillo, Louis Andrew. *The Golden Dial: Temporal Configuration in "Don Quijote."* Oxford: Dolphin Book Co., 1975. 178 p.

Myers, Catherine (Rodgers). *Time in the Narrative of "The Faerie Queen."* Salzburg: Inst. f. Engl. Sprache u. Literatur, Univ. Salzburg, 1973. 128 p.

Niglia, Giuseppe. *Le porte del tempo.* A cura di Onofrio Brindisi. Reggio Calabria: Parallelo 38, 1978. 250 p.

Popper, Frank. *Die kinetische Kunst.* Köln, 1975.

Quinones, Ricardo J. *The Renaissance Discovery of Time.* Cambridge, Mass.: Harvard University Press, 1972. 549 p.

Romilly, Jacqueline de. *Time in Greek Tragedy*. Ithaca, N.Y.: Cornell University Press, 1968. 180 p.

Ruberg, Uwe. *Raum und Zeit im Prosa-Lancelot*. München: Wilhelm Fink, 1966. 193 p.

Sant'Anna, Affonso Romano de. *Carlos Drummond de Andrade, o poeta "Gauche," no tiempo & espaco*. Bélo Horizonte, 1969. 284 p.

Schwarz, Peter Paul. *Aurora: zur romantischen Zeitstruktur bei Eichendorff*. Bad Homburg v. d. H.: Verlag Gehlen, 1970. 236 p.

Spencer, Sharon. *Space, Time and Structure in the Modern Novel*. New York: New York University Press, 1971. 251 p.

Stelzig, Eugene L. *All Shades of Consciousness: Wordsworth's Poetry and the Self in Time*. The Hague: Mouton, 1975. 212 p.

Sypher, Wylie. *The Ethic of Time: Structures of Experience in Shakespeare*. New York: Seabury Press, 1976. 216 p.

Tobin, Patricia Dreschler. *Time and the Novel*. Princeton: Princeton University Press, 1978.

Turner, Frederick. *Shakespeare and the Nature of Time: Moral and Philosophical Themes in Some Plays and Poems of William Shakespeare*. Oxford: Clarendon Press, 1971. 193 p.

Walker, G. *The Strong Necessity of Time: the Philosophy of Time in Shakespeare and in Elizabethan Literature*. The Hague: Mouton, 1976.

Weinrich, Harald. *Tempus*. Stuttgart, 1971.

6. *Education, Juvenile Literature, Time as Subject of Fiction*

Adler, Irving. *Time in Your Life*. Rev. ed. New York: John Day Co., 1969. 130 p.

Baumann, Hans. *Wieviel Uhr ist's anderswo?* Bilder von Antoni Boratynski. Stuttgart: K. Thienemanns, 1972. 30 p.

Bendick, Jeanne. *Space and Time*. Illustrated by the author. New York: F. Watts, 1967.

Bethancourt, T. Ernesto. *Tune In Yesterday*. New York: Holiday House, 1978. 156 p.

Bollnow, Otto Friedrich. *Das Verhältnis zur Zeit· ein Beitrag zur pädagogischen Anthropologie*. Heidelberg: Quelle & Meyer, 1972. 112 p.

Boston, Lucy Maria. *Treasure of Green Knowe*. Illustrated by Peter Boston. New York: Harcourt, Brace, Jovanovich, 1978. 185 p.

Burns, Marilyn. *This Book Is About Time*. Illustrated by Martha Weston. Boston: Little, Brown, 1978. 127 p.

Conaway, Judith. *The Discovery Book of Time*. Pictures by Bruce Witty. Milwaukee: Raintree Editions; Chicago: distributed by Childrens Press, 1977. 31 p.

Ecker, Ernst. *Über Zeit und Zeitverhältnisse in der Erzienhung*. Frieburg im Breisgau: Gutenbergdruckerei R. Oberkirch, 1966. 146 p.

Gibbons, Gail. *Clocks and How They Go*. New York: Crowell, 1979, 32 p.

Grée, Alain. *Petit Tom sait lire l'heure*. Imagé par Gérard Grée. Tournai: Casterman, 1977. 21 p.

Handford, Barbara. *Measuring Time*. Adapted from the text by Henri Michel. Illustrated by Raymond Renard. St. Louis: Webster Division, McGraw-Hill, 1968. 28 p.

Kajenn. *Vad är Klockan?* Bild: Fibben Hald. Stockholm: Natur och kultur, 1969. 24 p.

Lopez, Norbert C. *King Pancho and the First Clock*. Illustrated by Marianne Gutierrez. Mankato, Minn.: Oddo Publishing, 1968, 32 p.

Nestle, Werner. *Didaktik der Zeit und Zeitmessung: empir. Konstruktion e. Teilcurriculums z. Revision d. Lehrplans im Sachunterricht d. Grundschule*. Stuttgart: Klett, 1973. 228 p.

North, Joan. *The Light Maze*. New York: Farrar, Straus & Giroux, 1971. 185 p.

Shibles, Warren A. *Time: A Critical Analysis for Children*. Whitewater, Wis.: Language Press, 1978. 79 p.

Timmons, Christine. *The Time Book*. Chicago: Encyclopaedia Britannica, 1974. 29 p.

Wade, Harlan. *Time*. Concept and illustrations by Denis Wrigley. Milwaukee: Raintree Childrens Books, 1977. 32 p.

Zöpfl, Helmut. *Zeit und Geschichtlichkeit in pädagogischer Sicht*. München: Ehrenwirth, 1968. 89 p.

7. Psychology

Association de psychologie scientifique de langue française. *Du temps biologique au temps psychologique: symposium de l'Association de psychologie scientifique de langue française, Poitiers, 1977*. Paris: Presses universitaires de France, 1979. 386 p.

Bovet, Magali et al. *Perception et notion du temps*. Paris: Presses universitaires de France, 1967. 187 p.

Cohen, John. *Psychological Time in Health and Disease*. Springfield, Ill.: C. C. Thomas, 1967. 103 p.

Cottle, Thomas J., and Klineberg, Stephen L. *The Present of Things Future: Explorations of Time in Human Experience*. New York: Free Press, 1974. 290 p.

Culbertson, James T. *Sensations, Memories, and the Flow of Time*. Santa Margarita, Calif.: Cromwell Press, 1976.

Engelbrecht, Frederik Jacobus. *Tyd en neurose by die Bantoe*. Pietersburg: Universiteit van die Noorde, 1972. 30 p.

Evans, Daryl. *Explorations in Time*. Monticello, Ill: Council of Planning Librarians, 1975. 103 p.

Gorman, Bernard S., and Wessman, Alden E., eds. *The Personal Experience of Time*. New York: Plenum Press, 1977. 296 p.

Hugenholtz, Paul Theodoor. *Tijd en creativiteit. Ontwerp van een structurele antropologie*. 2e herz. dr. Eerder uitgeg. door de Noord-Hollandsche U. M. Utrecht: Het Spectrum, 1972. 334 p.

Jensen, Olav Storm. *Om oplevelse af varighed. En eksperimentel undersøgelse*. København: Københavns universitets Fond til tilvejebringelse af laeremidler, 1969. 157 p.

Kristofferson, Alfred Boyd. *Further Experiments on Successiveness Discrimination*. Washington: National Aeronautics and Space Administration, 1966. 65 p.

Leonov, Aleksei Arkhipovich, and Lebedev, Vladimir. *Space and Time Perception by the Cosmonaut*. Translated by B. Belitsky. Moscow: Mir Publishers, 1971. 200 p.

Lévy, Jean Claude. *Le Temps psychologique*. Paris: Dunod, 1969. 106 p.

Meerloo, Joost Abraham Maurits. *Along the Fourth Dimension: Man's Sense of Time and History*. Illustrated by Carl Smith. New York: John Day Co., 1970. 278 p.

Meissner, Rudolf. *Über de Bergriff der erlebten Zeit bei E. Minkowski*. Tübingen, 1970.

Michon, John Albertus. *Timing in Temporal Tracking*. Assen: Vam Gorcum, 1967. 127 p.

Minkowski, Eugène. *Lived Time: Phenomenological and Psychopathological Studies*. Translated by Nancy Metzel. Evanston, Ill.: Northwestern University Press, 1970. 455 p.

Monge, Jean. *Temps et mémoire*. Roanne: Horvath, 1972. 199 p.

Montangero, Jacques. *La Notion de durée chez l'enfant de 5 a 9 ans*. Paris: Presses universitaires de France, 1977. 252 p.

Müller, Siegfried. *Untersuchungen zur Messung pessimistischer und optimistischer Zukunftserwartungen: eine Studie z. Psychologie d. Zukunftserlebens*. Köln, Bonn: Hanstein, 1973. 182 + 97 p.

Newman, L. F. *Time*. High Wycombe: Published by the author, 1977. 3 leaves.

Orientation spatiale et temporelle. Paris: Éditions ESF, 1972. 130 + 46 p.

Orme, John Edward. *Time, Experience and Behaviour*. London, Iliffe; New York, American Elsevier, 1969. 189 p.

Ornstein, Robert Evans. *On the Experience of Time.* Harmondsworth: Penguin, 1969. 126 p.

Pauleikhoff, Bernhard. *Person und Zeit. Im Brennpunkt seelischer Störungen.* Heidelberg: Hüthig, 1979. 254 p.

Payk, Theo Rudolf. *Mensch und Zeit. Chronopathologie im Grundriss.* Stuttgart: Hippokrates, 1979. 140 p.

Piaget, Jean. *The Child's Conception of Time.* Translated by A. J. Pomerans. New York: Basic Books, 1970. 285 p.

Pieters, Herman A. *A Psychologist Looks at Space, Motion and Time.* Utrecht: Oosthoek, 1972. 34 p.

Plate, Bernward. *Die Erfahrung, die Zeit und das Mit-Dasein.* München, 1966. 224 p.

Rama Rao, Palamand. *Studies in Time Perception.* Delhi: Concept Pub. Co., 1978. 127 p.

Thass-Thienemann, Theodore. *The Subconscious Language.* New York: Washington Square Press, 1967. 437 p.

Theau, Jean. *La Conscience de la durée et la concept de temps.* Toulouse: É. Privat, 1969. 315 p.

Trommsdorff, Gisella. *Gruppeneinflüsse auf Zukunftbeurteilungen.* Hain: Mesienheim/Glan, 1978.

Underwood, Benton J. *Temporal Codes for Memories: Issues and Problems.* Hillsdale, N.J.: L. Erlbaum Associates; New York: distributed by Halsted Press, 1976. 158 p.

Uslar, Detlev von. *Die Wirklichkeit des Psychischen: Leiblichkeit, Zeitlichkeit.* Pfullingen: Neske, 1969. 105 p.

Vicario, Giovanni. *Tempo psicologico ed eventi.* Firenze: Giunti; G. Barbèra, 1973. 284 p.

von Franz, Marie-Louise. *Number and Time: Reflections Leading toward a Unification of Depth Psychology and Physics.* Evanston, Ill.: Northwestern University Press, 1974. 332 p.

Vroon, Pieter A. *Enkele psychofysische en cognitieve aspecten van de tijdzin.* N.p., 1972(?). 262 p.

———. *Some Aspects of the Lengthening Effect in Sequential Time Estimations.* Utrecht: Psychological Laboratory, University of Utrecht, 1972. 16 leaves.

Warhadpande, N. R. *Time, Space, and Motion: A Logical Analysis with Special Reference to Psychology.* Nagpur: Nagpur University, 1969. 236 p.

Winnubst, J. A. M. *Het westerse tijdssyndroom: conceptuele integratie en eerste aanzet tot construct validatie van een reeks molaire tijdsvariabelen in de psychologie.* Amsterdam: Swets & Zeitlinger, 1975. 272 p.

Yaker, Henri Marc et al., eds. *The Future of Time: Man's Temporal Environment.* Garden City, N.Y.: Doubleday, 1971. 512 p.

Zern, David. "The Influence of Certain Child-Rearing Factors Upon the Development of a Structured and Salient Sense of Time." *Genetic Psychology Monogaphs* 81 (1970): 197–254.

8. Biology

Bünning, Erwin. *The Physiological Clock: Circadian Clocks and Biological Chronometry.* New York: Springer-Verlag, 1973. 259 p.

Chizhevskiĭ, Aleksandr Leonidovich. *V ritme solntsa.* Moskva: Nauka, 1969. 112 p.

Cloudsley-Thompson, John Leonard. *Biological Clocks: Their Functions in Nature.* London: Weidenfeld and Nicolson, 1980. 138 p.

Curtis, Howard James. *Biological Mechanisms of Aging.* Springfield, Ill.: C. C. Thomas, 1966. 133 p.

Dmitriev, A. S., ed. *Fiziologicheskie mekhanizmy vospriiatiia i otsenki vremeni.* Ufa: Bashkirskiĭ gosudarstvennyĭ universitet, 1969. 152 p.

Gauquelin, Michel. *The Cosmic Clocks: From Astrology to a Modern Science.* Chicago: H. Regnery Co., 1967. 250 p.

Guzzo, Roberto. *La legge universale della differenziazione e del funzionalismo nel tempo e nell'eternità*. Roma: Noi pubblicisti, 1972. 135 p.

Halberg, Franz et al., eds. *Glossary of Chronobiology*. Milano: Il Ponte, 1977. 189 p.

Hastings, J. Woodward et al., eds. *The Molecular Basis of Circadian Rhythms*. Berlin: Dahlem Konferenzen, 1976. 461 p.

Holubář, Josef. *The Sense of Time: An Electrophysiological Study of Its Mechanisms in Man*. Translated from the Czech by John S. Barlow. Cambridge, Mass.: M. I. T. Press, 1969. 122 p.

Luce, Gay Gear. *Body Time*. New York: Pantheon, 1974. 394 p.

Nasioutzik, A. K. *Biologia tou chronou*. Athēnai: Kedros, 1970. 227 p.

Phaure, Jean. *Le cycle de l'humanité adamique: introduction à l'étude de la cyclologie traditionnelle et de la fin des temps*. Paris: Dervy-Livres, 1973. 656 p.

Poirel, Christian. *Les rhythmes circadiens en psychopathologie: perspectives neurobiologique sur les structures de la temporalité*. Paris: Masson, 1975. 113 p.

Rensing, Ludger. *Biologische Rhythmen und Regulation*. Stuttgart: Gustav Fischer, 1973. 265 p.

Scharf, J. H., and von Mayersbach, H., eds. *Chronobiologie*. Halle an der Saale: Deutsche Academie der Naturforscher, 1977. 759 p.

Scheving, Lawrence E. et al, eds. *Chronobiology*. Tokyo: Igaku Shoin, 1974. 784 p.

Smith, Anthony. *The Seasons: Life and Its Rhythms*. New York: Harcourt Brace Jovanovich, 1970. 318 p.

Sullivan, Navin. *Animal Timekeepers*. Illustrated by Haris Petie. Englewood Cliffs, N.J.: Prentice-Hall, 1966. 64 p.

Thumshirn, Werner. *Unsere inner Uhr. Auf der Spur der biologischen Zeit*. Zürich: Schweizer Verlagshaus, 1975. 319 p.

Ward, Ritchie R. *The Living Clocks*. Drawings by Hollett Smith. New York: Knopf, 1971. 385 p.

Winfree, Arthur T. *The Geometry of Biological Time*. New York: Springer-Verlag, 1980. 530 p.

Zoppis, Giuseppe. *Protoritmo. 27,365 giorni*. 2. edizione riveduta e ampliata. Gubbio: Tip. eugubina, 1969. 61 p.

9. Geography, Geology, Geochronology

Berry, William B. N. *Growth of a Prehistoric Time Scale, Based on Organic Evolution*. San Francisco: W. H. Freeman, 1968. 158 p.

Carlstein, Tommy et al., eds. *Timing Space and Spacing Time*. 3 vols. London: E. Arnold. 1978.

Eicher, Don L. *Geologic Time*. 2d ed. Englewood Cliffs, N.J.: Prentice-Hall, 1976. 150 p.

Franke, Herbert W. *Methoden der Geochronologie: die Suche nach den Daten der Erdgeschichte*. Berlin and New York: Springer-Verlag, 1969. 132 p.

Geokhronologiiâ Vostochno-Evropeĭskoĭ platformy i sochleniiâ Kavkazsko-Karpatskoĭ sistemy. Moskva: Akademia nauk SSSR, 1975. 167 p.

Harper, Christopher T., ed. *Geochronology: Radiometric Dating of Rocks and Minerals*. Stroudsburg, Pa.: Dowden, Hutchinson & Ross, 1973. 469 p.

Kirkaldy, John F. *Geological Time*. Edinburgh: Oliver and Boyd, 1971. 133 p.

Lynch, Kevin. *What Time Is This Place?* Cambridge, Mass.: M. I. T. Press, 1972. 276 p.

Philip, Graeme Maxwell. *Time and Geology*. Armidale, N.S.W.: University of New England, 1967. 16 p.

Rosenberg, G. D. et al., eds. *Growth Rhythms and the History of the Earth's Rotation.* New York: Wiley, 1975. 559 p.

Silverberg, Robert. *Clocks for the Ages: How Scientists Date the Past.* New York: Macmillan, 1971. 238 p.

Thornes, John B., and Brundsen, D. *Geomorphology and Time.* New York: Wiley, 1977. 208 p.

Willkomm, Horst. *Altersbestimmungen im Quartär: Datierungen mit Radiokohlenstoff und anderen kernphysikalischen Methoden.* München: K. Thiemig, 1976. 276 p.

York, Derek. *The Earth's Age and Geochronology.* Oxford and New York: Pergamon Press, 1972. 178 p.

10. Calendars, Chronologies

Achelis, Elisabeth. *The World Calendar: Addresses and Occasional Papers Chronologically Arranged on the Progress of Calendar Reform since 1930.* Ann Arbor, Mich.: Gryphon Books, 1971. 189 p.

Alavi, Is'haqunnabi. *The Arab Calendar Prevalent during the Lifetime of Muhammad: An Indian Discovery.* Compiled and translated by Abid Raza Bedar. Delhi: Rampur Institute of Oriental Studies, 1968. 48 p.

al-Bīrūnī. *The Chronology of Ancient Nations.* An English Version of the Arabic Text of the Athâr-ul-bâkiya of Albîrûnî or "Vestiges of the Past." Translated and ed. by C. Edward Sachau. (Unveränd. Nachdr. d. Ausg. London, 1879). Frankfurt: Minerva-Verlag, 1969. 464 p.

Alver, Brynjulf. *Dag og merke. Folkeleg tidsrekning og merkedagstradisjon.* Oslo: Universitetsforlaget, 1970. 176 p.

Barros, Maria de Nazareth Moreira Martins de. *Cronologia e historia.* Belém: Universidade Federal do Pará, Curso de Biblioteconomia, 1970. 46 leaves.

Bond, John James. *Handy-Book of Rules and Tables for Verifying Dates with the Christian Era, Giving an Account of the Chief Eras, and Systems Used by Various Nations.* Reprint of the 1889 edition. New York: Russell & Russell, 1966. 465 p.

Casas, Johanna Broda de. *The Mexican Calendar as Compared to Other Mesoamerican Systems.* Wien: Engelbert Stiglmayr, 1969. 105 p.

Chevallier, R. *Aiôn: le temps chez les Romains.* Paris: A. & J. Picard, 1976. 370 p.

Collison, Robert Lewis. *Hamlyn Dictionary of Dates and Anniversaries.* 2d rev. ed. J. M. Bailie, general ed. London and New York: Hamlyn, 1978. 415 p.

Colson, Francis Henry. *The Week: An Essay on the Origin & Development of the Seven-Day Cycle.* Westport, Conn.: Greenwood Press, 1974. 126 p.

Donaldson, Elizabeth & Gerald. *The Book of Days.* New York: A&W Publishers, 1979. ca. 400 p.

Donato, Hernâni. *História do calendário.* São Paulo: Edições Melhoramentos, 1976. 158 p.

Dressler, Adolf. *Kalender-Kunde. Eine Kulturhistorische.Studie.* München: 1972.

Ekrutt, Joachim W. *Der Kalendar im Wandel der Zeiten.* Stuttgart: 1972.

Evenson, A. E. *About the History of the Calendar.* Chicago: Children's Press, 1972. 41 p.

Gahlot, Sukhvir Singh. *Historians' Calendar: Giving Equivalent Dates of Christian, Vikram, Shaka, Hijari, Bengali, and Kollam Eras.* Vol. 1: 1544 A.D.–1643 A.D. Jodhpur: Hindi Sahitya Mandir, 1979.

García Larragueta, Santos Agustín. *Cronología (Edad Media).* Pamplona: Ediciones Universidad de Navarra, 1976. 106 p.

Gossling, Winifred. *A Time Chart of Social History.* London: Lutterworth Press, 1970. 55 p.

Grotefend, Hermann. *Taschenbuch der Zeitrechnung des deutschen Mittelalters und der Neuzeit*. Hannover: Verlag Hahnsche Buchhandlung, 1971. 224 p.

Gumpach, Johannes von. *Die Zeitrechnung der Babylonier und Assyrer*. (Unveränd. Neudr. d. Ausg. Heidelberg 1852). Walluf b. Wiesbaden: Dr. Martin Sändig oHG, 1972. 170 p.

Hammerschmidt, Ernst. *Äthiopische Kalendertafeln*. Wiesbaden: Steiner, 1977. 21 p.

Harvey, O. L. *The Chinese Calendar and the Julian Day Number*. Silver Spring, Md.: Harvey, 1977. 30 p.

———. *Miscellaneous Calendars*. Silver Spring, Md.: Harvey, 1977. 31 p.

———. *Time Shaper, Day Counter: Dionysius and Scaliger*. Silver Spring, Md.: Harvey, 1976. 54 p.

Hellyer, Brian. *Man the Timekeeper*. London: Priory Press, 1974. 96 p.

Hull, Terence H. *Almanak ˜penanggalan Jawa-Masehi untuk penelitian sosialekonomi*. Yogyakarta: Lembaga Kependudukan, Universitas Gadjah Mada, 1976. 55 leaves.

Kaletsch, Hans. *Tag und Jahr: die Geschichte unseres Kalenders*. Zurich: Artemis, 1970. 96 p.

Kasher, Menachem Mendel. *Kay ha-ta'arikh ha-yiśre-eli*. Jerusalem: Ho-tse-os base torah Schlomo, 1977. 329 p.

Kern, Hermann. *Kalenderbauten. Frühe astronomische Grossgeräte aus Indien, Mexiko und Peru*. München, 1976.

Lentz, Wolfgang. *Zeitrechnung in Nuristan und am Pamir*. Graz: Akadem. Druck- u. Verlagsanst, 1978. 52 + 211 p.

Mahler, Eduard. *Handbuch der jüdischen Chronologie*. (Reprografischer Nachdruck der Ausg. Frankfurt a. M., 1916). Hildesheim: Gg Olms, 1967. 635 p.

Mamedbeĭli, G. D. *Sinkhronicheskie tabilitŝy dlia perevoda dat*. Baku: Akademia nauk Azerbaijan, 1972. 100 p.

Miceli, Jean de. *Nouveaux instruments pour le calcul du temps: chronologie de la XVIIIe dynastie et du temps de l'Exode, Thoutmois II pharaon de l'Exode*. Tulle: Éditions du livre, 1974. 114 p.

1977 Date Book, Almanac & Authentic Calendar of Saints, Hoboes, Dervishes, Dancers, Lovers, Coevolutionaries, Healers, Helpers & Ordinary Holy Fools. Ojai, Calif.: Being Incorporated, 1976. ca. 250 p.

O'Neil, William Matthew. *Time and the Calendars*. Sydney: Sydney University Press, 1975. 138 p.

Piazzo, Marcello del. *Manuale di cronologia*. Roma: A. N. A. I., 1969. 148 p.

Piper, Ferdinand. *Karls des Grossen Kalendarium und Ostertafel: aus d. Pariser Urschrift hrsg. u. erl., nebst e. Abh. über d. latein. u. griech. Oxtercyklen d. Mittelalters*. Unverand. Neudr. d. Ausg. von 1858. Walluf bei Wiesbaden: Sändig, 1974. 168 p.

Rohner, Ludwig. *Kalendergeschichte und Kalender*. Wiesbaden: Akademische Verlagsgesellschaft Athenaion, 1978. 552 p.

Samuel, Alan Edourd. *Greek and Roman Chronology: Calendars and Years in Classical Antiquity*. München: Beck, 1972. 307 p.

Sar-Shalom, Rahamim H. *Yesodot ha-luah ha-'ivri*. Tel A'viv: Y'sod, 1967. 42 p.

Schmidt-Liebich, Jochen. *Daten englischer Geschichte: von d. Anfängen bis zur Gegenwart*. München: Deutscher Taschenbuch-Verlag, 1977. 348 p.

Seder 'Olam zuta. Tel A'viv: Tsi-on, 1969. 112 p.

Seleshnikov, Semen Isakovich. *Istoria kalendaria i khronologia*. Pod red. P. G. Kulikovskova. Moskva: Nauka, 1977. 224 p.

Stein, Werner. *Kulturfahrplan: die wichtigsten Daten der Kulturgeschichte von Anbeginn bis 1969*. München: F. A. Herbig, 1970. 1,564 p.

Steinberg, Sigfrid Henry. *Historical Tables, 58 B.C.—A.D. 1972*. 9th ed. New York: St. Martin's Press, 1973. 269 p.

TŠybul'skiĭ, Vladimir Vasil'evich. *Calendars of Middle East Countries: Conversion Tables and Explanatory Notes.* Moscow: Nauka Publishing House, Central Dept. of Oriental Literature, 1979. 255 p.

Tunnicliffe, K. C. *Aztec Astrology.* Romford, England: L. N. Fowler, 1979. 98 p.

Ungnad, Arthur. *Die Venustafeln und das 9. Jahr Samsuilunas (1741 v. Chr.).* Neudr. d. Ausg. Leipzig, Harrosowitz, 1940. Osnabrück: Zeller, 1972. 25 p.

Vom Aderlassen und Bräute machen: der älteste Basler Buchkalender und das Regiment der Gesundheit von 1513 als Ratgeber. Faksimile mit Übersetzung. Hrsg., Beat Trachsler. Basel: Gute Schriften, 1974. 123 p.

Xántus, János. *A természet kalendáriuma.* 2. böv. kiad. A rajzokat Unipán Helga készítette. Bukarest: Kriterion, 1972. 528 p.

11. Time Measurement: Clocks and Watches, Dials, Systems and Standards

Abeler, Jürgen. *Ullstein-Uhrenbuch: e. Kulturgeschichte d. Zeitmessung.* Zeichn., Dieter Messerschmidt. 2., neu bearb. Aufl. Berlin, Frankfurt/M, Wein: Ullstein, 1979. 324 p.

Aksel'rod, Zakhar Markovich. *Teoriia i proektirovanie priborov vremeni.* Uchebnik dlia spetsial'nosti "Pribory tochnoĭ mekhaniki" vuzov. Leningrad: Mashinostroenie, 1969. 480 p.

Andersen, David Yde. *Bornholmere og andre gamle ure.* 2. forøgede udg. København: Borgen, 1968. 205 p.

Becker, Karl-Ernst. *Uhren.* München: Battenberg, 1978. 184 p.

Berman, Gideon. *An Essay on Church Clocks, with Reference to the Clock from the Hospital Santa Cruz.* London: Turner and Devereux, 1974. 58 p.

Bilfinger, Gustav. *Die mittelalterlichen Horen und die modernen Stunden: ein Beitrag zur Kulturgeschichte.* Reprint of 1892 edition. Wiesbaden: M. Sändig, 1969. 279 p.

Blair, Byron Emerson. *Time and Frequency: A Bibliography of NBS Literature Published July 1955 – December 1970.* Washington: U.S. National Bureau of Standards, 1971. 50 p.

Briatore, Luigi. *Cronologia e tecniche della misura del tempo.* Firenze, Ciunti: G. Barbèra, 1976. 159 p.

Britten, Frederick James. *Britten's Watch & Clock Maker's Handbook: Dictionary and Guide.* Illustrated by John W. Wood. 16th ed., rev. by Richard Good. New York: Arco Pub. Co., 1978. 459 p.

Brusa, Giuseppe. *L'arte dell'orologeria in Europa: sette secoli di orologi meccanici.* Busto Arsizio: Bramante, 1978. 491 p.

Bruton, Eric. *The History of Clocks and Watches.* New York: Rizzoli, 1979. 288 p.

Buenos Aires. Museo Nacional de Arte Decorativo. *Historia del reloj: exposición.* Catalogación y textos: Maud De Ridder de Zemborain, Jaime Llavallol y Carlos Alfredo Zemborain. Buenos Aires: Asociación Amigos del Museo Nacional de Arte Devorativo, 1967. 86 p.

Burlingame, Roger. *Dictator Clock: 5,000 Years of Telling Time.* New York: Macmillan, 1966. 166 p.

Cipolla, Carlo M. *Clocks and Culture, 1300–1700.* London: Collins, 1967, 192 p.

Chodakowski, Joseph. *Wadjet, Essay in the Ancient Astronomy.* Maffra, Vic.: J. Chodakowski, 1973. 15 p.

Clutton, Cecil, and Daniels, George. *Watches: A Complete History of the Technical and Decorative Development of the Watch.* Rev. and enl. ed. London: Published for Sotheby Parke Bernet Publications by P. Wilson; Totowa, N.J.: Biblio Distribution Center, 1979. 312 p.

Coleman, Lesley. *A Book of Time.* New York: T. Nelson, 1971. 144 p.

Cousins, Frank W. *Sundials: A Simplified Approach by Means of the Equatorial Dial.* Illustrated by Malcolm Chandler. London: J. Baker, 1969. 247 p.
Cunynghame, Sir Henry Hardinge Samuel. *Time and Clocks: A Description of Ancient and Modern Methods of Measuring Time.* Detroit: Singing Tree Press, 1970. 200 p.

De Carle, Donald. *Complicated Watches and Their Repair.* Illustrated by E. A. Ayres. New York: Bonanza Book, 1979. 174 p.
Due Rojo, Antonio. *Edades y tiempos en el universo: conquistas de la cronometría moderna.* Madrid: Editorial Razón y Fe, 1966. 188 p.

Elliott, Douglas J. *Shropshire Clock and Watchmakers.* London: Phillimore, 1979. 172 p.
Erbich, Klaus. *Präzisionspendeluhren: von Graham bis Riefler.* München: Callwey, 1978. 254 p.

Franklin Institute, Philadelphia. *Horological Books and Pamphlets in the Franklin Institute Library.* Compiled by Walter A. R. Pertuch and Emerson W. Hilker. 2d ed. Philadelphia, 1968. 109 leaves.

Gibbs, James W. *Buckeye Horology: A Review of Ohio Clock and Watch Makers.* Columbia, Pa.: Art Crafters, 1971. 128 p.
————. *Dixie Clockmakers.* Gretna, La.: Pelican Pub. Co., 1979. 191 p.
Good, Richard. *Watches in Colour.* Poole: Blandford Press, 1978. 218 p.
Great Britain. Home Dept. *Review of British Standard Time.* London: H.M.S.O., 1970. 84 p.
Guye, Samuel. *Time & Space: Measuring Instruments from the 15th to the 19th Century.* Translated from the French by Diana Dolan. London: Pall Mall Press, 1970. 289 p.
Guyot, Edmond. *Histoire de la détermination de l'heure.* La Chaux-de-Fonds: Chambre suisse de l'horlogerie, 1969. 247 p.

Hood, Peter. *How Time is Measured.* 2nd ed. London: Oxford University Press, 1969. 64 p.
Horamatic: montres à remontage automatique de 1770 à 1978. Textes, J. C. Beuchat et al. Le Locle: Édition du Château des Monts, Musée d'horlogerie, 1978. 100 p.
Hüttenhain, Trude. *Die astronomische Uhr im Dom zu Münster.* Münster/Westf.: Aschendorff, 1970. 31 p.

Jespersen, James, and Fitz-Randolph, Jane. *From Sundials to Atomic Clocks: Understanding Time and Frequency.* Illustrated by John Robb. Washington: National Bureau of Standards, U.S. Dept. of Commerce, 1977. 175 p.
Jüttermann, Herbert. *Die Schwarzwalduhr.* Braunschweig, 1972.

King, Henry C. *Geared to the Stars: The Evolution of Planetariums, Orreries, and Astronomical Clocks.* Toronto: University of Toronto Press, 1978. 422 p.
Kochmann, Karl. *The Black Forest Cuckoo Clock: European Clockmaking.* Concord, Calif.: Antique Clock Pub., 1976. 240 p.

Lietuvos TSR Istorijos-etnografijos muziejus. *XVI—XIX A. laikrodžiai: katalogas.* Kataloga sudarė O. Mažeikienė; fotografavo V. Bortkevičius: atsakingas redaktorius A. Jankevičienė. Vilnius: Lietuvos TSR Istorijos-etnografijos muziejus, 1976. 26 p.
Lloyd, Herbert Alan. *Old Clocks.* 4th ed., rev. and enl. London: Benn; New York: Dover, 1970.
Lodén, Lars Olof. *Tid. En bok om tideräkning och kalenderväsen.* Stockholm: Bonnier, 1968. 210 p.

Loomes, Brian. *Complete British Clocks*. Newton Abbot, Eng.; and North Pomfret, Vt: David & Charles, 1978. 256 p.

Loske, Lothar M. *Die Sonnenuhren: Kunstwerke der Zeitmessung und ihre Geheimnisse*. 2., erg. Aufl. Berlin, New York: Springer-Verlag, 1970. 101 p.

Lunardi, Heinrich et al. *Bein dritten Ton war es genau . . . Die Ahnen unserer Uhr*. Basel: GS-Verlag, 1976. 168 p.

Magyar, László. *Óraipari zsebkönyv*. Budapest: Müszaki Kiadó, 1979. 319 p.

Mallet, Georges. *Étude de la mesure des temps*. Paris: Édition d'Organisation, 1971. 127 p.

Maurice, Klaus. *Die französische Pendule des 18. Jahrhunderts. Ein Beitrag zu ihrer Ikonologie*. Berlin: Gruyter, 1967. 124 p.

———. *Von Uhren und Automaten: das Messen der Zeit*. München: Prestel Verlag, 1968. 93 p.

Maurice, Klaus, and Mayr, Otto, eds. *The Clockwork Universe: German Clocks and Automata 1550–1650*. New York: Neal Watson Academic Publications, 1980. 314 p.

Meis, Reinhard. *Die alte Uhr: Geschichte, Technik, Stil: ein Handbuch für Sammler und Liebhaber*. Braunschweig: Klinkhardt & Biermann, 1978. 2 v.

Michel, Henri. *Les cadrans solaires de Max Elskamp*. Liège: Éditions du Musée wallon, 1966. 68 p.

New Haven Clock Company. *Illustrated Catalogue of Clocks Manufactured by the New Haven Clock Company*. Bristol, Conn.: American Clock and Watch Museum, 1976. 80 p.

Old Sturbridge Village, Sturbridge, Mass. *New England Clocks at Old Sturbridge Village: The J. Cheney Wells Collection*. By Amos G. Avery. 2d ed. Sturbridge, 1966.

Padelt, Erna. *Menschen messen Zeit und Raum*. Berlin: Verlag Technik, 1971. 168 p.

Pioneers of Precision Timekeeping: A Symposium. London: Antiquarian Horological Society, 1966(?). 117 p.

Pomella, Fulgido. *L'orologio da portare addosso: arte e tecnica nell'orologio tascabile dalle origini al 1820–30: ad uso di amatori e collezionisti*. Ivrea: Priuli & Verlucca, 1978. 289 p.

Price, Derek John de Solla. *Gears from the Greeks: The Antikythera Mechanism, a Calendar Computer from Ca. 80 B.C.* Philadelphia: American Philosophical Society, 1974. 70 p.

Proverbio, Edoardo. *Il problema della misura del temp e l'importanza dei servizi orari*. (Conferenza nel ciclo "Astronomia" tenuta presso L'Osservatorio astronomico di Brera in Milano). Milano, 1966. 59 p.

Risley, Allan S. *The National Measurement System for Time and Frequency*. Washington: U.S. Dept. of Commerce, National Bureau of Standards, 1976. 64 p.

Schmitt, Gustav. *Die Comtoiser Uhr*. Villingen/Schwarzwald: Müller, 1977. 400 p.

Shpoliânskiĭ, Vladimir Aleksandrovich. *Khronometriâ*. Moskva: Mashinostroenie, 1974. 654 p.

Synchronisation de stations éloignées par simple survol. Par Jean Besson et al. Châtillon: Office national d'études de recherches aérospatiales, 1970. 31 p.

Thomson, Malcolm M. *The Beginning of the Long Dash: A History of Timekeeping in Canada*. Toronto and Buffalo: University of Toronto Press, 1978. 190 p.

U.S. Dept. of Transportation. Office of Assistant General Counsel for Regulation. *Standard*

Time in the United States: A History of Standard and Daylight Saving Time in the United States and an Analysis of the Related Laws. Washington, 1970. 27 p.

Weinfeld, Stefan. *Czas.* Ilustrowal Zdzislaw Milach. Wyd 2., rozsz. Warszawa: Nasza Ksiegarnia, 1977. 77 p.

Weiss, Rolf. *Uhrensammlung Kellenberger, Winterthur.* Photos, Michael Speich. Basel: Gesellschaft für Schweizerische Kunstgeschichte, 1974. 35 p.

Welch, Kenneth F. *Time Measurement: An Introductory History.* Newton Abbot: David and Charles, 1972. 120 p.

Wieschebrink, Theodor. *Die astronomische Uhr im Dom zu Münster.* Hrsg. von Erich Hüttenhain. Münster/Westf.: Aschendorff, 1968. 83 p.

Zavel' skiĭ, Fridrikh Samuilovich. *Vremîâ i ego izmerenie. Ot billionnykh doleĭ sekundy do milliardov let.* Izd. 3-e, pererab. Moskva: Nauka, 1972. 272 p.

Zeeman, J. *De Nederlandse stoelklok.* 2e, bijgewerkte en herz. dr. Assen: Gorcum, 1978. 264 p.

12. Physical Science

Augustynek, Zdzislaw. *Natura czasu.* Warszawa: Państwowe Wydawn. Naukowe, 1975. 257 p.

———. *Wlasności czasu.* (Wyd. 2.) Warszawa: Państwowe Wydawn. Naukowe, 1972. 186 p.

Barbier, Osvaldo. *Tempo e relatività: il significato del tempo nella concezione relativistica.* Roma: Bizzarri, 1976. 151 p.

Basri, Saul A. *A Deductive Theory of Space and Time.* Amsterdam: North-Holland Pub. Co., 1966. 165 p.

Becker, Oskar. *Beiträge zue phänomenologischen Begründung der Geometrie und ihrer physikalischen Anwendung.* Tübingen: Niemeyer, 1973. 176 p.

Bergson, Henri Louis. *Durée et simultanéité. A propos de la théorie d'Einstein.* 7. éd. Paris: Presses universitaires de France, 1968. 216 p.

Blokhintŝev, Dmitriĭ Ivanovich. *Space and Time in the Microworld.* Translated by Zdenka Smith. Dordrecht, Holland; and Boston: Reidel, 1973. 330 p.

Böhme, Gernot. *Über die Zeitmodi. Eine Untersuchung über das Verstehen von Zeit als Gegenwart, Vergangenheit and Zukunft mit besonderer Berücksichtigung der Beziehungen zum zweiten Hauptsatz der Thermodynamik.* Göttingen: Vandenhoeck u. Ruprecht, 1966. 121 p.

Bunge, Mario Augusto. *The Furniture of the World.* Dordrecht and Boston: Reidel, 1977. 352 p.

Casanova, Gaston. *Renversement du temps et nouvelle théorie de l'électron.* Bologna: Azzoguidi, 1972. 38 p.

Chamberlain, Joseph Miles. *Time and the Stars.* Garden City, N.Y.: Natural History Press, 1964. 32 p.

Chudinov, Éngel's Matveevich. *Prostranstvo i vremîâ v sovremennoi fizike.* Moskva: Znanie, 1969. 47 p.

Conference on Quantum Theory and the Structures of Time and Space, Feldafing, Germany, 1974. *Quantum Theory and the Structures of Time and Space: Papers Presented at a Conference Held in Feldafing, July 1974.* Edited by L. Castell et al. München: C. Hanser, 1975. 252 p.

Conference on the Foundations of Space-Time Theories, Minneapolis, Minn., 1974. *Foundations of Space-Time Theories*. Edited by John Earman et al. for the Minnesota Center for Philosophy of Science. Minneapolis: University of Minnesota Press, 1977. 459 p.

Davies, P. C. W. *The Physics of Time Assymetry*. Berkeley: University of California Press, 1974. 214 p.
————. *Space and Time in the Modern Universe*. Cambridge and New York: Cambridge University Press, 1977. 232 p.
Denbigh, Kenneth George. *An Inventive Universe*. New York: G. Braziller, 1975. 219 p.

Egorov, Andreĭ Andreevich. *Dialekticheskoe otnoshenie prostranstva-vremeni k materiaľnomu dvizheniiŭ*. Leningrad: Izd-vo Leningradskovo universiteta, 1976. 126 p.
Esposito, Paul, and Witten, Louis, eds. *Asymptotic Structures of Space-Time*. New York: Plenum Press, 1977. 442 p.

Fahr, Hans Jörg. *Raumzeitdenken, Zwangsvorstellung Unendlichkeit*. Osnabrück: Fromm, 1973. 82 p.
Fratangelo, M. *Ipotesi sulla teoria dei temponi*. Bologna: Istituto di fisica A. Righi, 1976. 99 p.

Galli, Mario Giuseppe. *Spazio e tempo nella scienza moderna*. 2 v. Firenze: Tip. Baccini & Chiappi, 1965–67.
Geach, Kenneth. *Gravity is Time*. Walton-on-Thames: Published by the author, 1975. 5 p.
Gold, Thomas, and Schumacher, D. L., eds. *The Nature of Time*. Ithaca, N.Y.: Cornell University Press, 1967. 248 p.

Haber, Heinz. *Gefangen in Raum und Zeit. Die Grenzen der Menschlichen Vorstellungskraft über das Wesen der Schöpfung*. Stuttgart: Deutsche Verlagsanstalt, 1975. 103 p.
Häussling, Ansgar. *Die Reichweite der Physik. Zur Ontologie von Natur und Zeit*. Meisenheim am Glan: Hain, 1969. 173 p.

Janich, Peter. *Die Protophysik der Zeit*. Mannheim, Wien, Zürich: Bibliographisches Inst., 1969. 177 p.

Kim Đinh. *Chũ'thò'i*. Saigon: Sáng, 1967. 387 p.

Laprus, Wlodzimierz. *Wspólczesna koncepcja przestrzeni i czasu*. Warszawa: Wiedza Powszechna, 1970. 187 p.
Lepenies, Wolf. *Das Ende der Naturgeschichte. Wandel kultureller Selbverständlichkeiten in den Wissenschaften des 18. und 19. Jahrhunderts*. München: 1976.
Lucas, John Randolph. *A Treatise on Time and Space*. London: Methuen, 1973. 321 p.

Machamer, Peter K., and Turnbull, Robert G., eds. *Motion and Time, Space and Matter: Interrelations in the History of Philosophy and Science*. Columbus: Ohio State University Press, 1976. 559 p.
Marder, Leslie. *Time and the Space-Traveller*. Philadelphia: University of Pennsylvania Press, 1971. 208 p.
Merleau-Ponty, Jacques, and Morando, Bruno. *The Rebirth of Cosmology*. New York: Knopf, 1976. 284 p.
Mittelstaedt, Peter. *Der Zeitbegriff in der Physik*. Mannheim: Bibliographisches Institut, 1980. 188 p.

Molchanov, IUriĭ Borisovich. *Chetyre kontŝeptŝii vremeni v filosofii i fizike.* AN SSSR, In-t filosofii. Moskva: Nauka, 1977. 192 p.

Neal, Harry Edward. *The Mystery of Time.* New York: J. Messner, 1966. 190 p.

Palágyi, Menyhért. *Neue Theorie des Raumes und der Zeit. Die Grundbegriffe einer Metageometrie.* (Unveränderter reprografischer Nachdruck der Ausg. Leipzig 1901). Sonderausg. Darmstadt: Wissenschaftliche Buchgesellschaft, 1967. 48 p.

Pavlov, Vasiliĭ Terent'evich. *Logicheskie funkĉii kategoriĭ prostranstva i vremeni.* Kiev: Izd-vo Kievskovo universiteta, 1966. 234 p.

Park, David. *The Image of Eternity: The Roots of Time in the Physical World.* Amherst: University of Massachusetts Press, 1980. 149 p.

Pascaru, Ion. *Secunda cît mileniul?* Bucureşti: Editura Ştiinţifică, 1966. 101 p.

Prieto y Delgado, Luis. *Espacio, tiempo y relatividad.* Madrid: Colegio de Ingenieros de Caminos, Canales y Puertos, 1974. 143 p.

Prigogine, Ilyia. *From Being to Becoming: Time and Complexity in the Physical Sciences.* San Francisco: W. H. Freeman, 1980. 272 p.

Radojčić, Miloš. *Une construction axiomatique de la théorie de l'espace-temps de la relativité restreinte.* Beograd: Académie serbe des sciences et des arts, 1973. 169 p.

Rigal, J. L. ed. *Le Temps et la pensée physique contemporaine.* Paris: Dunod, 1968. 150 p.

Salmon, Wesley C. *Space, Time, and Motion: A Philosophical Introduction.* Encino, Calif.: Dickenson Pub. Co., 1975. 147 p.

Schlegel, Richard. *Time and the Physical World.* New York: Dover Pub., 1968. 211 p.

Sklar, Lawrence. *Space, Time, and Spacetime.* Berkeley: University of California, 1974. 423 p.

Strong, Richard Allen. *Star-Tracker; An Aid for Use in Learning about Time, Space and Our Earth.* Rev. 2d ed. Dayton, Ohio: Star-Tracker Systems, 1973. 24 p.

Suppes, Patrick Colonel, ed. *Space, Time and Geometry.* Dordrecht and Boston: Reidel, 1973. 424 p.

Torao, Masahisa. *Toki to wa nani ka.* Tokyo: Kidansha, 1969. 222 p.

Wagn, Klaus. *What Time Does.* Translated by Sophie Wilkins. München: Caann-Verlag, 1976.

Wallis, Robert. *Le temps, quatriéme dimension de l'esprit.* Paris: Flammarion, 1966. 281 p.

Watanabe, Michael Satoshi. *The History of Time, Seen Through Physical Science.* In Japanese. Tokyo: Tosho, 1973. 246 p.

⸻. *Time.* In Japanese. Tokyo: Kawade Shoboh Shinsha, 1974. 338 p.

Wheeler, John Archibald. "Frontiers of Time." In *Proceedings of the International School of Physics "Enrico Fermi,"* Course 72, p. 395–496. Amsterdam: North Holland, 1979.

Whiteman, Michael. *Philosophy of Space and Time and the Inner Constitution of Nature: A Phenomenological Study.* London: Allen & Unwin; New York: Humanities Press., 1967. 436 p.

B. Works of Possible Interest, 1900–1965

Abbott, Edwin Abbott. *Flatland: A Romance of Many Dimensions, With the Author, a Square.* 5th ed., rev. New York: Barnes & Noble, 1963. 108 p.

Abeler, Jürgen. *Uhren im Wandel der Zeiten: eine Ausstellung im Deutschen Goldschmiedehaus Hanau vornehmlich aus Beständen des Wuppertaler Uhrenmuseums.* Sammlung: Georg Abeler. Hanau: Deutsches Goldschmiedehaus, 1964. 77 p.

Abad Carretero, Luis. *Una filosofia del instante.* Mexico: El Colegio de México, 1954. 258 p.

————. *Instante, querer y realidad.* México: Fondo de Cultura Económica, 1958. 455 p.

Academy of Time. Annual report. New York, 1947– .

Adler, Irving. *Time in Your Life.* Illustrated by Ruth Adler. New York: J. Day Co., 1955. 127 p.

Alexander, Hubert Griggs. *Time as Dimension and History.* Albuquerque: University of New Mexico Press, 1945. 134 p.

Alexander, Samuel. *Spinoza and Time.* London: G. Allen & Unwin, 1921. 80 p.

Alfa, Peninah, ed. *Mi-boker 'ad boker.* Merhavya: Seforim Poelim, 1952. 70 p.

Allen, Chalinder. *The Tyranny of Time.* New York: Philosophical Library, 1947. 275 p.

Amado, Eliane Lévy-Valensi. *Le temps dans la vie psychologique.* Paris: Flammarion, 1965. 202 p.

Anding, Ernst. "Über Koordinaten und Zeit." In *Encyklopädie der mathematischen Wissenschaften,* bd. VI–2, hft. 1, p. 3–15. Leipzig, 1905.

Andrade, Jules Frédéric Charles. *Hologerie et chronométrie.* Paris: J.-B. Baillière et fils, 1924. 582 p.

————. *Le mouvement, mesures de l'étendue et mesures du temps.* Paris: F. Alcan, 1911. 328 p.

Arbuthnot, Forster Fitzgerald. *The Mysteries of Chronology, with Proposal for a New English Era, to Be Called the Victorian.* London: W. Heinemann, 1900. 239 p.

Argentiere, Romulo. *Aventura humana no espaço e no tempo.* São Paulo: Fulgor, 1962. 190 p.

Arnot, F. L. *Time and the Universe: A New Basis for Cosmology.* Sydney: Australasian Publishing Co., 1941. 73 p.

Arvanitakēs, G. L. *Chronologia tōn archaiōn kai neōterōn Hellēnon.* 'Athēnai: P. Sabbidē & N. Karagiannachē, 1940. 49 p.

Aschoff, Jürgen, ed. *Circadian Clocks: Proceedings. Feldafing Summer School.* Amsterdam: North-Holland Pub. Co., 1965. 479 p.

Asimov, Issac. *The Clock We Live On.* New rev. ed. New York: Collier Books, 1963. 158 p.

Aspecten van de tijd; een bundel wijsgerige studies (ter gelegenheid van zijn 25-jarig bestaan uitg. door het Genootschap voor Wetenschappelijke Philosophie). Assen: Van Gorcum, 1950. 307 p.

Axel, Robert. *Estimation of Time.* New York, 1924. 77 p.

Aznar, José Camón. *El tiempo en el arte.* Madrid: Sociedad de Estudios y Publicaciones, 1958. 381 p.

Bachelard, Gaston. *La dialectique de la durée.* Nouv. éd. Paris: Presses universitaires de France, 1950. 150 p.

————. *L'intuition de l'instant, étude sur la Siloë de Gaston Roupnel.* Paris: Stock, Delamain et Boutelleau, 1932. 128 p.

————. *The Poetics of Space.* Translated by Maria Jolas. New York: Orion Press, 1964. 240 p.

Backman, Gaston Viktor. *Wachstum und organische Zeit.* Leipzig: J. A. Barth, 1943. 195 p.

Baker, John Tull. *An Historical and Critical Examination of English Space and Time Theories from Henry More to Bishop Berkeley.* Bronxville, N.Y.: Sarah Lawrence College, 1930. 90 p.

Barahona, Maria Alzira. *Para um estudo da expressão do tempo no romance português contemporâneo.* Lisboa, 1968. 207 p.

Barr, James. *Biblical Words for Time.* Naperville, Ill.: A. R. Allenson, 1962. 174 p.

Barthe, Engelhard. *Takt und Tempo: Studien über die Zusammenhänge von Takt und Tempo.* Hamburg: Musikverlag H. Sikorski, 1960. 51 p.

Bauer, Edmond. *Critique des notions d'éther, d'espace et de temps cinématique de la relativité.* Paris: Hermann et Cie, 1932. 31 p.

Beemelmans, Friedrich. *Zeit und Ewigkeit nach Thomas von Aquino.* Münster: W. Aschendorff, 1914. 64 p.

Bell, Thelma Harrington. *The Riddle of Time.* New York: Viking Press, 1963. 160 p.

Benedicks, Carl Axel Fredrik. *Space and Time: An Experimental Physicist's Conception of These Ideas and of Their Alteration.* London: Methuen & Co., 1924. 98 p.

Benigar, Juan. *El concepto del Tiempo entre los Araucanos.* Academia nacional de la historia Buenos Aires, Bol., 1924. V. 1, p. 137–154.

Bergh, George van den. *De euro-klok: een eenheidsstelsel van klokhervorming voor geheel Europa ten westen ven het IJzeren Gordijn.* 4. druk. Haarlem: H. D. Tjeenk Willink, 1962. 114 p.

Bergsten, Staffan. *Time and Eternity: A Study in the Structure and Symbolism of T. S. Eliot's "Four Quartets."* Stockholm: Svenska bokförlaget, 1960. 258 p.

Bernea, Ernest. *Timpul la ţăranul român. Contributie la problema timpului in religie şi magie.* Bucuresti, 1941. 75 p.

Bertele, Hans von. *Globes and Spheres.* Lausanne: Swiss Watch and Jewelry Journal, 1961. 63 p.

Besançon, France. Université. Observatoire. *Règlement chronométrique.* Besançon: Impr. J. Millot & Cie, 1909. 20 p.

Black, Max. *Models and Metaphors: Studies in Language and Philosophy.* Ithaca, N.Y.: Cornell University Press, 1962. 267 p.

Blakely, William Arthur. *The Discrimination of Short Empty Temporal Intervals.* Urbana, Ill.: 1933. 6 p.

Blum, Harold Francis. *Time's Arrow and Evolution.* 2d ed. Princeton: Princeton University Press, 1955. 219 p.

Boas, George. *The Acceptance of Time.* University of California Publications in Philosophy, vol. 16, no. 12, p. 249–269. Berkeley: University of California Press, 1950.

Bolton, Lyndon. *Time Measurement: An Introduction to Means and Ways of Reckoning Physical and Civil Time.* London: G. Bell and Sons, 1924. 166 p.

Boodin, John Elof. *Time and Reality.* New York: Macmillan Co., 1904. 119 p.

Borel, Émile Félix Édouard Justin. *Space and Time.* Translated by Angelo S. Rappoport and John Dougall. New York: Dover Pub., 1960. 234 p.

Boston, Lucy Maria. *The Children of Green Knowe.* Illustrated by Peter Boston. New York: Harcourt, Brace, Jovanovich, 1955. 157 p.

Boucher, Maurice. *Essai sur l'hyperspace, le temps, la matière & l'énergie.* 3. éd. rev. et augm. Paris: Gauthier-Villars et Cie, 1927. 264 p.

Brandon, Samuel George Frederick. "Time as God and Devil." *John Rylands Library, Manchester, Bulletin* 47 (1964): 12–31.

Brearley, Harry Chase. *Time Telling through the Ages.* York: Doubleday, Page & Co., 1919. 294 p.

Buehr, Walter. *Keeping Time.* Illustrated by the author. New York: Putnam, 1960. 94 p.

Burger, Dionys. *Sphereland: A Fantasy about Curved Spaces and an Expanding Universe.* Translated by Cornelie J. Rhinboldt. New York: Crowell, 1965. 208 p.

Butler, S. T., and Messel, H., eds. *Time: Selected Lectures on Time and Relativity, the Arrow of Time and the Relation of Geological and Biological Time, and on Men of Science.* Oxford and New York: Pergamon Press, 1965. 315 p.

Calabresi, Renata. *La determinazione del presente psichico.* Firenze: R. Bemporad & figlio, 1930. 188 p.

Carr, Herbert Wildon. *'Time' and 'History' in Contemporary Philosophy: With Special Reference to Bergson and Croce*. London: Oxford University Press, 1918. 19 p.

Cherbonnier, Edmond La Beaume. *Freedom and Time: A Study in Some Recent Contributions to the Problem*. Ann Arbor: University Microfilms, 1952. 350 leaves.

Cherniev, Leonid Fedorovich. *Sluzhba vremeni na morskikh sudakh*. Moskva: Morskoĭ transport, 1963. 54 p.

Christoff, Daniel. *Le temps et les valeurs, essai sur l'idée de finalité et son usage en philosophie morale*. Neuchâtel: Éditions de la Baconnière, 1945. 222 p.

Church, Margaret. *Time and Reality: Studies in Contemporary Fiction*. Chapel Hill: University of North Carolina Press, 1963. 302 p.

Cittanova, Marie. *Analyse psychologique des notions d'espace, de temps et de relativité: essai de philosophie scientifique sur l'espace et temps*. Paris: Les Éditions Adyar, 1932. 368 p.

Congrès international de chronométrie, Paris, 1954. Procès-verbaux et mémoires, recueillis par Paul Libessart et présentés par René Baillaud. 3 v. Besançon: Observation national de Besançon, 1955–56. 1,436 p.

Conrad, Hedwig Martius. *Die Zeit*. München: Kösel-Verlag, 1954. 306 p.

Conti Rossini, Carlo. *Tabelle comparative del calendario etiopico col calendario romano*. Roma: Istituto per l'Oriente, 1948. 47 p.

Cook, Melvin Alonzo. *Geological Chronometry*. Salt Lake City: University of Utah, Dept. of Metallurgy, 1956. 59 p.

Coomaraswany, Ananda. *Time and Eternity*. Ascona: Arbitus Asiae, 1947. 140 p.

Cooper, Grosvenor, and Meyer, Leonard B. *The Rhythmic Structure of Music*. Chicago: University of Chicago, 1960. 212 p.

Cowan, Harrison J. *Time and Its Measurement: From the Stone Age to the Nuclear Age*. Cleveland: World Pub. Co., 1958. 159 p.

Cullmann, Oscar. *Christ and Time: The Primitive Christian Conception of Time and History*. Rev. (3d) ed. Translated by Floyd V. Filson. London: SCM Press, 1962. 253 p.

Defossez, Léopold. *Les savants du XVIIe siècle et la mesure du temps*. Lausanne. Édition du Journal suisse d'horlogerie et de bijourterie, 1946. 341 p.

Dinkler, Erich, ed. *Zeit und Geschichte*. Tübingen: JCB Mohr, 1964.

Dolgov, Pëtr Nikolaevich. *Opredelenie vremeni passazhnym instrumentom v meridiane*. Moskva: Gos. izd-vo tekniko-teoret. lit-ry, 1952. 396 p.

Dubé, Arthur Joseph. *The General Principles for the Reckoning of Time in Canon Law: An Historical Synopsis and Commentary*. Washington, D.C.: The Catholic University of America Press, 1941. 299 p.

Dupuis, Nathan Fellowes. *The Measures and the Measurement of Time*. Kingston, Ont.: The Jackson Press, 1940(?). 147 p.

Eggenspieler, Alfred. *Durée et instant: essai sur le caractère analogique de l'être*. Paris: Librarie philosophique J. Vrin, 1933. 146 p.

Elias bar Shīnāya. *La Chronographie d'Élie bar Šinaya, métropolitain de Nisibe*. Tr. par L.-J. Delaporte. Paris: H. Champion, 1910. 409 p.

Esclangon, Ernest. *La notion de temps, temps physique et relativité, la dynamique du point matériel*. Paris: Gauthier-Villars, 1938. 76 p.

Evans-Pritchard, E. E. *The Nuer*. London: Oxford University Press, 1940.

Fernández y Rodríquez, Obdulio. *El ritmo en la naturaleza*. Madrid: Academia de Ciencias Exactas, Físicas y Naturales, 1962. 161 p.

Fernández Suárez, Alvaro. *El tiempo y el "hay."* Madrid: Indice, 1955. 91 p.

Ferrall, Sarah Catherine. *The Absolute Judgment of Temporal Intervals*. Urbana, Ill.: 1935. 6 p.

Filgueira Valverde, José. *La Cántiga CIII: noción del tiempo y gozo eterno en la narrativa medieval*. Compostela: Universidad de Santiago de Compostela, Instituto de Estudios Regionales, 1936. 225 p.

Fliess, Wilhelm. *Das Jahr im Lebendigen*. Jena: E. Diederichs, 1924. 310 p.

———. *Vom Leben und vom Tod*; Jena: E. Diederichs, 1924. 139 p.

———. *Zur Periodenlehre*. Jena: E. Diederichs, 1925. 257 p.

Flynn, Harry Eugene. *Tick Tock: A Story of Time*. Illustrated by Herman Fay, Jr. Boston and New York: D. C. Heath and Co., 1938. 234 p.

Frankenhaeuser, Marianne. *Estimation of Time: An Experimental Study*. Stockholm: Almqvist & Wiksell, 1959. 135 p.

Freiberg, Stanley Kenneth. *The Artist's Year: A Study of the Meaning of Time in the Life and Works of William Blake*. Ann Arbor: University Microfilms, 1957. 155 leaves.

Freund, Walter. *Modernus und andere Zeitbegriffe des Mittelalters*. Köln, Graz, 1957. 114 p.

Friedman, Lopple Clayton. *The Growth of Time Concepts*. N.p., 1944. 31 p.

Fruin, Robert. *Handboek der chronologie, voornamelijk van Nederland*. Alphen aan den Rijn: N. Samsom n. v., 1934. 136 p.

Fuligatti, Giulio. *Degli horivoli a sole*. Ferrara: V. Baldini, 1616. 248 p.

Gaos, José. *Introducción a El ser y el tiempo de Martin Heidegger.* México: Fondo de Cultura Económica, 1951. 112 p.

Garriott, Christopher T. *Making the Most of the Time*. St. Louis: Bethany Press, 1959. 160 p.

Geach, Peter Thomas. *Some Problems about Time*. British Academy, London, Proceedings 51 (1965): 321–336.

Gent, Werner. *Das Problem der Zeit: eine historische und systematische Untersuchung*. Frankfurt a. M.: G. Schulte-Bulmke, 1934. 187 p.

Gentile, Giovanni. *Giordano Bruno e il pensiero del rinascimento*. 2. ed. riordinata. Firenze: Vallecchi, 1925. 303 p.

Ginzel, Friedrich Karl. *Handbuch der mathematischen und technischen Chronologie das Zeitrechnungswesen der Völker*. 3 v. Leipzig: J. C. Heinrichs, 1906–14.

Giorgi, G. *L'evoluzione della nozione di Tempo*. **Scientia**, Ser. 3, 55 (1934): 89–102.

Gladwin, Harold Sterling. *Tree-Ring Analysis: Problems of Dating*. Globe, Ariz.: Gila Pueblo, 1944.

Goeje, Claudius Henricus de. *What is Time?* Leiden: Brill, 1949. 51 p.

Goodennough, E. R. *Evaluation of Symbols Recurrent in Time, As Illustrated in Judaism*. Zürich: Rhine Verlag, 1952.

Goodwin, Brian C. *Temporal Organization in Cells: A Dynamic Theory of Cellular Control Processes*. London and New York: Academic Press, 1963. 163 p.

Goudriaan, J. C. *Over de schatting van het tempo van periodische geluidsprikkels en hare afhankelijkheid van physiologische factoren*. Amsterdam: Koninklijke akademie van wetenschappen, 1922. 160 p.

Gould, Rupert Thomas. *The Marine Chronometer: Its History and Development*. London: J. D. Potter, 1923. 17 p.

Granier, Jean. *La mesure du temps*. Paris: Presses universitaires de France, 1943. 118 p.

Granville, William Anthony. *The Fourth Dimension and the Bible*. Boston: R. G. Badger, 1922. 119 p.

Grize, Jean Blaise. *Essai sur le rôle du temps en analyse mathématique classique*. Neuchâtel, 1954. 104 p.

Grote, Louis Ruyter Radcliffe. *Das Zeitgesetz in Biologie und Pathologie*. Bremen, NS. Gauverlag Weser-Ems, 1942. 51 p.

Gschwind, Karl Heinz. *Raum, Zeit, Ablauf: Definitionen*. Würzburg: Vogel Verlag, 1958. 137 p.

Gurvitch, Georges. *The Spectrum of Social Time.* Translated by Myrtle Korenbaum, assisted by Phillip Bosserman. Dordrecht: D. Reidel Pub. Co., 1964. 152 p.

Guyot, Edmond. *Dictionnaire des termes utilisés dans la mesure du temps.* La Chaux-de-Fonds, Chambre suisse de l'horlogerie, 1953. 123 p.

Haeberlin, Carl. *Lebensrhythmen und Heilkunde: Entwurf einer biozentrischen ärztlichen Betrachtung,* Stuttgart, Leipzig: Hippokrates-verlag g. m. b. h., 1935. 74 p.

Hallett, Harold Foster. *Aeternitas: A Spinozistic Study.* Oxford: Clarendon Press, 1930. 344 p.

Hammerschmidt, William W. *Whitehead's Philosophy of Time.* New York: King's Crown Press, 1947. 108 p.

Haswell, J. Eric. *Horology: The Science of Time Measurement and the Construction of Clocks, Watches and Chronometers.* London: Chapman and Hall, 1928. 267 p.

Hatano, Seiichi. *Time and Eternity.* Translated by Ichiro Suzuki. Tokyo(?): Print. Bureau, Japanese Govt., 1963. 181 p.

Havet, Jacques. *Kant et le problème du temps.* Paris: Gallimard, 1946. 230 p.

Haya de la Torre, Víctor Raúl. *Espacio-tiempo-histórico: cinco ensayos y tres diálagos.* Lima, 1948. 189 p.

Heckert, Hilmar. *Lunationsrhythmen des menschlichen Organismus, Methodisches und Ergebnisse.* Leipzig: Akademische Verlagsgesellschaft Geest & Portig, 1961. 126 p.

Heidemann, Ingeborg. *Spontaneität und Zeitlichkeit. Ein Problem der Kritik der reinen Vernunft.* Köln, 1958. 275 p.

Hernández Perera, Jesús. *La pintura española y el reloj.* Madrid: R. Carbonell, 1958. 153 p.

Herzog, Oswald. *Zeit und Raum: das Absolute in Kunst und Natur.* Berlin, Frohnau: J. J. Otten, 1928. 62 p.

Heschel, Abraham Joshua. *The Sabbath: Its Meaning for Modern Man.* New York: Farrar, Straus and Young, 1951.

Hewitt, James Francis Katherinus. *Primitive Traditional History. The Primitive History and Chronology of India, South-eastern and South-western Asia, Egypt, and Europe, and the Colonies Thence Sent Forth.* 2 vols. London: J. Parker and Co., 1907.

Hinman, Edgar Lenderson. *On Time as an Absolute Principle of Negativity.* Lincoln, Nebraska, 1906. 20 p.

Hinton, Charles Howard. *The Fourth Dimension.* London: G. Allen & Unwin, 1934. 270 p.

Hochkeppel, W. *Die Veränderung des Zeitbewusstsein in modernen Theater.* München, 1957.

Hofland, H. J. A. *Geen tijd: op zoek naar oorzaken en gevlogen van het moderne tijdgebrek.* Amsterdam: Scheltema & Holkema, 1955. 147 p.

Hood, Peter. *How Time Is Measured.* London: Oxford University Press, 1955. 64 p.

Hooker, Charles William Ross. *What is the Fourth Dimension? Reflections Inspired by a Pair of Gloves.* London: A. & C. Black, 1934. 110 p.

Howeler, Casper. *Tijd en muziek.* Amsterdam: H. J. Paris, 1946. 149 p.

Hüttner, Max. *Zur Psychologie des Zeitbewusstseins bei kontinuierlichen Lichtreizen.* Leipzig: W. Engelmann, 1902. 48 p.

Hurst, Martha. *Some Aspects of the Problems of Time.* Chapel Hill: University of Carolina, 1934. 102 p.

Hutchinson, William M. *A Maxton Book about Time.* New York: Maxton Publishers, 1959.

Hyde, Margaret Oldroyd. *Animal Clocks and Compasses: From Animal Migration to Space Travel.* Illustrated by P. A. Hutchison. New York: Whittlesey House, 1960.

Ibérico, Mariano. *Perspectivas sobre el tema del tiempo.* Lima: Universidad Nacional Mayor de San Marcos, 1958. 195 p.

Ingarden, Roman. *Time and Modes of Being.* Translated by Helen R. Michejda. Springfield, Ill.: Thomas, 1964. 170 p.

Israeli, Nathan. *Abnormal Personality and Time.* New York, 1936. 123 p.

Jack, Butler. *New Light on Ancient Chronological Records in the Hebrew Scriptures: Showing, among Other Features, an Accurate Explanation of Daniel's Great Prophecy of the "Seventy Weeks."* Washington: Neale Publishing Co., 1902. 62 p.

Jacobowitz, Abraham Leib. *ha-Luah veshimusho ba-khronologiya.* Jerusalem: Ha-Universita ha-ivrit, 1953. 182 p.

Jaffé, George. *Drei Dialoge über Raum, Zeit und Kausalität.* Berlin: Springer, 1954. 211 p.

Jakubisiak, Augustin. *Essai sur les limites de l'espace et du temps.* Paris: F. Alcan, 1927. 196 p.

Johnson, Martin Christopher. *Time and Universe for the Scientific Conscience.* Cambridge University Press, 1952. 41 p.

————. *Time, Knowledge, and the Nebulae: An Introduction to the Meanings of Time in Physics, Astronomy, and Philosophy, and the Relativities of Einstein and of Milne.* New York: Dover Pub., 1947. 189 p.

Jünger, Ernst. *Das Sanduhrbuch.* Illustrierte Sonderausg. Frankfurt am Main: V. Klostermann, 1957. 261 p.

Jusué, Eduardo. *Tablas abreviadas para la reducción del cómputo árebe y del hebraico al cristiano y viceversa.* Madris: Establecimiento tipográfico de Fortanet, 1917. 427 p.

Kannegiesser, Karlheinz. *Raum, Zeit, Unendlichkeit.* Berlin: Deutscher Verlag der Wissenschaften, 1964. 137 p.

Kaufman, Gerald Lynton. *The Book of Time: A Review of the Many Fascinating Aspects of Time, the Unknown, Presented in Popular Style for the Intellectually Curious.* New York: J. Messner, Inc., 1938. 287 p.

Ketkar, Venkatesh Bapuji. *Indian and Foreign Chronology, with Theory, Practice and Tables, B.C. 3102 to 2100 A.D. and Notices of the Vedic, the Ancient Indian, the Chinese, the Jewish, the Ecclesiastical and the Coptic Calendars.* Bombay: British Indian Press, 1923. 214 p.

Khare, Gaṇeśa Sakhārāma. *Śivakālīna sampūrṇa śakavalī.* In Marathi. Bombay, 1923. ca. 450 p.

Klugmann, Friedhelm. *Die Kategorie der Zeit in der Musik.* Kassel-Wilhelmshöhe: Baerenreiter, 1961.

Kockelmans, A. *Tijd en ruimte.* Haarlem, 1958. 88 p.

Koşay, Hâmit Zübeyr. *Tarihöncesi ve tarih çağlarinin mukayeseli zaman tablosu.* İstanbul: Millî Eğitim Bakanliği, 1946.

Krauss, Joseph. *Vom Messen der Zeit im Wandel der Zeiten.* Wolfshagen-Scharbeutz: F. Westphal, 1950. 106 p.

Krusch, Bruno. *Studien zur christlich-mittelalterlichen Chronologie: die Enstehung unserer heutigen Zeitrechnung.* Berlin: Akademie der Wissenschaften, 1938. 87 p.

Kubitschek, Wilhelm. *Grundriss der antiken Zeitrechnung.* München: Beck, 1928. 241 p.

Kursus Kader Katolik. *Waktu.* Djakarta: K.K.K., 196–(?). 35 p.

Kurz, Peter. *200 Jahre Schwenninger Uhren. 1765–1965.* Schwenningen a. N.: Stadtverwaltung, 1965. 334 p.

Kuz'min, B. S. *Osnovy astronomicheskogo metoda izmereniiă vremeni.* Moskva: Gos. izd-vo tekhniko-teoret. lit-ry, 1954. 60 p.

Lacape. R. S. *A la recherche du temps vécu.* Paris: Hermann & Cᵢᵉ, 1935. 55 p.

La Harpe, Jean de. *Genèse et mesure du temps.* Neuchâtel: Université, 1941. 180 p.

Lamorte, André. *Le problème du temps dans le prophétisme biblique.* Beatenberg (Suisse) Éditions École biblique; Paris: Librairie protestante, 1960. 195 p.

Lautman, Albert. *Symétrie et dissymétrie en mathématiques et en physique: Le problème du temps.* Paris: Hermann, 1946. 50 p.

Lavelle, Louis. *Du temps et de l'éternité*. Paris, Aubier: Éditions Montaigne, 1945. 446 p.

Lecomte du Noüy, Pierre. *Biological Time*. New York: Macmillan Co., 1937. 180 p.

Leisegang, Hans. *Die Begriffe der Zeit und Ewigkeit im späteren Platonismus*. Münster i. W.: Aschendorff, 1913. 59 p.

Le Lionnais, François. *The Orion Book of Time*. Translated by William D. O'Gorman, Jr. New York: Orion Press, 1960. 110 p.

Lewis, Wyndham. *Time and Western Man*. Boston: Beacon Press, 1957.

Leohlin, John C. *The Influence of Different Activities on the Apparent Length of Time*. Washington: American Psychological Association, 1959. 27 p.

Lietmann, Hans. *Zeitrechnung der römischen Kaiserzeit, des Mittelalters und der Neuzeit für die Jahre 1–2000 nach Christus*. Berlin, Leipzig: W. de Gruyter & Co., 1934. 127 p.

Loske, Lothar M. *Die Sonnenuhren: Kunstwerke der Zeitmessung und ihre Geheimnisse*. Berlin: Springer, 1959. 88 p.

Lovejoy, Arthur Oncken. *The Reason, the Understanding, and Time*. Baltimore: Johns Hopkins Press, 1961. 210 p.

Lubac, Émile. *Présent conscient et cycles de durée: le rôle du corps à la venue sur le présent conscient*. Paris: F. Alcan, 1936. 200 p.

MacIver, Robert Morrison. *The Challenge of the Passing Years: My Encounter with Time*. New York: Simon and Schuster, 1962. 133 p.

Madeira, José Antonio. *Considerações sôbre a perceptibilidade auditiva dos sinais horários rítmicos radio-telegráficos no processo das coincidências por extinção de sinais*. Coimbra: Tip. da "Atlântida," 1942. 11 p.

Maeterlinck, Maurice. *The Life of Space*. Translated by Bernard Miall. New York: Dodd, Mead & Co., 1928. 194 p.

Malrieu, Philippe. *Les origines de la conscience du temps. Les attitudes temporelles de l'enfant*. Paris: Presses universitaires de France, 1953. 158 p.

Manning, Henry Parker, ed. *The Fourth Dimension Simply Explained: A Collection of Essays Selected from Those Submitted in the Scientific American's Prize Competition*. New York: Dover Pub., 1960. 251 p.

Marsden, Dora. *The Philosophy of Time*. Oxford: Holywell Press, 1955. 34 p.

Marshall, Roy Kenneth. *Sundials*. Illustrated by Jerry Cailor. New York: Macmillan, 1963. 126 p.

Martin, Alfred Wilhelm Otto von. *Sociology of the Renaissance*. New York: Oxford University Press, 1944. 100 p.

Martynov, Dmitriĭ IAkovlevich. *Veka i mgnoven'ia*. Moskva: Izd-vo Moskovskogo universiteta, 1961. 85 p.

Matamek Conference on Biological Cycles, Labrador, 1931. *Abstract of Papers and Discussions*. Prepared by Charles Elton et al. Matamek Factory, Canadian Labrador, 1933. 50 p.

Medina Peralta, Manuel, and Alonso Lerch, Federico. *Neuvas tablas para la aplicación del método de Díaz Covarrubias en la determinación de la hora*. Mexico: Universidad Nacional Autónoma de México, Departamento de Geodesia, 1963. 8 p.

Mendilow, Adam Abraham. *Time and the Novel*. London and New York: P. Nevill, 1952. 245 p.

Mercier, A. "Petits prolégomènes à une étude sur le temps." *Theoria* 29 (1963): 277–282.

Metz, André. *Temps, espace, relativité*. Paris: G. Beauchesne, 1928. 211 p.

Meyer, R. W., ed. *Das Zeitproblem in 20. Jahrhundert*. Bern: Francke Verlag, 1964. 363 p.

Miasnikov, Lev Leonidovich. *Atomnye chasy*. Leningrad, 1962. 53 p.

Millikan, Robert Andrews. *Time, Matter, and Values*. Chapel Hill: University of North Carolina Press, 1932.

Milne, Edward Arthur. "Some Points in the Philosophy of Physics: Time, Evolution, and Creation," in *Smithsonian Institution. Annual Report, 1933* (Washington, 1935). p. 219–238.

Moore, Wilbert Ellis. *Man, Time, and Society.* New York: Wiley, 1963. 163 p.

Moreau, Joseph. *L'espace et le temps selon Aristote.* Padova: Antenore, 1965. 212 p.

Morgenstern, Irvin. *The Dimensional Structure of Time, Together with the Drama and Its Timing.* New York: Philosophical Library, 1960. 174 p.

Murray, Rosalind. *Time and the Timeless.* London: Centenary Press, 1942. 126 p.

Nappi, Rosa. *Lo spazio, il tempo e la matematica nell'empirismo inglese.* Napoli: Libreria scientifica editrice, 1951. 72 p.

Needham, Joseph. *Time: The Refreshing River (Essays and Addresses, 1932–1942).* London: G. Allen & Unwin, 1943. 280 p.

Neugebauer, Paul Viktor. *Astronomische Chronologie.* 2 v. Berlin und Leipzig: W. de Gruyter & Co., 1929.

Nicoll, Maurice. *Living Time and the Integration of the Life.* London: V. Stuart, 1952. 252 p.

Nieden, Ernst zur. *...und trotzdem Zeit.* Baden-Baden: A. Lutzeyer, 1960. 64 p.

Nordmann, Charles. *The Tyranny of Time: Einstein or Bergson?* Translated by E. E. Fournier d'Albe. London: T. F. Unwin, 1925. 216 p.

Olston, Albert B. *The Story of Time.* Chicago: Jarvis Universal Clock Company, 1915. 112 p.

Palmer, Lionel Stanley. *Man's Journey through Time: A First Step in Physical and Cultural Anthropochronology.* New York: Philosophical Library, 1959. 184 p.

Parsons, Edmund. *Time Devoured: A Materialistic Discussion of Duration.* London: Allen & Unwin, 1964. 132 p.

Pascual Castán, Santiago. *Los términos judiciales: índice de los que comprende la vigente Ley de enjuiciamiento civil, con texto articulado y notas útiles de la jurisprudencia del Tribunal Supremo.* Barcelona, 1949. 111 p.

Pearce, Josephine Anna. *The Manipulations of Time in Shakespeare's English History Plays.* Ann Arbor: University Microfilms, 1955. 229 leaves.

Pemartín, José. *Introducción a una filosofía de lo temporal: doce lecciones sobre espacio, tiempo, causalidad.* Madrid: Espasa-Calpe, 1941. 196 p.

Polak, Isoif Fedorovich. *Vremía i kalendar'.* Izd. 4. Moskva: Gos. izd-vo fiziko-matematicheskoĭ: lit-ry, 1959. 46 p.

Poole, Reginald Lane. *The Beginnings of the Year in the Middle Ages.* London: H. Milford, 1921. 25 p.

———. *Medieval Reckonings of Time.* London: Society for Promoting Christian Knowledge, 1948. 47 p.

Priestley, John Boynton. *Man and Time.* Garden City, N.Y.: Doubleday, 1964. 319 p.

The Problem of Time. Lectures delivered before the Philosophical Union of the University of California, 1934. Berkeley: University of California Press, 1935. 225 p.

Quinn, John M. *The Doctrine of Time in St. Thomas: Some Aspects and Applications.* Washington: Catholic University of America Press, 1960. 54 p.

Rabello, Sylvio. *A representação do tempo na criança.* São Paulo: Companhia editora nacional, 1938. 182 p.

Rădulescu-Motru, Constantin. *Zeit und Schicksal.* Jena: W. Gronau, 1943. 190 p.

Ramige, Eldon Albert. *Contemporary Concepts of Time and the Idea of God.* Boston: Stratford Co., 1935. 132 p.

Reck, Alma Kehoe. *Clocks Tell the Time*. Illustrated by Janina Domanska. New York: Scribner, 1960. 48 p.

Reinberg, Alain, and Ghata, Jean. *Biological Rhythms*. Translated by C. J. Cameron. New York: Walker, 1964. 138 p.

Restrepo, Félix. *Entre el tiempo y la eternidad*. Bogotá: Voluntad, 1960. 170 p.

Reunión de Aproximación Filosófico-Científica. El tiempo. Institución "Fernando el Católico" (C. S. I. C.) de la Excma. Diputación Provincial de Zaragoza, 1958. 310 p.

Richter, Curt Paul. *Biological Clocks in Medicine and Psychiatry*. Springfield, Ill.: C. C. Thomas, 1965. 109 p.

Rotenstreich, Nathan, "History and Time: A Critical Examination of R. G. Collingwood's Doctrine." Offprint from *Studies in Philosophy, Scripta Hierosolymitana*. Jerusalem, 1960.

Ruch, E. A. *Space and Time: A Comparative Study of the Theories of Aristotle and A. Einstein*. Pretoria, 1958. 62 p.

Runia, Klaas. *De theologische tijd bij Karl Barth, met name in zijn anthropologie*. Franeker: T. Wever, 1955. 257 p.

Ryzewski, S. *Time Units in the Solar System: Geometrisation of Time*. 2 vols. Southampton, Eng., 1962–64.

Sabarini, Raniero. *Il tempo in G. B. Vico*. Milano: Bocca, 1954. 89 p.

Sachs, Curt. *Rhythm and Tempo: A Study in Music History*. New York: Norton, 1953. 391 p.

Sánchez de Bustamante, Teodoro. *Reflexiones sobre el espacio y el tiempo*. Buenos Aires, 1953. 20 p.

Saunier, Claudius. *Treatise on Modern Horology in Theory and Practice*. Translated by Julieu Tripplin and Edward Rigg. London: W. & G. Foyle, 1952. 844 p.

Schaeffler, Richard. *Die Struktur der Geschichtszeit*. Frankfurt/Main: V. Klostermann, 1963. 571 p.

Schaltenbrand, Georges, ed. *Zeit in nervenärztlicher Sicht*. Stuttgart: Ferdinand Enke, 1963. 135 p.

Schayer, Stanislaw. *Contributions to the Problem of Time in Indian Philosophy*. Krakow: Polski Buletyn Orientalistyczny, 1938.

Schram, Robert Gustav. *Kalendariographische und chronologische Tafeln*. Leipzig: J. C. Hinrichs, 1908. 368 p.

Schwieger, Heinz Gerhard. *Des Menschen Engel ist die Zeit*. Wien: P. Neff, 1959. 64 p.

Sen Gupta, Subodh Chandra. *The Whirligig of Time: The Problem of Duration in Shakespeare's Plays*. Bombay: Orient Longmans, 1961. 201 p.

Shackle, George Lennox Sharman. *Time in Economics*. Amsterdam: North Holland Pub. Co., 1958. 111 p.

Shapiro, Herman. *Motion, Time and Place According to William Ockham*. St. Bonaventure, N.Y.: Franciscan Institute, 1957. 151 p.

Smits, Jacobus Aloysius. *Bijdrage tot het onderzoek van het 24-uursrhytme bij de mens*. Hilversum: Drukkerij "Erica," 1948. 102 p.

Smythe, Elizabeth Jane. *A Normative, Genetic Study of the Development of Time Perception*. Ann Arbor: University Microfilms, 1956. 80 leaves.

Sollberger, Arne. *Biological Rhythm Research*. Amsterdam, New York: Elsevier Pub. Co., 1965. 461 p.

Sonnen- und Turmuhren. Dresden: Sachsenverlag, 1959.

Sonnet, André. *The Twilight Zone of Dreams*. Translated by J. T. Fraser. Philadelphia: Chilton Co., 1961. 230 p.

Sorokin, Pitirim Aleksandrovich. *Sociocultural Causality, Space, Time: A Study of Referential Principles of Sociology and Social Science*. Durham, N.C.: Duke University Press, 1943. 246 p.

Spier, J. M. *Tijd en eeuwigheid: een wijsgerig onderzoek bij het licht van Gods Woord.* Kampen: J. H. Kok, 1953. 230 p.

Spierdijk, C. *Klokken en Klokkenmakers, 1300–1900.* Amsterdam: DeBussy, 1962. 256 p.

Springer, Max. *Mensch, Zeit, Uhr: zur Geschichte der Zeitmessung.* Berlin: Ullstein, 1927. 150 p.

Stace, Walter Terence. *Time and Eterniity: An Essay in the Philosophy of Religion.* Princeton: Princeton University Press, 1952. 169 p.

Staiger, Emil. *Die Zeit als Einbildungskraft des Dichters: Untersuchungen zu Gedichten von Brentano, Goethe und Keller.* Zürich und Leipzig: M. Niehans, 1939. 203 p.

Stambaugh, Joan. *Untersuchungen zum Problem der Zeit bei Nietzsche.* Den Haag: M. Nijhoff, 1959. 235 p.

Stearns, Frank Preston. *Space and Time: A Critique on Herbert Spencer.* New York: Knickerbocker Press, 190–(?). 17 p.

Steffens, Joseph. *Entwicklung des Zeitbegriffs in der griechischen Philosophie bis Plato nebst einer einleitenden Gründung auf die vorphilosophische Anschauung.* Berlin: R. Trenkel, 1911. 58 p.

Stock, Augustine. *Kingdom of Heaven: The Good Tidings of the Gospel.* New York: Herder and Herder, 1964. 191 p.

Stocks, John Leofric. *Time, Cause and Eternity.* London: Macmillan and Co., 1938. 163 p.

Strehler, Bernard Louis et al., eds. *The Biology of Aging.* Washington: American Institute of Biological Sciences, 1960. 364 p.

Strzeszewski, C. *Problem czasu w ekonomice.* Lublin: KUL, 1959.

Stuart, Mary. *The Psychology of Time.* London: K. Paul, Trench, Trubner & Co.; New York: Harcourt, Brace & Co., 1925. 152 p.

Szumilewicz, Irena. *O kierunku uplywu czasu.* Warszawa: Panstwowe Wydawn. Naukowe, 1964. 127 p.

Tannenbaum, Beulah, and Stillman, Myra. *Understanding Time: The Science of Clocks and Calendars.* Illustrated by William D. Hayes. New York: Whittlesey House, 1958. 143 p.

Tardy (publisher). *Bibliographie générale de la mesure du temps, suivie d'un essai de'classification technique et géographique.* Paris, 1947. 352 p.

Tebaldeschi, Ivanhoe. *Tempo e coscienza.* Roma: Editoriale Arte e Storia, 195–(?). 132 p.

Temmer, Mark J. *Time in Rousseau and Kant: An Essay on French Preromanticism.* Genève: E. Droz, 1958. 79 p.

Theodoricus, Teutonicus, de Vriberg. "Meister Deitrich von Freiberg über die Zeit und das Sein." [The text of "De tempore" and "De mensuris durationis," with commentary by F. Stegmüller.] In *Archives d'histoire doctrinale et littéraire du moyen âge,* 1942.

Thieberger, Richard. *Der Begriff der Zeit bei Thomas Mann, vom Zauberg zum Joseph.* Baden-Baden: Verlag für Kunst und Wissenschaft, 1952. 162 p.

Time and Its Mysteries: Eight Lectures Given on the James Arthur Foundation, New York University. New York: Collier Books, 1962. 186 p.

Tlustý, Vojtěch. *Prostor a čas: studie z marxistické filosofie.* 2., přepracované vyd. Phara: Nakl. politické literatury, 1963. 176 p.

Tomb, John Walker. *An Essay on Time.* Princeton, N. J.: Dept. of Philosophy, Princeton University, 1953. 14 p.

Toulmin, Stephen Edelston, and Goodfield, June. *The Discovery of Time.* New York: Harper & Row, 1965. 280 p.

Treisman, Michel. *Temporal Discrimination and the Indifference Interval Implications for a Model of the "Internal Clock."* Washington: American Psychological Association, 1963. 31 p.

Trejos, Juan. *La doctrina del eterno retorno y los avances de la ciencia: ensayo.* Madrid: Ediciones Iberoamericanas, 1964. 54 p.

Ubbelohde, Alfred René. *Time and Thermodynamics.* London: Oxford University Press, 1947. 110 p.

Uka, Ngwobia. *The Development of Time Concepts in African Children of Primary School Age.* Ibadan: Institute of Education, University College, 1962. 25 p.

Ungerer, Emil. *Zeit-ordnungsformen des organischen Lebens.* Leipzig: J. A. Barth, 1936. 72 p.

Ushenko, Andrew Paul. *The Logic of Events: An Introduction to a Philosophy of Time.* Berkeley: University of California Press, 1929. 180 p.

Vachon, André. *Le temps et l'espace dans l'oeuvre de Paul Claudel: expérience chrétienne et imagination poétique.* Paris: Éditions du Seuil, 1965. 435 p.

Van Riper, Benjamin Whitman. *Some Views of the Time Problem.* Menasha, Wis.: George Banta, 1916. 99 p.

Vera, Francisco. *Espacio, hiperespacio y tiempo.* Madrid: Editorial Páez, 1928. 192 p.

Verwiebe, Walter. *Welt und Zeit bei Augustin.* Leipzig: F. Meiner, 1933. 86 p.

Vestdijk, Simon. *Het eeuwige telaat: dialogen over den tijd.* Amsterdam: Uitgeverij Contact, 1946. 131 p.

Villanueva Mejia, Demetrio. *La duración supra-oposicional: ensayo de una metafísica de la duración.* Lima, 1956. 338 p.

Vil' nitskiĭ, Moiseĭ Borisovich. *K istorii razvitiiâ predstavleniĭ o prostranstve i vremeni v klassicheskoĭ fizike.* Kiev: Izd-vo Akademii nauk Ukr. CCR, 1955. 234 p.

Vogel, Karl Max. *What is Time?* Boston: Club of Odd Volumes, 1948. 33 p.

Volkelt, Johannes Immanuel. *Phänomenologie und Metaphysik der Zeit.* München: Beck, 1925. 200 p.

Wahl, Jean André. *Du rôle de l'idée de l'instant dans la philosophie de Descartes.* Paris: F. Alcan, 1920. 48 p.

Waldeck, Hans. *Ebbe und Flut im Menschen: die neue Biorhythmik, von der Einfachen zur höheren Rhythmuslehre.* Büdingen: Lebensweiser-Verlag, 1952. 67 p.

———. *Der Rhythmus deines Blutes: die Biorhythmik als Natur-gesetz.* 2. Aufl. Büdingen: Lebensweiser-Verlag, 1953. 121 p.

Walder, Francis. *L'existence Profonde: le souvenir, l'instant, l'espérance.* Paris: Aubier, Éditions Montaigne, 1953. 158 p.

Walter, Johnston Estep. *Nature and Cognition of Space and Time.* West Newton, Pa.: Johnston and Penney, 1914. 186 p.

Wang, Tsu-hua. *Shih chien ti chêng fu.* Taiwan, 1953. 84 p.

Ward, F. A. B. *London Science Museum: Handbook of the Collection Illustrating Time Measurement.* 4th ed. London: H.M. Stationery Office., 1958 –

Watanabe, Satoshi. *Jikan.* Tokyo, 1948. 3ōō p.

Wax, Murray. "Time, Magic, and Asceticism: A Comparative Study of Time Perspectives." Photocopy. Chicago: University of Chicago Library, 1959. 299 leaves.

Weber, Christian Oliver. *The Concept of Duration as Key to the Logical Forms of Reason and Their Psychological Processes.* Lincoln, Neb., 1925. 109 p.

Weber, Jean Paul. "Edgar Poe and the Theme of the Clock." *La Nouvelle Revue Francaise.* **68** and **69** (1958).

Weinreich, Otto. *Phöbus, Aurora, Kalender und Uhr: über eine Doppelform der epischen Zeitbestimmung in der Erzählkunst der Antike und Neuzeit.* Stuttgart: W. Kohlhammer, 1937. 42 p.

Weishaupt, Adam. *Zweifel über die Kantischen Begriffe von Zeit und Raum.* Nürnberg, 1788. Bruxelles: Culture et Civilisation, 1968. 120 p.

Wells, Herbert George. *The Time Machine: An Invention.* New York: H. Holt and Co., 1922. 216 p.

Werfel, Franz V. *Star of the Unborn.* Translated by Gustave O. Arlt. New York: Viking Press, 1946. 645 p.

Wernli, Hans J. *Biorhythm: A Scientific Exploration into the Life Cycles of the Individual.* Translated by Rosemary Colmers. New York: Crown Pub., 1961. 128 p.

Weyl, Hermann. *Space—Time—Matter.* Translated by Henry L. Brose. London: Methuen & Co., 1922. 330 p.

White, Carroll T. *Temporal Numerosity and the Psychological Unit of Duration.* Washington: American Psychological Association, 1963. 37 p.

Widenmeyer, Walter Harold. *Über das Zeiterlebnis bei Geisteskranken mit ungestörten Merk- und Gedächtnisfunktionen.* Bonn: Buchdruckerei P. Kubens, 1931. 25 p.

Wieland, Wolfgang. *Schellings Lehre von der Zeit: Grundlagen und Voraussetzungen der Weltalterphilosophie.* Heidelberg: C. Winter Universitätsverlag, 1956. 100 p.

Wijk, Walter Emile van. *Het eerste leerboek der technische tijdrekenkunde (Scaliger's Isagogici chronologiae canones, 1606).* 's-Gravenhage: M. Nijhoff, 1954. 40 p.

Winniczuk, Lidia. *Kalendarz starożytnych Greków i Rzymian.* Warszawa, 1951. 37 p.

Wright, David Jay. *Time Estimation Measurements and Schizophrenia.* Ann Arbor, Mich.: University Microfilms, 1958. 75 leaves.

Wullimann, Franz. *Tageszeit und psychischer Allgemeinzustand.* Winterthur: P. G. Keller, 1955. 78 p.

Zaunick, Rudolph, Hrsg. *Das Zeit-Problem.* Bericht über die Jahresversammlung der Deutschen Akademie der Naturforscher Leopoldina. Leipzig, 1959. 313 p.

Zeuner, Friedrich Eberhard. *Dating the Past: An Introduction to Geochronology.* 4th ed. Darien, Conn.: Hafner Pub. Co., 1958. 516 p.

Ziner, Feenie, and Thompson, Elizabeth. *The Junior True Book of Time.* Illustrations by Katherine Evans. London: Frederick Muller, 1959.

Zuckerkandl, Victor. *Sound and Symbol: Music and the External World.* Translated by Willard R. Trask. New York: Pantheon Books, 1956. 399 p.

Zwicky, Laurie Bowmann. *Milton's Use of Time: Image and Principle.* Ann Arbor, Mich.: University Microfilms, 1959. 122 leaves.

2. Periodical Literature—a Statistical Survey

A computer survey was conducted to determine, or at least suggest, the number of articles published in professional periodicals, which, in the collective approach of these four volumes, would be classified as useful and relevant to the study of time. Fifteen databases of an educational information retrieval system were searched covering, more or less, the period 1965–1980 and containing almost 15 million records.

The databases searched, the subjects sought and the strategy of the search are described below. The number of entries identified is a few short of 18,500. A sampling suggests that an entry-by-entry check would reduce this figure to, perhaps,

12,000, because of multiple listings. On the other hand, the indexers of the 15 million records must be regarded as untutored in the study of time. Many papers which, by the standards of *The Study of Time* volumes would be classified under time, were surely not so recognized. Again, a random, though rather extensive sampling suggests that a careful reading of the literature would more than triple this number, taking us up to about 40,000. It is a further estimate that an extension of the survey to include the first 65 years of the century would less than double the total count. We may, therefore, settle on the hypothetical figure of 65,000 as the number of periodical articles in this century potentially related to a systematic study of time

This is not necessarily the number of citations which would have to be included in an authoritative bibliography, but it is the number of papers which the editors of such a work would have to examine, accept, reject, classify, annotate and cross-index by subject, author, and source. The scope of such an enterprise may be comparable to that of a well-known reference work in the history of science, the *Isis Cumulative Bibliography**. It is a four volume work of over 2500 pages. The first 1300 pages contain about 40,000 entries classified by personalities significant in the history of science. There seems to be sufficient material in the periodical literature to supply a comparable number of entries for a multidisciplinary bibliography of time. However, in view of the rapidly increasing time-related literature in many fields of intellectual and empirical discipline, it is not at all certain that a bibliography, published in a traditional form, would be practical to compile or useful to employ. Yet, a multidisciplinary work conceived and executed on a lesser scale, is likely to remain unauthentic.

Modern technology does, however, offer a way for accomplishing the task earlier envisaged for printed bibliographies. Accordingly, the feasibility of an open database, under the name TIMELINE, is now being studied.

Table I lists those sources of the DIALOG Information Retrieval System which were searched, courtesy of the Librarian and Staff of the Public Library, Westport, Connecticut.

In addition to the sources listed in Table I, the following indeces were consulted: *Social Sciences and Humanities Index*, 1966–1974; *Social Sciences Index*, 1975–1980; *Humanities Index*, 1975–1980; *Index Medicus*, 1960–1979.

In the computer search, only descriptors and identifiers were used. Descriptors are controlled terms, equivalent to subject headings. Identifiers are uncontrolled phrases, assigned by the indexers. Since neither lists of descriptors nor lists of identifiers were available, terms were selected by sampling the categories of classifications of known papers. The subjects listed in Table II are not necessarily either the identifiers or the descriptors used. They are, instead, the most general headings under which the descriptors and identifiers could be combined. Within each subject, the number of entries were weighted for duplication, as estimated through sampling. No attempt was made, however, to adjust for duplication between different subjects.

The uncertainty due to the absence of rigorous vocabularies, the many different jargons involved and the fact that the indexing was done by specialists naive in the

---

*A bibliography formed from *Isis* critical bibliographies, 1913–1965. Ed. by Magda Whitrow. London: Mansell, 1971–1981.

Table I. List of sources searched, periods covered, and of the approximate total number of entries available in each source.

Database	Period Covered	Approximate Number of Records
ABC-Clio, Inc., Art Literature	1974–1980	28,000
American Institute of Physics	1975–1980	114,000
Biological Abstracts and Bioresearch Index	1969–1980	2,600,000
Chemical Abstracts	1967–1980	4,362,000
Comprehensive Dissertation Abstracts	1861–1980	648,000
Excerpta Medica	1974–1980	1,160,000
Historical Abstracts	1973–1980	54,000
Institute for Scientific Information (Science and Technology)	1974–1980	2,970,000
Institute for Scientific Information (Social Sciences)	1972–1980	765,000
Language and Language Behavior Abstracts	1973–1980	33,000
Modern Language Association Bibliography	1976–1980	121,500
Philosopher's Index	1940–1980	81,300
Public Affairs Information Service International	1972–1980	107,000
Psychological Abstracts	1967–1980	305,500
Science Abstracts	1969–1980	1,404,000

Table II. An initial inventory of the number of time-related citations in the sources listed in Table I.

General subject	Number of citations
AGING	6606
ART AND TIME	54
BIOLOGICAL RHYTHM	4322
CALENDAR	79
COSMOLOGY AND TIME	466
DRUGS AND TIME	235
ENTROPY AND TIME	2351
INTELLECTUAL AND SOCIAL HISTORY AND TIME	151
ORGANIC EVOLUTION AND TIME	247
QUANTUM THEORY AND TIME	144
TIME ALLOCATION	43
TIME AND ECONOMIC REACTIONS	119
TIME AND RELATIVITY THEORY	872
TIME IN LANGUAGE AND LITERATURE	225
TIME IN PHILOSOPHY	726
TIME IN PSYCHOLOGY	864
TIME MEASUREMENT	902

study of time makes Table II useful for what it is intended to be: An initial inventory.

3. A Guide to the Time-related Subject Headings of the Library of Congress

Those who wish to pursue the bibliographic background to the study of time could do no better than to begin by examining that extraordinary collection known as the Library of Congress Card Catalog System.

Table III lists some ninety Library of Congress subject headings. Class numbers, where shown, represent the most common aspects of the subjects. Two vast and important fields in the study of time, philosophy and history, are represented only by their most general class designations.

The entries in Table III are not subjects one would construct for a card catalog specially designed for the study of time. Such subject headings are more likely to resemble the twelve major divisions of Part A of the bibliography of books on page 235. Instead, the subjects in Table III are those under which the Library of Congress has found it appropriate to classify various books which deal with time. The guide does not say what one is to look for. But it will be helpful in identifying suitable entries if one begins with a plan.

(following page: Table III)

Table III. Library of Congress subject headings most frequently called for in the study of time (See page 269)

Membership List

STUART ALBERT, University of Pennsylvania, Philadelphia, Pennsylvania, U.S.A.

*GONZALO ALCAINO, Instituto para la Investigation Astronomica, Santiago, Chile.

*PAUL K. ALKON, University of Southern California, Los Angeles, California, U.S.A.

JACOB A. ARLOW, New York, New York, U.S.A.

SETH G. ATWOOD, The Time Museum, Rockford, Illinois, U.S.A.

MARK AULTMAN, Yellow Springs, Ohio, U.S.A.

*PETER H. BARNETT, City University of New York, New York, New York, U.S.A.

H. BARREAU, Université Louis Pasteur, Strasbourg, France

JUDITH BECKER, Univ. of Michigan, Ann Arbor, Michigan, U.S.A.

JEAN BERNABE, Ottawa, Ontario, Canada

H. G. BEVANS, St. James' University Hospital, Leeds, England

LUDWIK BIELAWSKI, Polish Academy of Sciences, Warsaw, Poland

*RICHARD BLOCK, Montana State University, Bozeman, Montana, U.S.A.

*DAVID BOYER, Washington, D.C., U.S.A

GARRY L. BRODHEAD, Ithaca College, Ithaca, New York, U.S.A.

ROBERT S. BRUMBAUGH, Yale University, New Haven, Connecticut, U.S.A.

THOMAS J. BRUNEAU, University of Guam, Agana, Guam

A. G. CAIRNS-SMITH, University of Glasgow, Scotland

---

*Corresponding Member.

GILBERT CANTOR, Philadelphia, Pennsylvania, U.S.A.

*SARAH M. CARRIG, Chicago, Illlinois, U.S.A.

EVA CASSIRER, Berlin, West Germany

*PHILIP CHAPNIK, Glen Ellen, California, U.S.A.

FERREL CHRISTENSEN, University of Alberta, Canada

TIM CLOUDSLEY, Glasgow College of Technology, Glasgow, Scotland

DENIS CORISH, Bowdoin College, Brunswick, Maine, U.S.A.

OLIVIER COSTA DE BEAUREGARD, Bourron Marlotte, France

K. G. DENBIGH, F.R.S., Council for Science and Society, London, England

ROBERT EFRON, San Francisco, California, U.S.A

CHARLES EHRET, Argonne National Laboratory, Argonne, Illinois, U.S.A

DAVID EPSTEIN, Massachusetts Institute of Technology, Cambridge, Massachusetts, U.S.A.

JOHANNES FABIAN, University of Amsterdam, Amsterdam, Netherlands

*LAWRENCE W. FAGG, Catholic University of America, Washington, D.C., U.S.A.

*ROBERT A. FAGUET, Los Angeles, California, U.S.A

*STUART FEDER, New York, New York, U.S.A.

*HERBERT H. FINE, Middletown, New Jersey, U.S.A.

DAVID FINKELSTEIN, Georgia Institute of Technology, Atlanta, Georgia, U.S.A.

GEORGE FORD, University of Rochester, Rochester, New York, U.S.A.

MARIE-LOUISE VON FRANZ, Küsnacht, Switzerland

J. T. FRASER, Westport, Connecticut, U.S.A.

*DOUGLAS R. GIVENS, Saint Louis Community College, St. Louis, Missouri, U.S.A.

BARRY GLASSNER, Syracuse University, Syracuse, New York, U.S.A.

STEVEN GOLDMAN, Lehigh University, Bethlehem, Pennsylvania, U.S.A.

WILLIAM GOODY, London, England

B. C. GOODWIN, University of Sussex, Falmer Brighton, Sussex, England

JOSEPH B. GOODWIN, Smithsonian Institution, Washington, District of Columbia, U.S.A.

FRANCIS HABER, University of Maryland, Silver Springs, Maryland, U.S.A

*EDWARD T. HALL, Santa Fe, New Mexico, U.S.A.

MICHAEL HELLER, Pontifical Faculty, Cracow, Poland

*BERTRAND P. HELM, Southwest Missouri State University, Springfield, Missouri, U.S.A.

*SALLY CARR HELTON, Bloomington, Indiana, U.S.A.

*DAVID LEON HIGDON, Texas Tech University, Lubbock, Texas, U.S.A.

MAE WAN HO, Open University, Milton Keynes, England

OLIVER W. HOLMES, Wesleyan University, Middletown, Connecticut, U.S.A.

*PHILIP JAFFRAY HUGHES, Victoria, British Columbia, Canada

*ELLIOT JACQUES, Brunel University, Uxbridge, Middlesex, England

*DEVORAH KALEKIN-FISHMAN, University of Haifa, Haifa, Israel

HANS KALMUS, University College London, London, England

SUSUMU KAMEFUCHI, University of Tsukuba, Ibaraki, Japan

JONATHAN D. KRAMER, University of Cincinnati, Cincinnati, Ohio, U.S.A.

*I. LACHMAN-SZUMILEWICZ, London, England

I. P. LALÈYÊ, Kinshasa, Republique du Zaire

*ROBERT H. LAUER, Southern Illinois University, Edwardsville, Illinois, U.S.A.

NATHANIEL LAWRENCE, Williams College, Williamstown, Massachusetts, U.S.A.

*JEAN-MARC LEVY-LEBLOND, Université de Nice, Nice, France

SVERRE LYNGSTAD, New Jersey Institute of Technology, Newark, New Jersey, U.S.A.

SAMUEL L. MACEY, University of Victoria, Victoria, Canada

*CARLOS ALBERTO MALLMANN, Fundacion Bariloche, San Carlos de Bariloche, Argentina

*RICHARD MARTIN, Amherst, Massachusetts, U.S.A.

ERKKA MAULA, Hauho, Finland

ALBERT MAYR, Troghi, Italy

WOLFE MAYS, University of Manchester, England

MURRAY MELBIN, Boston University, Bostom, Massachusetts, U.S.A.

A. A. MENDILOW, The Hebrew University, Jerusalem, Israel

JACQUE MERLEAU-PONTY, Paris, France

JOHN A. MICHON, Rijksuniversiteit Groningen, Netherlands

*SUCHOON S. MO, University of Southern Colorado, Pueblo, Colorado, U.S.A.

*ADRIAN C. MOULYN, Stamford, Connecticut, U.S.A.

GERT H. MÜLLER, University of Heidelberg, West Germany

J. D. NORTH, Oxford, England

HELGA NOWOTNY, University of Vienna, Vienna, Austria

*OSCAR NUDLER, Fundacion Bariloche, San Carlos de Bariloche, Argentina

*L. NATHAN OAKLANDER, University of Michigan, Flint, Michigan, U.S.A.

*MARION A. O'CONNOR, The United Nations, New York, New York, U.S.A.

HIDEO OGAWA, Keio University, Tokyo, Japan

KEN-ICHI ONO, University of Tokyo, Tokyo, Japan

*HENRY FRAGER ORLOV, Wesleyan University, Middletown, Connecticut, U.S.A.

*WALTER OTT, University of New Brunswick, Fredericton, New Brunswick, Canada

*ADOLPHE PACAULT, Centre de Researche Paul Pascal, Talence, France

RAIMUNDO PANIKKAR, University of California, Santa Barbara, California, U.S.A.

DAVID PARK, Williams College, Williamstown, Massachusetts, U.S.A.

MARLENE PILARCIK, Penn State University, Dunmore, Pennsylvania, U.S.A.

ERNST PÖPPEL, Institut für Medizinische Psychologie, München, West Germany

*HARI SHANKAR PRASAD, Australian National University, Canberra, Australia

ALBERT I. RABIN, Michigan State Univerrsity, East Lansing, Michigan, U.S.A.

PAOLA REALE, Parma, Italy

*MARC RICHELLE, Université Liège, Liège, Belgium

GEORGE ROCHBERG, University of Pennsylvania, Philadelphia, Pennsylvania, U.S.A.

*GILBERT J. ROSE, Rowayton, Connecticut, U.S.A.

LEWIS . ROWELL, Indiana University, Bloomington, Indiana, U.S.A.

UNNI ROWELL, Bloomington, Indiana, U.S.A.

P. T. SAUNDERS, Universitty of London, London, England

ALBERT . SCHMIDT, Fairfield, Connecticut, U.S.A.

AGNES C. SCHULDT, Moscow, Idaho, U.S.A.

BARRY SCHWARTZ, University of Georgia, U.S.A.

JANET SCHWARTZ, Athens, Georgia, U.S.A.

SHARON C. SCHWARZE, Cabrini College, Radnor, Pennsylvania, U.S.A.

CHARLES M. SHEROVER, Hunter College, New York, New York, U.S.A.

PAUL SMYTH, Paralia Aegina, Greece

MICHAEL SPANGLER, Washington, District of Columbia, U.S.A.

*RUTH H. STONE, Indiana University, Bloomington, Indiana, U.S.A.

*RUDOLF S. STRAUSS, London, England

*THOMAS M. TERRY, University of Connecticut, Storrs, Connecticut, U.S.A.

DARLENE THOMAS, Lock Haven State College, Lock Haven, Pennsylvania, U.S.A.

MASANAO TODA, Hokkaido University, Sapporo, Japan

MARIANNA TORGOVNICK, Williams College, Williamstown, Massachusetts, U.S.A.

*ERIK TRELL, University of Lund, Malmö, Sweden

GISELA TROMMSDORFF, Universität Mannheim, West Germany

FREDERICK TURNER, Kenyon College, Gambier, Ohio, U.S.A.

* KIMON VALASKAKIS, University of Montreal, Montreal, Quebec, Canada

WALDMAR VOISÉ, Warsaw, Poland

* FRIEDRICH WALLNER, Theresienfeld, Austria

MICHAEL WATANABE, Honolulu, Hawaii, U.S.A.

MOGENS WEGENER, University of Aarhus, Aarhus, Denmark

JOHN G. WEIHAUPT, San Jose State University, San Jose, California, U.S.A.

EUGENE WEINER, Haifa University, Haifa, Israel

* RUDOLF WENDORFF, Gütersloh, West Germany

LEE F. WERTH, Cleveland State University, Cleveland, Ohio, U.S.A.

G. J. WHITROW, University of London, London, England

GORDON C. WINSTON, Williams College, Williamstown, Massachusetts, U.S.A.

MAKOTO YAMAMOTO, University of Tokyo, Tokyo, Japan

MICHAEL YANASE, Sophia University, Tokyo, Japan

Cumulative Index and Tables of Contents for
The Study of Time I–IV

In the Cumulative Index, Roman numerals identify the volume, Arabic numerals the first page of the article. For instance, Automatons are discussed in the papers that begin on p. 399 and p. 451 in *The Study of Time II*. Reference to the Cumulative Tables of Contents will identify the authors and titles of the articles.

There are many concepts which recur in almost every article. They are indexed only if they are critically examined rather than merely mentioned. Only the most significant names are indexed.

All entries are to be understood in their relation to time. Only where convention or clarity required it was Time made part of the entry.

J. T. F.

Cumulative Tables of Contents
The Study of Time I-IV

The Study of Time II

The Study of Time III

The Study of Time IV

The Study of Time

Edited by

J.T. Fraser, N. Lawrence, and D. Park

The Study of Time

Proceedings of the First Conference of the
International Society for the Study of Time
Oberwolfach (Black Forest) — West Germany

1972. approx. 560p. 65 figures. cloth.
ISBN 3-540-05824-9

The Study of Time II

Proceedings of the Second Conference of the
International Society for the Study of Time
Lake Yamanaka — Japan

1975. 493p. 80 figures. cloth.
ISBN 3-540-07321-3

The Study of Time III

Proceedings of the Third Conference of the
International Society for the Study of Time
Alpbach — Austria

1978. 735p. 34 figures. cloth.
ISBN 0-387-90311-9